Functional Data Analysis with R

Emerging technologies generate data sets of increased size and complexity that require new or updated statistical inferential methods and scalable, reproducible software. These data sets often involve measurements of a continuous underlying process, and benefit from a functional data perspective. *Functional Data Analysis with R* presents many ideas for handling functional data including dimension reduction techniques, smoothing, functional regression, structured decompositions of curves, and clustering. The idea is for the reader to be able to immediately reproduce the results in the book, implement these methods, and potentially design new methods and software that may be inspired by these approaches.

Features:
- Functional regression models receive a modern treatment that allows extensions to many practical scenarios and development of state-of-the-art software.
- The connection between functional regression, penalized smoothing, and mixed effects models is used as the cornerstone for inference.
- Multilevel, longitudinal, and structured functional data are discussed with emphasis on emerging functional data structures.
- Methods for clustering functional data before and after smoothing are discussed.
- Multiple new functional data sets with dense and sparse sampling designs from various application areas are presented, including the NHANES linked accelerometry and mortality data, COVID-19 mortality data, CD4 counts data, and the CONTENT child growth study.
- Step-by-step software implementations are included, along with a supplementary website (www.FunctionalDataAnalysis.com) featuring software, data, and tutorials.
- More than 100 plots for visualization of functional data are presented.

Functional Data Analysis with R is primarily aimed at undergraduate, master's, and PhD students, as well as data scientists and researchers working on functional data analysis. The book can be read at different levels and combines state-of-the-art software, methods, and inference. It can be used for self-learning, teaching, and research, and will particularly appeal to anyone who is interested in practical methods for hands-on, problem-forward functional data analysis. The reader should have some basic coding experience, but expertise in R is not required.

Ciprian M. Crainiceanu is Professor of Biostatistics at Johns Hopkins University working on wearable and implantable technology (WIT), signal processing, and clinical neuroimaging. He has extensive experience in mixed effects modeling, semiparametric regression, and functional data analysis with application to data generated by emerging technologies.

Jeff Goldsmith is Associate Dean for Data Science and Associate Professor of Biostatistics at the Columbia University Mailman School of Public Health. His work in functional data analysis includes methodological and computational advances with applications in reaching kinematics, wearable devices, and neuroimaging.

Andrew Leroux is an Assistant Professor of Biostatistics and Informatics at the University of Colorado. His interests include the development of methodology in functional data analysis, particularly related to wearable technologies and intensive longitudinal data.

Erjia Cui is an Assistant Professor of Biostatistics at the University of Minnesota. His research interests include developing functional data analysis methods and semiparametric regression models with reproducible software, with applications in wearable devices, mobile health, and imaging.

MONOGRAPHS ON STATISTICS AND APPLIED PROBABILITY

Editors: F. Bunea, R. Henderson, L. Levina, N. Meinshausen, R. Smith,

Recently Published Titles

For more information about this series please visit: https://www.crcpress.com/Chapman--HallCRC-Monographs-on-Statistics--Applied-Probability/book-series/CHMONSTAAPP

Functional Data Analysis
with R

Ciprian M. Crainiceanu, Jeff Goldsmith, Andrew Leroux,
and Erjia Cui

CRC Press
Taylor & Francis Group
Boca Raton London New York

CRC Press is an imprint of the
Taylor & Francis Group, an **informa** business

A CHAPMAN & HALL BOOK

First edition published 2024
by CRC Press
2385 Executive Center Drive, Suite 320, Boca Raton, FL 33431, U.S.A.

and by CRC Press
4 Park Square, Milton Park, Abingdon, Oxon, OX14 4RN

CRC Press is an imprint of Taylor & Francis Group, LLC

Library of Congress Cataloging-in-Publication Data

Names: Crainiceanu, Ciprian, author. | Goldsmith, Jeff, author. | Leroux, Andrew, author. | Cui, Erjia, author.
Title: Functional data analysis with R / Ciprian Crainiceanu, Jeff Goldsmith, Andrew Leroux, and Erjia Cui.
Description: First edition. | Boca Raton : CRC Press, 2024. |
Series: CRC monographs on statistics and applied probability | Includes bibliographical references and index. | Summary: "Functional Data Analysis with R is primarily aimed at undergraduate, masters, and PhD students, as well as data scientists and researchers working on functional data analysis. The book can be read at different levels and combines state-of-the-art software, methods, and inference. It can be used for self-learning, teaching, and research, and will particularly appeal to anyone who is interested in practical methods for hands-on, problem-forward functional data analysis. The reader should have some basic coding experience, but expertise in R is not required"-- Provided by publisher.
Identifiers: LCCN 2023041843 (print) | LCCN 2023041844 (ebook) | ISBN 9781032244716 (hbk) | ISBN 9781032244723 (pbk) | ISBN 9781003278726 (ebk)
Subjects: LCSH: Multivariate analysis. | Statistical functionals. | Functional analysis. | R (Computer program language)
Classification: LCC QA278 .C73 2024 (print) | LCC QA278 (ebook) | DDC 519.5/35--dc23/eng/20231221
LC record available at https://lccn.loc.gov/2023041843
LC ebook record available at https://lccn.loc.gov/2023041844

ISBN: 978-1-032-24471-6 (hbk)
ISBN: 978-1-032-24472-3 (pbk)
ISBN: 978-1-003-27872-6 (ebk)

DOI: 10.1201/9781003278726

Typeset in CMR10
by KnowledgeWorks Global Ltd.

Publisher's note: This book has been prepared from camera-ready copy provided by the authors.

To Bianca, Julia, and Adina,
may your life be as beautiful as you made mine.

Ciprian

To my family and friends, for your unfailing support and encouragement.

Jeff

To Tushar, mom, and dad, thank you for all you do to keep me centered and sane.
To Sarina and Nikhil, you're all a parent could ever ask for. Never stop shining your light on the world.

Andrew

To my family, especially my mom, for your unconditional love.

Erjia

Contents

Preface

Around the year 2000, several major areas of statistics were witnessing rapid changes: functional data analysis, semiparametric regression, mixed effects models, and software development. While none of these areas was new, they were all becoming more mature, and their complementary ideas were setting the stage for new and rapid advancements. These developments were the result of the work of thousands of statisticians, whose collective achievements cannot be fully recognized in one monograph. We will try to describe some of the watershed moments that directly influenced our work and this book. We will also identify and contextualize our contributions to functional data analysis.

The Functional Data Analysis (FDA) book of Ramsay and Silverman [244, 245] was first published in 1997 and, without a doubt, defined the field. It considered functions as the basic unit of observation, and introduced new data structures, new methods, and new definitions. This amplified the interest in FDA, especially with the emergence of new, larger, and more complex data sets in the early 2000s. Around the same time, and largely independent of the FDA literature, nonparametric modeling was subject to massive structural changes. Starting in the early 1970s, the seminal papers of Grace Wahba and collaborators [54, 150, 303] were setting the stage for smoothing spline regression. Likely influenced by these ideas, in 1986, Finbarr O'Sullivan [221] published the first paper on penalized splines (B-splines with a smaller number of knots and a penalty on the roughness of the regression function). In 1996, Marx and Eilers [71] published a seminal paper on P-splines (similar to O'Sullivan's approach, but using a different penalty structure) and followed it up in 2002 by showing that ideas can be extended to Generalized Additive Models (GAM) [72]. In 1999, Brumback, Ruppert, and Wand [26] pointed out that regression models incorporating splines with coefficient penalties can be viewed as particular cases of Generalized Linear Mixed Models (GLMM). This idea was expanded upon in a series of papers that led to the highly influential *Semiparametric Regression* book by Ruppert, Wand, and Carroll [258], which was published in 2003. The book showed that semiparametric models could incorporate additional covariates, random effects, and nonparametric smoothing components in a unified mixed effects inferential framework. It also demonstrated how to implement these models in existing mixed effects software. Simon Wood and his collaborators, in a series of papers that culminated with the 2006 *Generalized Additive Models* book [315], set the current standards for methods and software integration for GAM. The substantially updated 2017 second edition of this book [319] is now a classic reference for GAM.

In early 2000s, the connection between functional data analysis, semiparametric regression, and mixed effects models was not yet apparent, though some early cross-pollination work was starting to appear. In 1999, Marx and Eilers [192] introduced the idea of P-splines for signal regression, which is closely related to the Functional Linear Model with a scalar outcome and functional predictors described by Ramsay and Silverman; see also extensions in the early 2000s [72, 171, 193]. In 2007, Reiss and Ogden [252] introduced a version of the method proposed by Marx and Eilers [192] using a different penalty structure, described methods for principal component regression (FPCR) and functional partial least squares (FPLS), and noted the connection with the mixed effects model representation of penalized splines described in [258]. In spite of these crucial advancements, by 2008 there was still no reliable FDA software for implementing these methods. In 2008, Wood gave a Royal Scientific Society (RSS) talk (https://rb.gy/o1zg5), where he showed how to use mgcv to fit scalar-on-function regression (SoFR) models using "linear functional terms." This talk clarified the conceptual and practical connections between functional and semiparametric regression; see pages 17–20 of his presentation. In a personal note, Wood mentioned that his work was influenced by that of Eilers, Marx, Reiss, and Ogden, though he points to Wahba's 1990 book [304] and Tikhonov, 1963 [294] as his primary sources of inspiration. In his words: "[Grace Wahba's equation] (8.1.4), from Tikhonov, 1963, is essentially the signal regression problem. It just took me a long time to think up the summation convention idea that mgcv uses to implement this." In 2011, Wood published the idea of penalized spline estimation for the functional coefficient in the SoFR context; see Section 5.2 in his paper, where methods are extended to incorporate non-Gaussian errors with multiple penalties.

Our methods and philosophy were also informed by many sources, including the now classical references discussed above. However, we were heavily influenced by the mixed effects representation of semiparametric models introduced by Ruppert, Wand, and Carroll [258]. Also, we were interested in the practical implementation and scalability of a variety of FDA models beyond the SoFR model. The 2010 paper by Crainiceanu and Goldsmith [48] and the 2011 paper led by Goldsmith and Bobb [102] outlined the philosophy and practice underlying much of the functional regression chapters of this book: (1) where necessary, project observed functions on a functional principal component basis to account for noise, irregular observation grids, and/or missing data; (2) use rich-basis spline expansions for functional coefficients and induce smoothing using penalties on the spline coefficients; (3) identify the mixed effects models that correspond to the specific functional regression; and (4) use existing mixed effects model software (in their case WinBUGS [187] and nlme [230], respectively) to fit the model and conduct inference. Regardless of the underlying software platform, one of our main contributions was to recognize the deep connections between functional regression, penalized spline smoothing, and mixed effects inference. This allowed extensions that incorporated multiple scalar covariates, random effects, and multiple functional observations with or without noise, with dense or sparse sampling patterns, and complete or missing data. Over time, the inferential approach was extended to scalar-on-function regression (SoFR), function-on-scalar regression (FoSR), and function-on-function regression (FoFR). We have also contributed to increasing awareness of new data structures and the need for validated and supported inferential software.

Around 2010–2011, Philip Reiss and Crainiceanu initiated a project to assemble existing R functions for FDA. It was started as the package refund [105] for "REgression with FUNctional Data," though it never provided any refund, it was not only about regression, and was not particularly easy to find on Google. However, it did bring together a group of statisticians who were passionate about developing FDA software for a wide audience. We would like to thank all of these contributors for their dedication and vision. The refund package is currently maintained by Julia Wrobel.

Fabian Scheipl, Sonja Greven, and collaborators have led a series of transformative papers [128, 262, 263] that started to appear in 2015 and expanded functional regression in many new directions. The 2015 paper by Ivanescu, Staicu, Scheipl, and Greven [128] showed how to conduct function-on-function regression (FoFR) using the philosophy outlined by Goldsmith, Bobb, and Crainiceanu [48, 102]. The paper made the connection to the "linear functional terms" implementation in mgcv, which merged previously disparate lines of work in FDA. This series of papers led to substantial revisions of the refund package and the addition of the powerful function pffr(), which provides a functional interface based on the mgcv package. The function pfr(), initially developed by Goldsmith, was updated to the same standard. Scheipl's contributions to refund were transformative and set a new bar for FDA software. Finally, the ideas came together and showed how functional regression can be modeled semiparametrically using splines, smoothness can be induced via specific penalties on parameters, and penalized models can be treated as mixed effects models, which can be fit using modern software. This body of work provides much of the infrastructure of Chapters 4, 5, and 6 of this book.

To address the larger and increasingly complex data applications, new methods were required for Functional Principal Components Analysis (FPCA). To the best of our knowledge, in 2010 there was no working software for smoothing covariance matrices for functional data with more than 300 observations per function. Luo Xiao was one of the main contributors who introduced the FAst Covariance Estimation (FACE), a principled method for nonparametric smoothing of covariance operators for high and ultra-high dimensional functional data. Methods use "sandwich estimators" of covariance matrices that are guaranteed to be symmetric and positive definite and were deployed in the `refund::fpca.face()` function [331]. Xiao's subsequent work on sparse and multivariate sparse FPCA was deployed as the standalone functions `face::face.sparse()` [328, 329] and `mfaces::mface.sparse()` [172, 173]. During the writing of this book, it became apparent that methods were also needed for FPCA-like methods for non-Gaussian functional data. Andrew Leroux and Wrobel led a paper on fast generalized FPCA (fastGFPCA) [167] using local mixed effects models and deployed the accompanying `fastGFPCA` package [324]. These developments are highlighted in Chapters 2 and 3 of this book.

Much less work has been dedicated to survival analysis with functional predictors and, especially, to extending the semiparametric regression ideas to this context. In 2015, Jonathan Gellar introduced the Penalized Functional Cox Regression [94], where the effect of the functional predictor on the log-hazard was modeled using penalized splines. However, methods were not immediately deployed in `mgcv` because this option only became available in 2016 [322]. In subsequent publications, Leroux [164, 166] and Erjia Cui [55, 56] made clear the connection to the "linear functional terms" in `mgcv` and substantially enlarged the range of applications of survival analysis with functional predictors. This work provides the infrastructure for Chapter 7 of this book.

In 2009, Chongzhi Di, Crainiceanu, and collaborators introduced the concept of Multilevel Functional Principal Component (MFPCA) for functional data observed at multiple visits (e.g., electroencephalograms at every 30 seconds during sleep at two visits several years apart). They developed and deployed the `refund::mfpca.sc()` function. A much improved version of the software was deployed recently in the `refund::mfpca.face()` function based on a paper led by Cui and Ruonan Li [58]. Much work has been dedicated to extending ideas to structured functional data [272, 273], led by Haochang Shou, longitudinal functional data [109], led by Greven, and ultra-high dimensional data [345, 346], led by Vadim Zipunnikov. Many others have provided contributions, including Ana-Maria Staicu, Goldsmith, and Lei Huang. Fast methods for fixed effects inference in this context were developed, among others, by Staicu [223] and Cui [57]. These methods required specialized software to deal with the size and complexity of new data sets. This work forms the basis of Chapter 8 of this book.

As we were writing this book we realized just how many open problems still remain. Some of these problems have been addressed along the way; some are still left open. In the end, we have tried to provide a set of coherent analytic tools based on statistical principled approaches. The core set of ideas is to model functional coefficients parametrically or nonparametrically using splines, penalize the spline coefficients, and conduct inference in the resulting mixed effects model. The book is accompanied by detailed software and a website http://www.FunctionalDataAnalysis.com that will continue to be updated.

We hope that you enjoy reading this book as much as we enjoyed writing it.

1

Basic Concepts

Our goal is to create the most useful book for the widest possible audience without theoretical, methodological, or computational compromise.

Our approach to statistics is to identify important scientific problems and meaningfully contribute to solving them through timely engagement with data. The development of general-purpose methodology is motivated by this process, and must be accompanied by computational tools that facilitate reproducibility and transparency. This "problem forward" approach is critical as technological advances rapidly increase the precision and volume of traditional measurements, produce completely new types of measurements, and open new areas of scientific research.

Our experience in public health and medical research provides numerous examples of new technologies that reshape scientific questions. For example, heart rate and blood pressure used to be measured once a year during an annual medical exam. Wearable devices can now measure them continuously, including during the night, for weeks or months at the time. The resulting data provide insights into blood pressure, hypertension, and health outcomes and open completely new areas of research. New types of measurements are continuously emerging, including physical activity measured by accelerometers, brain imaging, ecological momentary assessments (EMA) via smart phone apps, daily infection and mortality during the COVID-19 pandemic, or CD4 counts from the time of sero-conversion. These examples and many others involve measurements of a continuous underlying process, and benefit from a functional data perspective.

1.1   Introduction

Functional Data Analysis (FDA) provides a conceptual framework for analyzing functions instead of or in addition to scalar measurements. For example, physical activity is a continuous process over the course of the day and can be observed for each individual; FDA considers the complete physical activity trajectory in the analysis instead of reducing it to a single scalar summary, such as the total daily activity. In this book we denote the observed functions by $W_i : S \to \mathbb{R}$, where S is an interval (e.g., $[0,1]$ in $\mathbb{R}$ or $[0,1]^M$ in $\mathbb{R}^M$), i is the basic experimental unit (e.g., study participant), and $W_i(s)$ is the functional observation for unit i at $s \in S$. In general, the domain S does not need to be an interval, but for the purposes of this book we will work under this assumption.

We often assume that $W_i(s) = X_i(s) + \epsilon_i(s)$, where $X_i : S \to \mathbb{R}$ is the true functional process and $\epsilon_i(s)$ are independent noise variables. We will see various generalizations of this definition, but for illustration purposes we use this notation. We briefly summarize the properties of functional data that can be used to better target the associated analytic methods:

- Continuity is the property of the observed functions, $W_i(s)$, and true functional processes, $X_i(s)$, which allows it to be sampled at a higher or lower resolution within S.

- Ordering is the property of the functional domain, S, which can be ordered and has a distance.

- Self-consistency is the property of the observed functions, $W_i(s)$, and true functional processes, $X_i(s)$, to be on the same scale and have the same interpretation for all experimental units, i, and functional arguments, s.

- Smoothness is the property of the true functional process, $X_i(s)$, which is not expected to change substantially for small changes in the functional argument, s.

- Colocalization is the property of the functional argument, s, which has the same interpretation for all observed functions, $W_i(s)$, and true functional processes, $X_i(s)$.

These properties differentiate functional from multivariate data. As the functional argument, $s \in S$, is often time or space, FDA can be used for modeling temporal and/or spatial processes. However, there is a fundamental difference between FDA and spatio-temporal processes. Indeed, FDA assumes that the observed functions, $W_i(s)$, and true functional processes, $X_i(s)$, depend on and are indexed by the experimental unit i. This means that there are many repetitions of the time series or spatial processes, which is not the case for time series or spatial analysis.

The FDA framework serves to guide methods development, interpretation, and exploratory analysis. We emphasize that the concept of continuously observed functions differs from the practical reality that functions are observed over discrete grids that can be dense or sparse, regularly spaced or irregular, and common or unique across functional observations. Put differently, in practice, *functional data are multivariate data with specific properties*. Tools for understanding functional data must bridge the conceptual and practical to produce useful insights that reflect the data-generating and observation processes.

FDA has a long and rich tradition. Its beginnings can be traced at least to a paper by C.R. Rao [247], who proposed to use Principal Component Analysis (PCA), a multivariate method, to analyze growth curves. Several monographs on FDA already exist, including [86, 153, 242, 245]. In addition, several survey papers provide insights into current developments [154, 205, 250, 299, 307]. This book is designed to complement the existing literature by focusing on methods that (1) combine parametric, nonparametric, and mixed effects components; (2) provide statistically principled approaches for estimation and inference; (3) allow users to seamlessly add or remove model components; (4) are associated with high-quality, fast, and easy-to-modify R software; and (5) are intuitive and friendly to scientific applications.

This book provides an introduction to FDA with R [240]. Two packages will be used throughout the book: (1) **refund** [105], which contains a large number of FDA models and many of the data sets used for illustration in this book; and (2) **mgcv** [317, 319], a powerful inferential software developed for semiparametric inference. We will show how this software, originally developed for semiparametric regression, can be adapted to FDA. This is a crucial contribution of the book, which is built around the idea of providing tools that can be readily used in practice. The book is accompanied by the web page http://www.FunctionalDataAnalysis.com, which contains vignettes and R software for each chapter of this book. All vignettes use the **refund** and **mgcv** packages, which are available from CRAN and can be loaded into R [240] as follows.

```
library(refund)
library(mgcv)
```

General-purpose, stable, and fast software is the key to increasing the popularity of FDA methods. The book will present the current version of the software, while acknowledging that software is changing much faster than methodology. Thus, the book will change slowly, while the web page http://www.FunctionalDataAnalysis.com and accompanying vignettes will be adapted to the latest developments.

1.2   Examples

We now introduce several examples that illustrate the ubiquity and complexity of functional data in modern research, and that will be revisited throughout the book. These examples highlight various types of functional data sampling, including dense, regularly-spaced grids that are common across participants, and sparse, irregular observations for each participant.

1.2.1   NHANES 2011–2014 Accelerometry Data

The National Health and Nutrition Examination Survey (NHANES) is a large, ongoing, cross-sectional study of the non-institutionalized US population conducted by the Centers for Disease Control and Prevention (CDC) in two-year waves using a multi-stage stratified sampling scheme. NHANES collects a vast array of demographic, socioeconomic, lifestyle and medical data, though the exact data collected and population samples vary from year to year. The wrist-worn accelerometry data collected in the NHANES 2011–2012 and 2013–2014 waves were released in December 2020. This data set is of particular interest because (1) it is publicly available and linked to the National Death Index (NDI) by the National Center for Health Statistics (NCHS); (2) it was collected from the wrist and processed into "monitor-independent movement summary" (MIMS) units [138] using an open source, reproducible algorithm (`https://bit.ly/3cDnRBF`); and (3) the protocol required 24-hour continuous wear of the wrist accelerometers, including during sleep, for multiple days for each study participant.

In total there were 14,693 study participants who agreed to wear an accelerometer. To ensure the quality of the accelerometry data for each subject, we excluded study participants who had less than three good days (2,083 study participants excluded), where a good day is defined as having at least 95% "good data." "Good data" is defined as data that was flagged as "wear" (`PAXPREDM` $\in \{1, 2, 4\}$) and did not have a quality problem flag (`PAXFLGSM = ""`) in the NHANES data set. The final data set has 12,610 study participants with an average age of 36.90 years and 51.23% females. Note that the variable name "gender" used in the data set and elsewhere is taken directly from the framing of the questions in NHANES, and is not intended to conflate sex and gender. The proportions of Non-Hispanic White, Non-Hispanic Black, Non-Hispanic Asian, Mexican American, Other Hispanic and Other Race were 35.17%, 24.81%, 11.01%, 15.16%, 9.87%, and 3.98%, respectively. The data set is too large to be made available through `refund`, but it is available from the website http://www.FunctionalDataAnalysis.com associated with this book.

Figure 1.1 displays objective physical activity data measured in MIMS for three study participants in the NHANES 2011–2014. Each panel column corresponds to one study participant and each panel row corresponds to a day of the week. The first study participant (labeled `SEQN 75111`) had seven days of data labeled `Sunday` through `Saturday`. The second study participant (labeled `SEQN 77936`) had five days of data labeled `Tuesday` through `Saturday`. The third study participant (labeled `SEQN 82410`) had six days of data that included all days of the week except `Friday`. This happened because the data recorded on

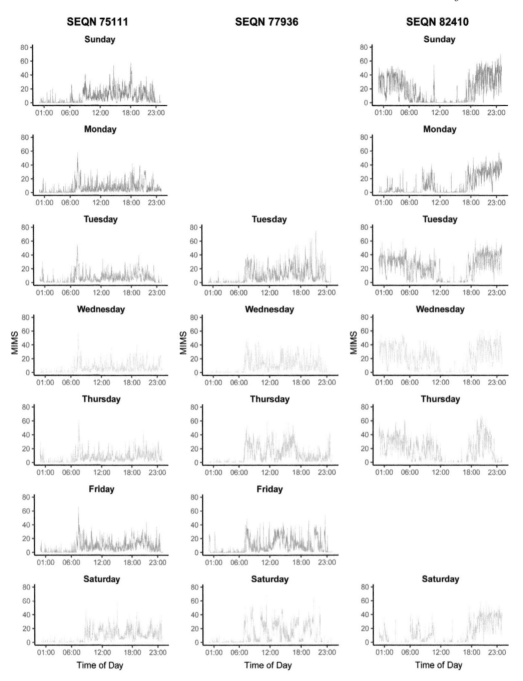

FIGURE 1.1: Physical activity data measured in MIMS for three study participants in the NHANES 2011–2014 summarized at every minute of the day. Each study participant is shown in one column and each row corresponds to a day of the week. The x-axis in each panel is time in one-minute increments from midnight to midnight.

Friday had less than 95% of "good data" and were therefore excluded. The x-axis for each panel is time in one-minute increments from midnight (beginning of the day) to midnight (end of the day). The y-axis is MIMS, a measure of physical activity intensity.

Some features of the data become apparent during visual inspection of Figure 1.1. First, activity during the night (0–6 AM) is reduced for the first two study participants, but not for the third. Indeed, study participant SEQN 82410 has clearly more activity during the night than during the day (note the consistent dip in activity between 12 PM and 6 PM). Second, there is substantial heterogeneity of the data from one minute to another both within and between days. Third, data are positive and exhibit substantial skewness. Fourth, the patterns of activity of study participant SEQN 75111 on Saturday and Sunday are quite different from their pattern of activity on the other days of the week. Fifth, there seems to be some day-to-day within-individual consistency of observations.

Having multiple days of minute-level physical activity for the same individual increases the complexity and size of the data. A potential solution is to take averages at the same time of the day within study participants. This is equivalent to averaging the curves in Figure 1.1 by column at the same time of the day. This reduces the data to one function per study participant, but ignores the visit-to-visit variability around the person-specific mean.

To illustrate the population-level data structure, Figure 1.2 displays the smooth means of several groups within NHANES. Data were smoothed for visualization purposes; technical details on smoothing are discussed in Section 2.3. The left panel displays the average physical activity data for individuals who died (blue line) and survived (red line). Mortality indicators were based on the NHANES mortality release file that included events up to December 31, 2019. Mortality information was available for 8,713 of the 12,610 study participants. There were 832 deceased individuals and 7,881 who were still alive on December 31, 2019. The plot indicates that individuals who did not die had, on average, higher physical activity throughout the day, with larger differences between 8 AM and 11 PM. This result is consistent with the published literature on the association between physical activity and mortality; see, for example, [64, 65, 136, 170, 259, 275, 292].

The right panel in Figure 1.2 displays the smooth average curves for groups stratified by age and gender. For illustration purposes, four age groups (in years) were used and identified

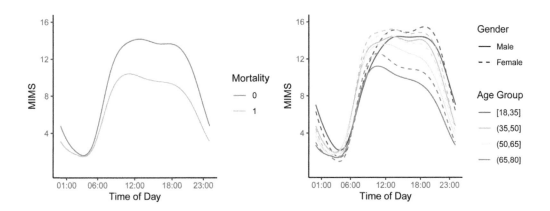

FIGURE 1.2: Average physical activity data (expressed in MIMS) in NHANES 2011–2014 as a function of the minute of the day in different groups. Left panel: deceased (blue line) and alive individuals (red line) as of December 31, 2019. Right panel: females (dashed lines) and males (solid lines) within age groups [18, 35] (red), (35, 50] (orange), (50, 65] (light blue), and (65, 80] (dark blue).

by a different color: $[18, 35]$ (red), $(35, 50]$ (orange), $(50, 65]$ (light blue), and $(65, 80]$ (dark blue). Within each age group, data for females is shown as dashed lines and for males as solid lines. In all subgroups physical activity averages are lower at night, increase sharply in the morning and remain high during the day. The average for the $(50, 65]$ and $(65, 80]$ age groups exhibit a steady decrease during the day. This pattern is not apparent in the younger age groups. These findings are consistent with the activity patterns described in [265, 327]. In addition, for every age group, the average activity during the day is higher for females compared to males. During the night, females have the same or slightly lower activity than males. These results contradict the widely cited literature [296] which indicated that "Males are more physically active than females." However, they are consistent with [327], which found that women are more active than men, especially among older individuals.

Rich, complex data as displayed in Figures 1.1 and 1.2 suggest multiple scientific problems, including (1) quantifying the association between physical activity patterns and health outcomes (e.g., prevalent diabetes or stroke) with or without adjustment for other covariates (e.g., age, gender, body mass index); (2) identifying which specific components of physical activity data are most predictive of future health outcomes (e.g., incident mortality or cardiovascular events); (3) visualizing the directions of variation in the data; (4) investigating whether clusters exist and if they are scientifically meaningful; (5) evaluating transformations of the data that may provide complementary information; (6) developing prediction methods for missing observations (e.g., one hour of missing data for a person); (7) quantifying whether the timing or fragmentation of physical activity provides additional information above and beyond summary statistics (e.g., mean, standard deviation over the day); (8) studying how much data are needed to identify a particular study participant; (9) predicting the activity for the rest of the day given data up to a particular time and day (e.g., 12 PM on `Sunday`); (10) determining what levels of data aggregation (e.g., minute, hour, day) may be most useful for specific scientific questions; and (11) proposing data generating mechanisms that could produce data similar to the observed data.

The daily physical activity curves have all the properties that define functional data: continuity, ordering, self-consistency, smoothness, and colocalization. The measured process is continuous, as physical activity is continuous. While MIMS were summarized at the minute level, data aggregation could have been done at a finer (e.g., ten-, or one-second intervals) or coarser (e.g., one- or two-hour intervals) scale. The functional data have the ordering property, because the functional argument is time during the day, which is both ordered and has a well-defined distance. The data and the measured process have the self-consistency property because all observations are expressed in MIMS at the minute level. The true functional process can be assumed to have the smoothness property, as one does not expect physical activity to change substantially over short periods of time (e.g., one second). The functional argument has the colocalization property, as the time when physical activity is measured (e.g., 12:00 PM) has the same interpretation for every study participant and day of measurement.

The observed data can be denoted as a function $W_{im} : S \to \mathbb{R}_+$, where $W_{im}(s)$ is the MIMS measurement at minute $s \in S = \{1, \dots, 1440\}$ and day $m = 1, \dots, M_i$, where M_i is the number of days with high-quality physical activity data for study participant i. Data complexity could be reduced by taking the average $W_i(s) = \frac{1}{M_i} \sum_{m=1}^{M_i} W_{im}(s)$ at every minute s or the average over days and minutes $\overline{W}_i = \frac{1}{M_i|S|} \sum_{s=1}^{|S|} \sum_{m=1}^{M_i} W_{im}(s)$, where $|S|$ denotes the number of elements in the domain S. Such reductions in complexity improve interpretability and make analyses easier, though some information may be lost. Deciding at what level to summarize the data without losing crucial information is an important goal of FDA.

Here we have identified the domain of the functions $W_{im}(\cdot)$ as $S = \{1, \ldots, 1440\}$, which is a finite set in $\mathbb{R}$ and does not satisfy the basic requirement that S is an interval. This could be a major limitation as basic concepts such as continuity or smoothness of the functions cannot be defined on the sampling domain $S = \{1, \ldots, 1440\}$. This is due to practical limitations of sampling that can only be done at a finite number of points. Here the theoretical domain is $[0, 1440]$ minutes, or $[0, 24]$ hours, or $[0, 1]$ days, depending on how we normalize the domain. Recall that the functions have the continuity property, which assumes that the function could be measured anywhere within this theoretical domain. While not formally correct, we will refer to both of these domains as S to simplify the exposition; whenever necessary we will indicate more precisely when we refer to the theoretical (e.g., $S = [0, 1440]$) or sampling (e.g., $S = \{1, \ldots, 1440\}$) domain. This slight abuse of notation will be used throughout the book and clarifications will be added, as needed.

1.2.2   COVID-19 US Mortality Data

COVID-19 is an infectious disease caused by the SARS-Cov-2 virus that was first identified in Wuhan, China in 2019. The virus spreads primarily via airborne mechanisms. In COVID-19, "CO" stands for corona, "VI" for virus, "D" for disease, and 19 for 2019, the first year the virus was identified in humans. According to the World Health Organization, COVID-19 has become a world pandemic with more than 767 million confirmed infections and almost 7 million confirmed deaths in virtually every country of the world by June 6, 2023 (`https://covid19.who.int/`). Here we focus on mortality data collected in the US before and during the pandemic. The COVID-19 data used in this book can be loaded using the following lines of code.

```
#Load refund
library(refund)
#Load the COVID-19 data
data(COVID19)
```

Among other variables, this data set contains the US weekly number of all-cause deaths, weekly number of deaths due to COVID-19 (as assessed on the death certificate), and population size in the 50 US states plus Puerto Rico and District of Columbia as of July 1, 2020. Figure 1.3 displays the total weekly number of deaths in the US between the week ending on January 14, 2017 and the week ending on December 12, 2020 for a total of 205 weeks. The original data source is the National Center for Health Statistics (NCHS) and the data set link is called National and State Estimates of Excess Deaths. It can be accessed from `https://bit.ly/3wjMQBY`. The file can be downloaded directly from `https://bit.ly/3pMAAaA`. The data stored in the `COVID19` data set in the `refund` package contains an analytic version of these data as the variable `US_weekly_mort`.

In Figure 1.3, each dot corresponds to one week and the number of deaths is expressed in thousands. For example, there were 61,114 deaths in the US in the week ending on January 14, 2017. Here we are interested in excess mortality in the first 52 weeks of 2020 compared to the first 52 weeks of 2019. The first week of 2020 is the one ending on January 4, 2020 and the 52nd week is the one ending on December 26, 2020. There were 3,348,951 total deaths in the US in the first 52 weeks of 2020 (red shaded area in Figure 1.3) and 2,852,747 deaths in the first 52 weeks of 2019 (blue shaded area in Figure 1.3). Thus, there were 496,204 more deaths in the US in the first 52 weeks of 2020 than in the first 52 weeks of 2019. This is called the (raw) excess mortality in the first 52 weeks of the year. Here we use this intuitive definition (number of deaths in 2020 minus the number of deaths in 2019), though slightly different definitions can be used. Indeed, note that the population size increases from 2019 to 2020 and some additional deaths can be due to the increase in

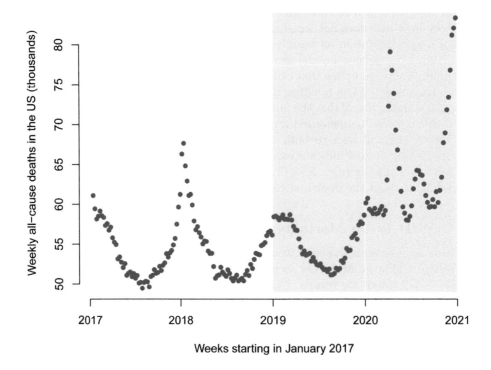

FIGURE 1.3: Total weekly number of deaths in the US between January 14, 2017 and December 12, 2020. The COVID-19 epidemic is thought to have started in the US sometime between January and March 2020.

population. For example, the US population was 330,024,493 on December 26, 2020 and 329,147,064 on December 26, 2019 for an increase of 877,429. Using the mortality rate in 2019 of 0.0087 (number of deaths divided by the total population), the expected increase in number of deaths due to increase in the population would be 7,634. Thus, the number of deaths associated with the natural increase in population is about 1.5% of the total excess all-cause deaths in 2020 compared to 2019.

Figure 1.3 displays a higher mortality peak at the end of 2017 and beginning of 2018, which is likely due to a severe flu season. The CDC estimates that in the 2017–2018 flu season in the US there were "an estimated 35.5 million people getting sick with influenza, 16.5 million people going to a health care provider for their illness, 490,600 hospitalizations, and 34,200 deaths from influenza" (https://bit.ly/3H8fa1b).

As indicated in Figure 1.3, the excess mortality can be calculated for every week from the beginning of 2020. The blue dots in Figure 1.4 display this weekly excess all-cause mortality as a function of time from January 2020. Excess mortality is positive in every week with an average of 9,542 excess deaths per week for a total of 496,204 excess deaths in the first 52 weeks. Excess mortality is not a constant function over the year. For example, there were an average of 1,066 all-cause excess deaths per week between January 1, 2020 and March 14, 2020. In contrast, there were an average of 14,948 all-cause excess deaths per week between March 28, 2020 and June 23, 2020.

One of the most watched indicators of the severity of the pandemic in the US was the number of deaths attributed to COVID-19. The data is made available by the US Center for Disease Control and Prevention (CDC) and can be downloaded directly from

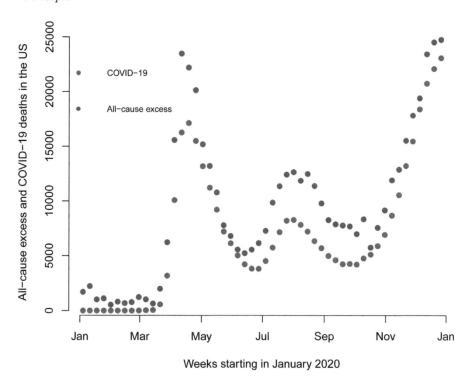

FIGURE 1.4: Total weekly number of deaths attributed to COVID-19 and excess mortality in the US. The x-axis is time expressed in weeks from the first week in 2020. Red dots correspond to weekly number of deaths attributed to COVID-19. Blue dots indicate the difference in the total number of deaths between a particular week in 2020 and the corresponding week in 2019.

https://bit.ly/3iE2xjo. The data stored in the COVID19 data set in the refund package contains an analytic version of these data as the variable US_weekly_mort_CV19. The red dots in Figure 1.4 represent the weekly mortality attributed to COVID-19 according to the death certificate. Visually, COVID-19 and all-cause excess mortality have a similar pattern during the year with some important differences: (1) all-cause excess mortality is larger than COVID-19 mortality every week; (2) the main association does not seem to be delayed (lagged) in either direction; and (3) the difference between all-cause excess and COVID-19 mortality as a proportion of COVID-19 mortality is highest in the summer.

Figure 1.4 indicates that there were more excess deaths than COVID-19 attributed deaths in each week of 2020. In fact, the total US all-cause excess deaths in the first 52 weeks of 2020 was 496,204 compared to 365,122 deaths attributed to COVID-19. The difference is 131,082 deaths, or 35.9% more excess deaths than COVID-19 attributed deaths. So, what are some potential sources for this discrepancy? In some cases, viral infection did occur and caused death, though the primary cause of death was recorded as something else (e.g., cardiac or pulmonary failure). This could happen if death occurred after the infection had already passed, infection was present and not detected, or infection was present but not adjudicated as the primary cause of death. In other cases, viral infection did not occur, but the person died due to mental or physical health stresses, isolation, or deferred health care. There could also be other reasons that are not immediately apparent.

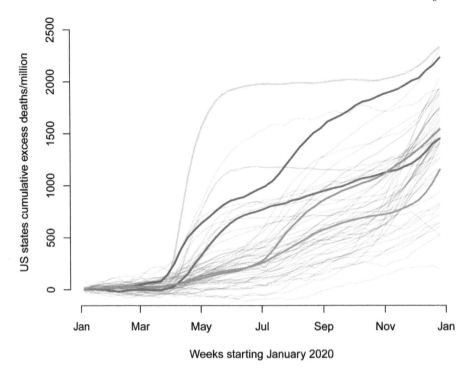

FIGURE 1.5: Each line represents the cumulative weekly all-cause excess mortality per million for each US state plus Puerto Rico and District of Columbia. Five states are emphasized: New Jersey (green), Louisiana (red), Maryland (blue), Texas (salmon), and California (plum).

In addition to data aggregated at the US national level, the `COVID19` data contains similar data for each state and two territories, Puerto Rico and Washington DC. The all-cause weekly excess mortality data for each state in the US is stored as the variable `States_excess_mortality` in the `COVID19` data set.

Figure 1.5 displays the total cumulative all-cause excess mortality per million in every state in the US, Puerto Rico and District of Columbia. For each state, the weekly excess mortality was obtained as described for the US in Figures 1.3 and 1.4. For every week, the cumulative excess mortality was calculated by adding the excess mortality for every week up to and including the current week. To make data comparable across states, cumulative excess mortality was then divided by the estimated population of the state or territory on July 1, 2020 and multiplied by 1,000,000. Every line represents a state or territory with the trajectory for five states being emphasized: New Jersey (green), Louisiana (red), Maryland (blue), Texas (salmon), and California (plum). For example, New Jersey had 1,916 excess all-cause deaths per one million residents by April 30, 2020. This corresponds to a total of 17,019 excess all-cause deaths by April 30, 2020 because the population of New Jersey was 8,882,371 as of July 1, 2020 (the reference date for the population size).

The trajectories for individual states exhibit substantial heterogeneity. For example, New Jersey had the largest number of excess deaths per million in the US. Most of these excess deaths were accumulated in the April–June period, with fewer between June and November, and another increase in December. In contrast, California had a much lower cumulative excess number of deaths per million, with a roughly constant increase during

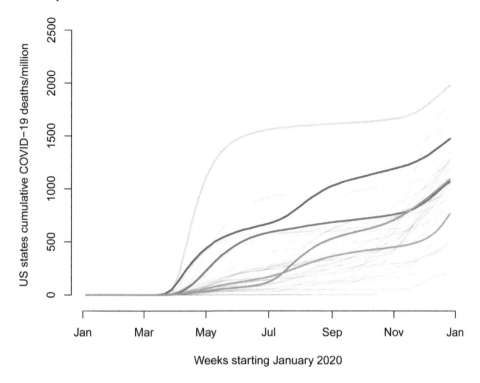

FIGURE 1.6: Each line represents the cumulative COVID-19 mortality for each US state plus Puerto Rico and District of Columbia in 2020. Cumulative means that numbers are added as weeks go by. Five states are emphasized: New Jersey (green), Louisiana (red), Maryland (blue), Texas (salmon), and California (plum).

2020. Maryland had about a third the number of excess deaths per million of New Jersey at the end of June and about half by the end of December.

We now investigate the number of weekly deaths attributed to COVID-19 for each state in the US, which is stored as the variable States_CV19_mortality in the COVID19 data set. Figure 1.6 is similar to Figure 1.5, but it displays the cumulative number of deaths attributed to COVID-19 for each state per million residents. Each line corresponds to a state and a few states are emphasized using the same color scheme as in Figure 1.5. The y-axis was kept the same as in Figure 1.5 to illustrate that, in general, the number of cumulative COVID-19 deaths tends to be lower than the excess all-cause mortality. However, the main patterns exhibit substantial similarities.

There are many scientific and methodological problems that occur from such a data set. Here are a few examples: (1) quantifying the all-cause and COVID-19 mortality at the state level as a function of time; (2) identifying whether the observed trajectories are affected by geography, population characteristics, weather, mitigating interventions, or intervention compliance; (3) investigating whether the strength of the association between reported COVID-19 and all-cause excess mortality varies with time; (4) identifying which states are the largest contributors to the observed excess mortality in the January–March period; (5) quantifying the main directions of variation and clusters of state-specific mortality patterns; (6) evaluating the distribution of the difference between all-cause excess and COVID-19 deaths as a function of state and time; (7) predicting the number of COVID-19 deaths and infections based on the excess number of deaths; (8) evaluating dynamic prediction models for mortality trajectories; (9) comparing different data transformations for analysis, visualization, and communication of results; and (10) using data from countries with good health statistics systems to estimate the burden of COVID-19 in other countries using all-cause excess mortality.

In the COVID-19 example it is not immediately clear that data could be viewed as functional. However, the partitioning of the data by state suggests that such an approach could be useful, at least for visualization purposes. Note that data in Figures 1.5 and 1.6 are curves evaluated at every week of 2020. Thus, the measured process is continuous, as observations could have been taken at a much finer (e.g., days or hours) or coarser (e.g., every month) time scale. Data are ordered by calendar time and is self-consistent because the number or proportion of deaths has the same interpretation for each state and every week. Moreover, one can assume that the true number of deaths is a smooth process as the number of deaths is not expected to change substantially for small changes in time (e.g., one hour). Data are also colocalized, as calendar time has the same interpretation for each state and territory.

The observed data can be denoted as functions $W_{im} : S \to \mathbb{R}_+$, where $W_{im}(s)$ is the number or cumulative number of deaths in state i per one million residents at time $s \in S = \{1, \ldots, 52\}$. Here $m \in 1, 2$ denotes all-cause excess mortality ($m = 1$) and COVID-19 attributed mortality ($m = 2$), respectively. Because each m refers to different types of measurements on the same unit (in this case, US state), this type of data is referred to as "multivariate" functional data. Observations can be modeled as scalars by focusing, for example, on $W_{im}(s)$ at one s at a time or on the average of $W_{im}(s)$ over s for one m. FDA focuses on analyzing the entire function or combination of functions, extracting information using fewer assumptions, and suggesting functional summaries that may not be immediately evident. Most importantly, FDA provides techniques for data visualization and exploratory data analysis (EDA) in the original or a transformed data space.

Just as in the case of NHANES physical activity data, the domain of the functions $W_{im}(\cdot)$ is $S = \{1, \ldots, 52\}$ expressed in weeks, which is a finite set that is not an interval. This is due to practical limitations of sampling that can only be done at a finite number of points. Here the theoretical domain is $[0, 52]$ weeks, or $[0, 12]$ months, or $[0, 1]$ years, depending on how we normalize the domain. Recall that the functions have the continuity property, which assumes that the function could be measured anywhere within this theoretical domain. While not formally correct, we will refer to both of these domains as S to simplify the exposition.

1.2.3 CD4 Counts Data

Human immune deficiency virus (HIV) attacks CD4 cells, which are an essential part of the human immune system. This reduces the concentration of CD4 cells in the blood, which affects their ability to signal other types of immune cells. Ultimately, this compromises the immune system and substantially reduces the human body's ability to fight off infections. Therefore, the CD4 cell count per milliliter of blood is a widely used measure of HIV progression. The CD4 counts data used in this book can be loaded as follows.

```
#Load refund
library(refund)
#Load the CD4 data
data(cd4)
```

This data contains the CD4 cell counts for 366 HIV infected individuals from the Multicenter AIDS Cohort Study (MACS) [66, 144]. We would like to thank Professor Peter Diggle for making this important de-identified data publicly available on his website and for giving

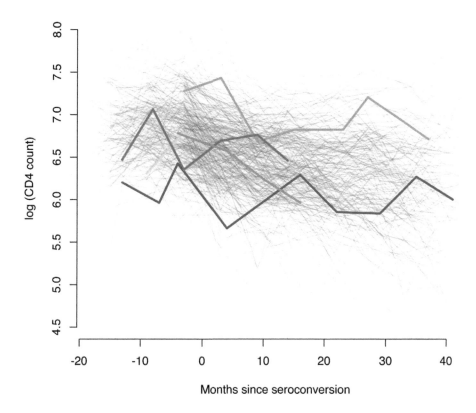

FIGURE 1.7: Each line represents the log CD4 count as a function of month, where month zero corresponds to seroconversion. Five study participants are identified using colors: green, red, blue, salmon, and plum.

us the permission to use it in this book. We would also like to thank the participants in this MACS sub-study. Figure 1.7 displays the log CD4 count for up to 18 months before and 42 months after sero-conversion. Each line represents the log CD4 count for one study participant as a function of month, where month zero corresponds to sero-conversion.

There are a total of 1,888 data points, with between 1 and 11 (median 5) observations per study participant. Five study participants are highlighted using colors: green, red, blue, salmon, and plum. Some of the characteristics of these data include (1) there are few observations per curve; (2) the time of observations is not synchronized across individuals; and (3) there is substantial visit-to-visit variation in log CD4 counts before and after seroconversion.

Figure 1.8 displays the same data as Figure 1.7 together with the raw (cyan dots) and smooth (dark red line) estimator of the mean. The raw mean is the average of log CD4 counts of study participants who had a visit at that time. The raw mean exhibits substantial variation and has a missing observation at time $t = 0$. The smooth mean estimator captures the general shape of the raw estimator, but provides a more interpretable summary. For example, the smooth estimator is relatively constant before seroconversion, declines rapidly in the first 10–15 months after seroconversion, and continues to decline, but much slower after month 15. These characteristics are not immediately apparent in the raw mean or in the person-specific log CD4 trajectories displayed in Figure 1.6.

There are many scientific and methodological problems suggested by the CD4 data set. Here we identify a few: (1) estimating the time-varying mean, standard deviation and

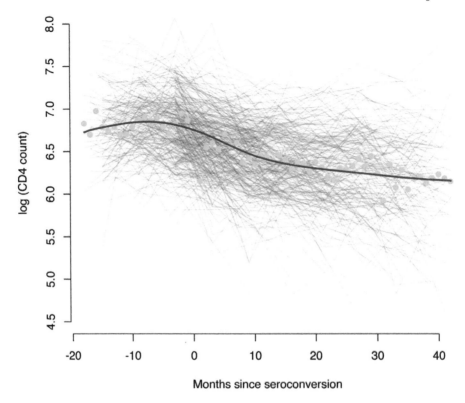

FIGURE 1.8: Each gray line represents the log CD4 count as a function of month, where month zero corresponds to seroconversion. The point-wise raw mean is shown as cyan dots. The smooth estimator of the mean is shown as a dark red line.

quantiles of the log CD4 counts as a function of time; (2) producing confidence intervals for these time-varying population parameters; (3) identifying whether there are specific sub-groups that have different patterns over time; (4) designing analytic methods that work with sparse data (few observations per curve that are not synchronized across individuals); (5) predicting log CD4 observations for each individual at months when measurements were not taken; (6) predicting the future observations for one individual given observations up to a certain point (e.g., 10 months after seroconversion); (7) constructing confidence intervals for these predictions; (8) quantifying the month-to-month measurement error (fluctuations along the long-term trend); (9) studying whether the month-to-month measurement error depends on person-specific characteristics, including average log CD4 count; and (10) de-signing realistic simulation studies that mimic the observed data structure to evaluate the performance of analytic methods.

Data displayed in Figures 1.7 and 1.8 are observed at discrete time points and with substantial visit-to-visit variability. We leave it as an exercise to argue that the CD4 data has the characteristics of functional data: continuity, ordering, self-consistency, smoothness, and colocalization.

The observed data has the structure $\{s_{ij}, W_i(s_{ij})\}$, where $W_i(s_{ij})$ is the log CD4 count at time $s_{ij} \in S = \{-18, -17, \ldots, 42\}$. Here $i = 1, \ldots, n$ is study participant, $j = 1, \ldots, p_i$ is the observation number, and p_i is the number of observations for study participant i. In statistics, this data structure is often encountered in longitudinal studies and is typically modeled using linear mixed effects (LME) models [66, 87, 161, 196]. LMEs use a pre-specified, typically parsimonious, structure of random effects (e.g., random intercepts and slopes) to capture the person-specific curves. Functional data analysis complements LMEs by considering more complex and/or data-dependent designs of random effects [134, 254, 255, 283, 328, 334, 336]. It is worth noting that this data structure and problem are equivalent to the matrix completion problem [29, 30, 214, 312]. Statistical approaches can handle different levels of measurement error in the matrix entries, and produce both point estimators and the associated uncertainty for each matrix entry.

In this example, one could think about the sampling domain as being $S = \{-18, -17, \ldots, 42\}$ expressed in months. This is a finite set that is not an interval. The theoretical domain is $[-18, 42]$ in months from seroconversion, though the interval could be normalized to $[0, 1]$. The difference from the NHANES and COVID-19 data sets is that observations are not available at every point in $S = \{-18, -17, \ldots, 42\}$ for each individual. Indeed, the minimum number of observations per individual is 1 and the maximum is 11, with a median number of observations of 5, which is $100 \times 5/(42 + 19) = 8.2\%$ of the number of possible time points between -18 and 42. This type of data is referred to in statistics as "sparse functional data." In strict mathematical terms this is a misnomer, as the sampling domain $S = \{-18, -17, \ldots, 42\}$ is itself mathematically sparse in $\mathbb{R}$. Here we will use the definition that sparse functional data are observed functions $W_i(s_{ij})$ where $j = 1, \ldots, p_i$, p_i is small (at most 20) at sampling points s_{ij} that are not identical across study participants. Note that this is a property of the observed data $W_i(s_{ij})$ and not of the true underlying process, $X_i(s)$, which could be observed/sampled at any point in $[-18, 42]$. While this definition is imprecise, it should be intuitive enough for the intents and purposes of this book. We acknowledge that there may be other definitions and also that there is a continuum of scientific examples between "dense, equally spaced functional data" and "sparse, unequally spaced functional data."

1.2.4   The CONTENT Child Growth Study

The CONTENT child growth study (referred to in this book as the CONTENT study) was funded by the Sixth Framework Programme of the European Union, Project CONTENT (INCO-DEV-3-032136) and was led by Dr. William Checkley. The study was conducted between May 2007 and February 2011 in Las Pampas de San Juan Miraflores and Nuevo Paraíso, two peri-urban shanty towns with high population density located on the southern edge of Lima city in Peru. The shanty towns had approximately 40,000 residents with 25% of the population under the age of 5 [38, 39]. A simple census was conducted to identify pregnant women and children less than 3 months of age. Eligible newborns and pregnant women were randomly selected from the census and invited to participate in the study. Only one newborn was recruited per household. Written informed consent was required from parents or guardians before enrollment. The study design was that of a longitudinal cohort study with the primary objective to assess if infection with Helicobacter pylori (H. pylori) increases the risk of diarrhea, which, in turn, could adversely affect the growth in children less than 2 years of age [131].

Anthropometric data were obtained longitudinally on 197 children weekly until the child was 3 months of age, every two weeks between three and 11 months of age, and once monthly thereafter for the remainder of follow-up up to age 2. Here we will focus on child length and weight, both measured at the same visits. Even if visits were designed to be equally spaced, they were obtained within different days of each sampling period. For example, the observation on week four for a child could be on day 22 or 25, depending on the availability of the contact person, day of the week, or on the researchers who conducted the visit.

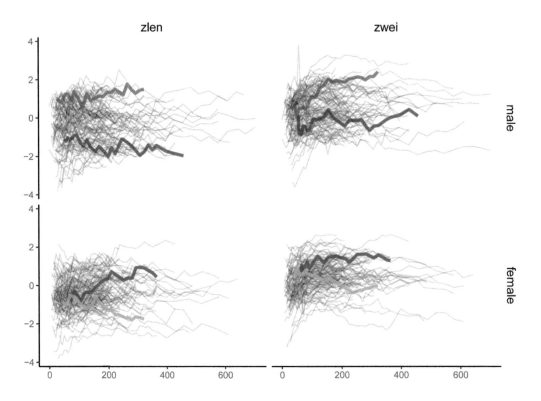

FIGURE 1.9: Longitudinal observations of z-score for length (zlen, first column) and z-score for weight (zwei, second column) shown for males (first row) and females (second row) as a function of day from birth. Data for two boys (shown as light and dark shades of red) and two girls (shown as light and dark shades of blue) are highlighted. The same shade of color identifies the same individual.

Moreover, not all planned visits were completed, which provided the data a quasi-sparse structure, as observations are not temporally synchronized across children.

We would like to thank Dr. William Checkley for making this important de-identified data publicly available and to the members of the communities of Pampas de San Juan de Miraflores and Nuevo Paraíso who participated in this study. The data can be loaded directly using the `refund` R package as follows.

```
#Load refund
library(refund)
#Load the CONTENT data
data(content)
```

Figure 1.9 provides an illustration of the z-score for length (zlen) and z-score for weight (zwei) variables collected in the CONTENT study. Data are also available on the original scale, though for illustration purposes here we display these normalized measures. For example, the zlen measurement is obtained by subtracting the mean and dividing by the standard deviation of height for a given age of children as provided by the World Health Organization (WHO) growth charts.

Even though the study was designed to collect data up to age 2 (24 months), for visualization purposes, observations are displayed only through day 600, as data become very

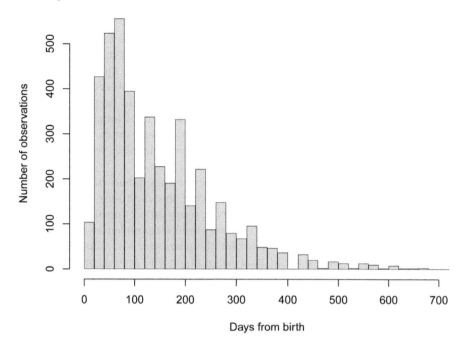

FIGURE 1.10: Histogram of the number of days from birth in the CONTENT study. There are a total of 4,405 observations for 197 children.

sparse thereafter. Data for every individual is shown as a light gray line and four different panels display the zlen (first column) and zwei (second column) variables as a function of day from birth separately for males (first row) and females (second row). Data for two boys is highlighted in the first row of panels in red. The lighter and darker shades of red are used to identify the same individual in the two panels. A similar strategy is used to highlight two girls using lighter and darker shades of blue. Note, for example, that both girls who are highlighted start at about the same length and weight z-score, but their trajectories are substantially different. The z-scores increase for both height and weight for the first girl (data shown in darker blue) and decrease for the second girl (data shown in light blue). Moreover, after day 250 the second girl seems to reverse the downward trend in the z-score for weight, though that does not happen with her z-score for height, which continues to decrease.

These data were analyzed in [127, 169] to dynamically predict the growth patterns of children at any time point given the data up to that particular time. Figure 1.10 displays the histogram of the number of days from birth in the CONTENT study. There are a total of 4,405 observations for 197 children, out of which 2006 (45.5% of total) are in the first 100 days and 3,299 (74.9% of total) are in the first 200 days from birth. Observations become sparser after that, which can also be observed in Figure 1.9.

There are several problems suggested by the CONTENT growth study including (1) estimating the marginal mean, standard deviation and quantiles of anthropometric measurements as a function of time; (2) producing pointwise and joint confidence intervals for these time-varying parameters; (3) identifying whether there are particular subgroups or individuals that have distinct patterns or individual observations; (4) conducting estimation and inference on the individual growth trajectories; (5) quantifying the contemporaneous and lagged correlations between various anthropometric measures; (6) estimating anthropometric measures when observations were missing; (7) predicting future observations for one individual given observations up to a certain point (e.g., 6 months after birth); (8) quantifying the month-to-month measurement error and study whether it is differential among children; (9) developing methods that are designed for multivariate sparse data (few observations per curve) with the amount of sparsity varying along the observation domain; (10) identifying outlying observations or patterns of growth that could be used as early warnings of growth stunting; (11) developing methods for studying the longitudinal association between multivariate growth outcomes and time-dependent exposures, such as infections; and (12) designing realistic simulation scenarios that mimic the observed data structure to evaluate the performance of analytic methods.

Data displayed in Figure 1.9 are observed at discrete time points and with substantial visit-to-visit and participant-to-participant variability. These data have all the characteristics of functional data: continuity, ordering, self-consistency, smoothness, and colocalization. Indeed, data are continuous because growth curves could be sampled at any time point at both higher and lower resolutions. The choice for the particular sampling resolution was a balance between available resources and knowledge about the growth process of humans. Data are also ordered as observations are sampled in time. That is, we know that a measurement at week 3 was taken before a measurement at month 5 and we know exactly how far apart the two measurements were taken. Also, the observed and true functional processes have the self-consistency property as they are expressed in the same units of measurement. For example, height is always measured in centimeters or is transformed into normalized measures, such as zlen. Data are also smooth, as the growth process is expected to be gradual and not have large minute-to-minute or even day-to-day fluctuations. Even potential growth spurts are smooth processes characterized by faster growth but small day-to-day variation. Observations are also colocalized, as the functional argument, time from birth, has the same interpretation for all functions. For example, one month from birth means the same thing for each baby.

The observed functional data in CONTENT has the structure $\{s_{ij}, W_{im}(s_{ij})\}$, where $W_{im} : S \to \mathbb{R}$ is the mth anthropometric measurement at time $s \in S \subset [0, 24]$ (expressed in months from birth) for study participant i. Here the time of the observations, s_{ij}, depends on the study participant, i, and visit number, j, but not the anthropometric measure, m. The reason is that if a visit was completed, all anthropometric measures were collected. However, this may not be the case for all studies and observations may depend on m in other studies. Each such variation on sampling requires special attention and methods development. In this example it is difficult to enumerate the entire sampling domain because it is too large and observations are not equally spaced. One way to obtain this space in R is using the function

```
#Find all unique observations
sampling_S <- sort(unique(content$agedays))
```

A similar notation, $W_{im}(s)$, was used to describe the NHANES data structure in Section 1.2.1. In NHANES m referred to the day number from initiating the accelerometry study. However, in the CONTENT study, m refers to the type of anthropometric measure. Thus, while in NHANES functions indexed by m measure the same thing every day (e.g., physical activity at 12 PM), in CONTENT each function measures something different (e.g., zlen and zwei at month 2). In FDA one typically refers to the NHANES structure as multilevel and to the CONTENT structure as multivariate functional data. Another difference is that data are not equally spaced within individuals and are not synchronized across individuals. Thus, the CONTENT data has a multivariate (multiple types of measurement), functional (has all characteristics of functional data), sparse (few observations per curve that are not synchronized across individuals), and unequally spaced (observations were not taken at equal intervals within study participants). The CONTENT data is highly complex and contains additional time invariant (e.g., sex) and time-varying observations (e.g., bacterial infections).

As the CD4 counts data presented in Section 1.2.3, the CONTENT data is at the interface between traditional linear mixed effects models (LME) and functional data. While both approaches can be used, this is an example when FDA approaches are more reasonable, at least as an exploratory tool to understand the potential hidden complexity of individual trajectories. In these situations, one also starts to question or even test the standard residual dependence assumptions in traditional LMEs. In the end, we will show that every FDA is a form of LME, but this will require some finesse and substantial methodological development.

1.3 Notation and Methodological Challenges

In all examples in Section 1.2, the data are comprised of functions $W_i : S \to \mathbb{R}$, though in the CONTENT example, one could argue that the vector $W_i(\cdot) = \{W_{i1}(\cdot), W_{i2}(\cdot)\}$, where $W_{i1}(\cdot)$ and $W_{i2}(\cdot)\}$ are the z-scores for length and weight, respectively, takes values in $\mathbb{R}^2$. Here, the conceptual and practical framing of functional data should be noted: conceptually, the theoretical domain S (where functional data could be observed) is an interval in $\mathbb{R}$ or $\mathbb{R}^M$; practically, the sampling domain S (where functional data is actually observed) is a finite subset of points of the theoretical domain. We will, at times, be specific about our use of a particular framing, but frequently the distinction can be elided (or at least inferred from context) without detracting from the clarity of our discussion.

Continuity is an important property of functional data, indicating that measurements could, in principle, have been taken at any point in the interval spanned by the sampling domain S. For example, in the NHANES study, data are summarized at every minute of the day, which results in 1,440 observations per day. However, data could be summarized at a much finer or coarser resolution. Thus, the domain of the function is considered to be an interval and, without loss of generality, the $[0, 1]$ interval. In NHANES the start of the day (midnight or 12:00 AM) would correspond to 0, the end of the day (11:59 PM) would correspond to 1 and minute s of the day would correspond to $(s - 1)/1439$.

Most common functional data are of the type $W_i : [0, 1] \to \mathbb{R}$, though many variations exist. An important assumption is that there exists an underlying, true process, $X_i : [0, 1] \to \mathbb{R}$, and $W_i(s)$ provides proxy measurements of $X_i(s)$ at the points where $W_i(\cdot)$ is observed. The observed function is $W_i(s) = X_i(s) + \epsilon_i(s)$, where $\epsilon_i(s)$ are independent noise variables, which could be Gaussian, but could refer to binary, Poisson, or other types of errors.

Thus, FDA assumes that there exists an infinite-dimensional data generating process, $X_i(\cdot)$, for every study participant, while information is accumulated at a finite number of points via the measured process, $W_i(s)$, where $s \in S$ and S is the sampling domain. This inferential problem is addressed by a combination of smoothing and simplifying (modeling) assumptions. The sampling location (s points where $W_i(\cdot)$ are measured), measurement type (exactly what is measured), and underlying signal structure (the distribution of $X_i(\cdot)$) raise important methodological problems that need to be addressed to bridge the theoretical assumption of continuity with the reality of sampling at a finite number of points.

First, connecting the continuity of $X_i(\cdot)$ to the discrete measurement $W_i(\cdot)$ needs to be done through explicit modeling and assumptions.

Second, the density and number of observations at the study participant level could vary substantially. Indeed, there could be as few as two or three to as many as hundreds of millions of observations per study participant. Moreover, observations can be equally or unequally spaced within and between study participants as well as when aggregated across study participants. Each of these scenarios raises its own specific set of challenges.

Third, the complexity of individual and population trajectories is a priori unknown. Extracting information is thus a balancing act between model assumptions and signal structure often in the presence of substantial noise. As shown in the examples in this chapter, functional data are seldom linear and often non-stationary.

Fourth, the covariance structure within experimental units (e.g., study participants) requires a new set of assumptions that cannot be directly extended from traditional statistical models. For example, the independence and exchangeability assumptions from longitudinal data analysis are, at best, suspect in many high-resolution FDA applications. The autoregressive assumption is probably way too restrictive, as well, because it implies stationarity of residuals and an exponential decrease of correlation as a function of distance. Moreover, as sampling points are getting closer together (higher resolution) the structure of correlation may change substantially. The unstructured correlation assumption is more appropriate for FDA, but it requires the estimation of a very large dimensional correlation matrix. This can raise computational challenges for moderate to high-dimensional functions.

Fifth, observed data may be non-Gaussian with high skewness and thicker than normal tails. While much is known about univariate modeling of such data, much more needs to be done when the marginal distributions of functional data exhibit such behavior. Binary or Poisson functional data raise their own specific sets of challenges.

To understand the richness of FDA, one could think of all problems in traditional data analysis where some of the scalar observations are replaced with functional observations. This requires new modeling and computational tools to accommodate the change of all or some measurements from scalars to high-dimensional, highly structured multivariate vectors, matrices or arrays. The goal of this book is to address these problems by providing a class of self-contained, coherent analytic methods that are computationally friendly. To achieve this goal, we need three important components: dimensionality reduction, penalized smoothing, and unified regression modeling via mixed effects models inference. Chapter 2 will introduce these ideas and principles.

1.4   R Data Structures for Functional Observations

As the preceding text makes clear, there is a contrast between the conceptual and practical formulations of functional data: conceptually, functional data are continuous and infinite dimensional, but practically they are observed over discrete grids. This book relies on both formulations to provide interpretable model structures with concrete software implementations. We will use a variety of data structures for the storage, manipulation, and use of functional observations, and discuss these briefly now.

In perhaps the simplest case, functional data are observed over the same equally spaced grid for each participant or unit of observation. Physical activity is measured at each minute of the day for each participant in the NHANES data set, and deaths due to COVID-19 are recorded weekly in each state in the COVID-19 US mortality data. A natural way of representing such data is a matrix in which rows correspond to participants and columns to the grid over which data are observed.

For illustration purposes, we display below the "wide format" data structure of the NHANES physical activity data. This is stored in the variable MIMS of the data nhanes_fda_with_r. This NHANES data consists of a 12,610 × 1,440 matrix, with columns containing MIMS measurements from 12:00 AM to 11:59 PM. Here we approximated the MIMS up to the second decimal for illustration purposes, so the actual data may vary slightly upon closer inspection. This data structure is familiar to many statisticians, and can be useful in the implementation of specific methods, such as Functional Principal Component Analysis (FPCA).

```
#Storage format for the accelerometry data in NHANES data set
nhanes_fda_with_r$MIMS
          MIN0001   MIN0002   MIN0003   MIN0004   ...   MIN1439   MIN1440
  62161    1.11      3.12      1.47      0.94      ...    1.38      1.53
  62163   25.15     19.16     17.84     20.33      ...    7.38     15.93
  62164    1.92      1.67      2.38      0.93      ...    3.03      4.46
  62165    3.98      3.00      1.91      0.89      ...    2.18      0.31
   ...     ...       ...       ...       ...       ...    ...       ...
  83730    1.50      2.11      1.34      0.16      ...    1.07      1.14
  83731    0.09      0.01      0.49      0.10      ...    0.86      0.46
```

It is possible to use matrices for data that are somewhat less simple, although care is required. When data can be observed over the same grid but are sparse for each subject, a matrix with missing entries can be used. For the CD4 data, observations are recorded at months before or after seroconversion. The observation grid is integers from -18 to 42, but any specific participant is measured only at a subset of these values. Data like these can be stored in a relatively sparse matrix, again with rows for study units and columns for elements of the observation grid. Our data examples focus on equally spaced grids, but this is not required for functional data in general or for the use of matrices to store these observations.

For illustration purposes, we display the CD4 count data in the same "wide format" used for NHANES. The structure is similar to that of NHANES data, where each row corresponds to an individual and each column corresponds to a potential sampling point, in this case a month from seroconversion. However, in the CD4 data example most observations are not available, as indicated by the NA fields. Indeed, as we discussed, only 1,888 data points are available out of the $366 \times 61 = 22{,}326$ entries of the matrix, or 8.5%. Having one look at the data matrix and knowing that less than 10% of the entries are known, immediately creates the idea that the matrix and the data are "sparse." Note, however, that this concept refers to the percent of non-missing entries into a matrix and not to the mathematical concept of sparsity. In most of the book, "sparsity" will refer to matrix sparsity and not to the mathematical concept of sparsity of a set.

```
#Storage format for CD4 data in refund
CD4
        -18  -17  -16  -15  -14  -13  -12  -11  -10   -9   ...   41   42
  [1,]   NA   NA   NA   NA   NA   NA   NA   NA   NA   548  ...   NA   NA
  [2,]   NA   NA   NA   NA   NA   NA   NA   NA   NA   NA   ...   NA   NA
  [3,]   NA   NA   NA  846   NA   NA   NA   NA   NA  1102  ...   NA   NA
   ...   ...  ...  ...  ...  ...  ...  ...  ...  ...  ...  ...  ...  ...
[363,]   NA   NA   NA   NA   NA   NA   NA   NA 1661   NA   ...   NA   NA
[364,]   NA   NA   NA  646   NA   NA   NA  882   NA   NA   ...   NA   NA
[365,]   NA   NA   NA   NA   NA   NA   NA   NA   NA   NA   ...  294   NA
[366,]   NA   NA   NA   NA   NA   NA   NA   NA   NA   NA   ...  462   NA
```

Storing the CD4 in wide format is not a problem because the matrix is relatively small and does not take that much memory. However, this format is not efficient and could be extremely cumbersome when data matrices increase both in terms of number of rows or columns. The number of columns can increase very quickly when the observations are irregular across subjects and the union of sampling point across study participants is very large. In the extreme, but commonly encountered, case when no two observations are taken at exactly the same time, the number of columns of the matrix would be equal to the total number of observations for all individuals. Additionally, observation grid values are not directly accessible, and must be stored as column names or in a separate vector.

Using the "long format" for sparse functional data can address some disadvantages that are associated with the "wide format." In particular, a data matrix or frame with columns for study unit ID, observation grid point, and measurement value can be used for dense or sparse data and for regular or irregular observation grids, and makes the observation grid explicit. Below we show the CD4 counts data in "long format," where all the missing data are no longer included. The price to pay is that we add the column ID, which contains many repetitions, while the column `time` also contains some repetitions to explicitly indicate the month where the sample was taken.

The long format of the data is much more memory efficient when data are sparse, though these advantages can disappear or become disadvantages when data become denser. For example, when the observation grid is common across subjects and there are many observations for each study participant, the `ID` and `time` column require substantial additional memory without providing additional information. Long format data may also repeat subject-level covariates for each element of the observation grid, which further exacerbates memory requirements. Moreover, complexity and memory allocation can increase substantially when multiple functional variables are observed on different observation grids. From a practical perspective, different software implementations require different data structures, which can be a reason for frustration. In general `refund` tends to use the wide format of the data, whereas our implementation of FDA in `mgcv` often uses the long format.

```
#CD4 count data in long format
CD4
CD4 count        time        ID
   548            -9          1
   ...            ...         ...
   846           -15          3
  1102            -9          3
   ...            ...         ...
  1661           -10         363
   ...            ...         ...
   646           -15         364
   882           -11         364
   ...            ...         ...
   294            41         365
   ...            ...         ...
   462            41         366
```

Given these considerations, we will use both the wide and long formats and we will discuss when and how we make the transition between these formats. We recognize the increased popularity of the `tidyverse` for visualization and exploratory data analysis, which prefers the long format of the data. Over the last several years, many R users have gravitated toward data frames for data storage. This shift has been facilitated by (and arguably is attributable to) the development of the `tidyverse` collection of packages, which implement general-purpose tools for data manipulation, visualization, and analysis.

The `tidyfun` [261] R package was developed to address issues that arise in the storage, manipulation, and visualization of functional data. Beginning from the conceptual perspective that a complete curve is the basic unit of analysis, `tidyfun` introduces a data type (`"tf"`) that represents and operates on functional data in a way that is analogous to numeric data. This allows functional data to easily sit alongside other (scalar or functional) observations in a data frame in a way that is integrated with a tidyverse-centric approach to manipulation, exploratory analysis, and visualization. Where possible, `tidyfun` conserves memory by avoiding data duplication.

We will use both the `tidyverse` and the `usualverse` (completely made up word) and we will point out the various approaches to handling the data. In the end, it is a personal choice of what tools to use, as long as the main inferential engine works.

One can reasonably ask why a book of methods places such an emphasis on data structures? The reason is that this is a book on "functional data analysis with R" and not a book on "functional data analysis without R." Thus, in addition to methods and inference we emphasize the practical implementation of methods and the combination of data structures, code, and methods that is amenable to software development.

1.5 Notation

Throughout the book we will attempt to use notation that is consistent across chapters. This will not be easy or perfect, as functional data analysis can test the limits of reasonable notation. Indeed, the Latin and Greek alphabet using lower- and uppercase, bold and regular font were heavily tested by the data structures discussed in this book. To provide some order ahead of starting the book in earnest we introduce the following notation.

- n: number of study participants

- i: the index for the study participant, $i = 1, \ldots, n$

- S: the sampling or theoretical domain of the observed functions; this will depend on the context

- Y_i: scalar outcome for study participant i

- $W_i(s_j)$: observed functional measurement for study participant i and location $s_j \in S$, for $j = 1, \ldots, p$ when data are observed on the same grid (dense, equal grid)

- $W_i(s_{ij})$: observed functional measurement for study participant i and location $s_{ij} \in S$, for $j = 1, \ldots, p_i$ when data are observed on different grids across study participants (sparse, different grid)

- $W_{im}(\cdot)$: observed functional measurement for multivariate or multilevel data. For multivariate data $m = 1, \ldots, M$, whereas for multivariate data $m = 1, \ldots, M_i$, though in some instances $M_i = M$ for all i

- $X_i(s_j)$, $X_i(s_{ij})$, $X_{im}(\cdot)$: same as $W_i(s_j)$, $W_i(s_{ij})$, $W_{im}(\cdot)$, but for the underlying, unobserved, functional process

- $\mathbf{Z}_i$: column vector of additional scalar covariates

- vectors: defined as columns and referred to using **bold**, typically lower case, font

- matrices: referred to using **bold**, typically upper case, font

2

Key Methodological Concepts

In this chapter we introduce some of the key methodological concepts that will be used extensively throughout the book. Each method is important in itself, but it is the specific combination of these methods that provides a coherent infrastructure for FDA inference and software development. Understanding the details of each approach is not essential for the application of these methods. Readers who are less interested in a deep dive into these methods and more interested in applying them can skip this chapter for now.

2.1 Dimension Reduction

Consider the case when functional data are of the form $W_{\mathrm{raw},i}(s)$ for $i = 1,\ldots,n$ and $s \in S = \{s_1,\ldots,s_p\}$, where $p = |S|$ is the number of observations in S. Assume that all functions are measured at the same values, s_j, $j = 1,\ldots,p$, and that there are no missing observations. The centered and normalized functional data is

$$W_i(s_j) = \frac{1}{\sqrt{np}}\{W_{\mathrm{raw},i}(s_j) - \overline{W}_{\mathrm{raw}}(s_j)\}\,,$$

where $\overline{W}_{\mathrm{raw}}(s_j) = \frac{1}{n}\sum_{i=1}^{n} W_{\mathrm{raw},i}(s_j)$ is the average of functional observations over study participants at s_j. This transformation is not strictly necessary, but will simplify the connection between the discrete observed measurement process and the theoretical underlying continuous process. In particular, dividing by $\sqrt{np}$ will keep measures of data variation comparable when the number of rows (study participants) or columns (data sampling resolution) change.

The data can be organized in an $n \times p$ dimensional matrix, $\mathbf{W}$, where the ith row contains the observations $\{W_i(s_j) : j = 1,\ldots,p\}$. Each row in $\mathbf{W}$ corresponds to a study participant and each column has mean zero. The dimension of the problem refers to p and dimension reduction refers to finding a smaller set of functions, $K_0 < p$, that contains most of the information in the functions $\{W_i(s_j) : j = 1,\ldots,p\}$.

There are many approaches to dimension reduction. Here we focus on two closely related techniques: Singular Value Decomposition (SVD) and Principal Component Analysis (PCA). While the linear algebra will get slightly involved, SVD and PCA are essential analytic tools for high-dimensional FDA. Moreover, the SVD and PCA of any $n \times p$ dimensional matrix can easily be computed in R [240] as described below.

```
#SVD of matrix W
SVD_of_W <- svd(W)
#PCA of matrix W
PCA_of_W <- princomp(W)
```

2.1.1 The Linear Algebra of SVD

The SVD of $\mathbf{W}$ is the decomposition $\mathbf{W} = \mathbf{U}\boldsymbol{\Sigma}\mathbf{V}^t$, where $\mathbf{U}$ is an $n \times n$ dimensional matrix with the property $\mathbf{U}^t\mathbf{U} = \mathbf{I}_n$, $\boldsymbol{\Sigma}$ is an $n \times p$ dimensional diagonal matrix, and $\mathbf{V}$ is a $p \times p$ dimensional matrix with the property $\mathbf{V}^t\mathbf{V} = \mathbf{I}_p$. Here $\mathbf{I}_n$ and $\mathbf{I}_p$ are the identity matrices of size n and p, respectively. The diagonal entries d_k of $\boldsymbol{\Sigma}$, $k = 1, \ldots, K = \min(n, p)$, are called the singular values of $\mathbf{W}$. The columns of $\mathbf{U}$, $\mathbf{u}_k = \{u_{ik} : i = 1, \ldots, n\}$, and $\mathbf{V}$, $\mathbf{v}_k = \{v_k(s_j) : j = 1, \ldots, p\}$, for $k = 1, \ldots, K$ are the left and right singular vectors of $\mathbf{W}$, respectively. The matrix form of the SVD decomposition can be written in entry-wise form for every $s \in S$ as

$$W_i(s) = \sum_{k=1}^{K} d_k u_{ik} v_k(s) \ . \tag{2.1}$$

This provides an explicit linear decomposition of the data in terms of the functions, $\{v_k(s_j) : j = 1, \ldots, p\}$, which are the columns of $\mathbf{V}$ and form an orthonormal basis in $\mathbb{R}^p$. These right singular vectors are often referred to as the main directions of variation in the functional space. Because $\mathbf{v}_k$ are orthonormal, the coefficients of this decomposition can be obtained as

$$d_k u_{ik} = \sum_{j=1}^{p} W_i(s_j) v_k(s_j) \ .$$

Thus, $d_k u_{ik}$ is the inner product between the ith row of $\mathbf{W}$ (the data for study participant i) and the kth column of $\mathbf{V}$ (the kth principal direction of variation in functional space).

We will show that $\{d_k^2 : k = 1, \ldots, K\}$ quantify the variability of the observed data explained by the vectors $\{v_k(s_j) : j = 1, \ldots, p\}$ for $k = 1, \ldots, K$. The total variance of the original data is

$$\frac{1}{np} \sum_{i=1}^{n} \sum_{j=1}^{p} \{W_{\mathrm{raw},i}(s_j) - \overline{W}_{\mathrm{raw}}(s_j)\}^2 = \sum_{i=1}^{n} \sum_{j=1}^{p} W_i^2(s_j) \ ,$$

which is equal to $\mathrm{tr}(\mathbf{W}^t\mathbf{W}) = \mathrm{tr}(\mathbf{V}\boldsymbol{\Sigma}^t\mathbf{U}^t\mathbf{U}\boldsymbol{\Sigma}\mathbf{V}^t)$, where $\mathrm{tr}(\mathbf{A})$ denotes the trace of matrix $\mathbf{A}$. As $\mathbf{U}^t\mathbf{U} = \mathbf{I}_n$, $\mathrm{tr}(\mathbf{W}^t\mathbf{W}) = \mathrm{tr}(\mathbf{V}\boldsymbol{\Sigma}^t\boldsymbol{\Sigma}\mathbf{V}^t) = \mathrm{tr}(\boldsymbol{\Sigma}^t\boldsymbol{\Sigma}\mathbf{V}^t\mathbf{V})$, where we used the property that $\mathrm{tr}(\mathbf{AB}) = \mathrm{tr}(\mathbf{BA})$ for $\mathbf{A} = \mathbf{V}$ and $\mathbf{B} = \boldsymbol{\Sigma}^t\boldsymbol{\Sigma}\mathbf{V}^t$. As $\mathbf{V}^t\mathbf{V} = \mathbf{I}_p$ and $\boldsymbol{\Sigma}^t\boldsymbol{\Sigma} = \sum_{k=1}^{K} d_k^2$, it follows that

$$\sum_{i=1}^{n} \sum_{j=1}^{p} W_i^2(s_j) = \sum_{k=1}^{K} d_k^2 \ , \tag{2.2}$$

indicating that the total variance is equal to the sum of squares of the singular values.

In practice, for every $s \in S$, $W_i(s)$ is often approximated by $\sum_{k=1}^{K_0} d_k u_{ik} v_k(s)$ that is, by the first K_0 right singular vectors, where $0 \leq K_0 \leq K$. We now quantify the variance explained by these K_0 right singular vectors. Denote by $\mathbf{V} = [\mathbf{V}_{K_0} | \mathbf{V}_{-K_0}]$ the partition of $\mathbf{V}$ in the $p \times K_0$ dimensional sub-matrix $\mathbf{V}_{K_0}$ and $p \times (p - K_0)$ dimensional sub-matrix $\mathbf{V}_{-K_0}$ containing the first K_0 and the last $(p - K_0)$ columns of $\mathbf{V}$, respectively. Similarly, denote by $\boldsymbol{\Sigma}_{K_0}$ and $\boldsymbol{\Sigma}_{-K_0}$ the sub-matrices of $\boldsymbol{\Sigma}$ that correspond to the first K_0 and last $(K - K_0)$ singular values, respectively. With this notation, $\mathbf{W} = \mathbf{U}\boldsymbol{\Sigma}_{K_0}\mathbf{V}_{K_0}^t + \mathbf{U}\boldsymbol{\Sigma}_{-K_0}\mathbf{V}_{-K_0}^t$ or, equivalently, $\mathbf{W} - \mathbf{U}\boldsymbol{\Sigma}_{K_0}\mathbf{V}_{K_0}^t = \mathbf{U}\boldsymbol{\Sigma}_{-K_0}\mathbf{V}_{-K_0}^t$. Using a similar argument to the one for the decomposition of the total variation, we obtain $\mathrm{tr}(\mathbf{V}_{-K_0}\boldsymbol{\Sigma}_{-K_0}^t\mathbf{U}^t\mathbf{U}\boldsymbol{\Sigma}_{-K_0}\mathbf{V}_{-K_0}^t) = \sum_{k=K_0+1}^{K} d_k^2$. Therefore,

$$\mathrm{tr}(\mathbf{W} - \mathbf{U}\boldsymbol{\Sigma}_{K_0}\mathbf{V}_{K_0}^t)^t(\mathbf{W} - \mathbf{U}\boldsymbol{\Sigma}_{K_0}\mathbf{V}_{K_0}^t) = \sum_{k=K_0+1}^{K} d_k^2 \ .$$

Changing from matrix to entry-wise notation this equality becomes

$$\sum_{i=1}^{n}\sum_{j=1}^{p}\{W_i(s_j) - \sum_{k=1}^{K_0} d_k u_{ik} v_k(s_j)\}^2 = \sum_{k=K_0+1}^{K} d_k^2 \, . \tag{2.3}$$

Equations (2.2) and (2.3) indicate that the first K_0 right singular vectors of $\mathbf{W}$ explain $\sum_{k=1}^{K_0} d_k^2$ of the total variance of the data, or a fraction equal to $\sum_{k=1}^{K_0} d_k^2 / \sum_{k=1}^{K} d_k^2$. In many applications d_k^2 decrease quickly with k indicating that only a few $v_k(\cdot)$ functions are enough to capture the variability in the observed data.

It can also be shown that for every $K_0 = 1, \ldots, K$

$$\sum_{i=1}^{n}\sum_{j=1}^{p}\{W_i(s_j) - \sum_{k \neq K_0} d_k u_{ik} v_k(s_j)\}^2 = d_{K_0}^2 \, , \tag{2.4}$$

where the sum over $k \neq K_0$ is over all $k = 1, \ldots, K$, except K_0. Thus, the K_0th right singular vector explains $d_{K_0}^2$ of the total variance, or a fraction equal to $d_{K_0}^2 / \sum_{k=1}^{K} d_k^2$. The proof is similar to the one for equation (2.3), but partitions the matrix $\mathbf{V}$ into a sub-matrix that contains its K_0 column vector and a sub-matrix that contains all its other columns.

In summary, equation (2.1) can be rewritten for every $s \in S$ as

$$W_i(s) = \sum_{k=1}^{K_0} d_k u_{ik} v_k(s) + \sum_{k=K_0+1}^{K} d_k u_{ik} v_k(s) \, , \tag{2.5}$$

where $\sum_{k=1}^{K_0} d_k u_{ik} v_k(s)$ is the approximation of $W_i(s)$ and $\sum_{k=K_0+1}^{K} d_k u_{ik} v_k(s)$ is the approximation error with variance equal to $\sum_{k=K_0+1}^{K} d_k^2$. The number K_0 is typically chosen to explain a given fraction of the total variance of the data, but other criteria could be used.

We now provide the matrix equivalent of the approximation in equation (2.5). Recall that $W_i(s_j)$ is the (i,j)th entry of the matrix $\mathbf{W}$. If $\mathbf{u}_k$ and $\mathbf{v}_k$ denote the left and right singular vectors of $\mathbf{W}$, the (i,j) entry of the matrix $\mathbf{u}_k \mathbf{v}_k^t$ is equal to $u_{ik} v_k(s_j)$. Therefore, the matrix format of equation (2.5) is

$$\mathbf{W} = \sum_{k=1}^{K_0} d_k \mathbf{u}_k \mathbf{v}_k^t + \sum_{k=K_0+1}^{K} d_k \mathbf{u}_k \mathbf{v}_k^t \, . \tag{2.6}$$

The matrix $\sum_{k=1}^{K_0} d_k \mathbf{u}_k \mathbf{v}_k^t$ is called the rank K_0 approximation of $\mathbf{W}$.

2.1.2   The Link between SVD and PCA

The PCA [140, 229] of $\mathbf{W}$ is the decomposition $\mathbf{W}^t \mathbf{W} = \mathbf{V} \mathbf{\Lambda} \mathbf{V}^t$, where $\mathbf{V}$ is the $p \times p$ dimensional matrix with the property $\mathbf{V}^t \mathbf{V} = \mathbf{I}_p$ and $\mathbf{\Lambda}$ is a $p \times p$ diagonal matrix with positive elements on the diagonal $\lambda_1 \geq \ldots \geq \lambda_p \geq 0$. PCA is also known as the discrete Karhunen-Loéve transform [143, 184]. Denote by $\mathbf{v}_k$, $k = 1, \ldots, K = \min(n,p)$, the $p \times 1$ dimensional column vectors of the matrix $\mathbf{V}$. The vector $\mathbf{v}_k$ is the kth eigenvector of the matrix $\mathbf{V}$, corresponds to the eigenvalue λ_k, and has the property that $\mathbf{W}^t \mathbf{W} \mathbf{v}_k = \lambda_k \mathbf{v}_k$. In FDA the $\mathbf{v}_k$ vectors are referred to as eigenfunctions. In image analysis the term eigenimages is used instead.

Just as with SVD, $\mathbf{v}_k$ form a set of orthonormal vectors in $\mathbb{R}^p$. It can be shown that every $\mathbf{v}_{k+1}$ explains the most residual variability in the data matrix, $\mathbf{W}$, after accounting for the eigenvectors $\mathbf{v}_1, \ldots, \mathbf{v}_k$. We will show this for $\mathbf{v}_1$ first. Note that $\mathbf{W}^t \mathbf{W} = \sum_{k=1}^{K} \lambda_k \mathbf{v}_k \mathbf{v}_k^t$. If

$\mathbf{v}$ is any $p \times 1$ dimensional vector such that $\mathbf{v}^t\mathbf{v} = 1$, the variance of $\mathbf{Wv}$ is $\mathbf{v}^t\mathbf{W}^t\mathbf{Wv} = \sum_{k=1}^{K} \lambda_k \mathbf{v}^t \mathbf{v}_k \mathbf{v}_k^t \mathbf{v}$. Denote by $\mathbf{v} = \sum_{l=1}^{K} a_l \mathbf{v}_l$ the expansion of $\mathbf{v}$ in the basis $\{\mathbf{v}_l : l = 1, \ldots, K\}$. Because $\mathbf{v}_k$ are orthonormal $\mathbf{v}_k^t \mathbf{v} = \sum_{l=1}^{K} a_l \mathbf{v}_k^t \mathbf{v}_l = a_k$ and $\mathbf{v}^t\mathbf{v} = \sum_{l=1}^{K} a_l^2 = 1$. Therefore, $\mathbf{v}^t\mathbf{W}^t\mathbf{Wv} = \sum_{k=1}^{K} \lambda_k a_k^2 \leq \lambda_1 \sum_{k=1}^{K} a_k^2 = \lambda_1$. Equality can be achieved only when $a_1 = 1$ and $a_k = 0$ for $k = 2, \ldots, K$, that is, when $\mathbf{v} = \mathbf{v}_1$. Thus, $\mathbf{v}_1$ is the solution to the problem

$$\mathbf{v}_1 = \arg\max_{||\mathbf{v}=1||} \mathbf{v}^t\mathbf{W}^t\mathbf{Wv} \,. \tag{2.7}$$

Once $\mathbf{v}_1$ is known, the projection of the data matrix on $\mathbf{v}_1$ is $\mathbf{A}_1\mathbf{v}_1$ and the residual variation in the data is $\mathbf{W} - \mathbf{A}_1\mathbf{v}_1^t$, where $\mathbf{A}_1$ is an $n \times p$ dimensional matrix. Because $\mathbf{v}_k$ are orthonormal, it can be shown that $\mathbf{A}_1 = \mathbf{Wv}_1$ and the unexplained variation is $\mathbf{W} - \mathbf{Wv}_1\mathbf{v}_1^t = \sum_{k=2}^{K} \lambda_k \mathbf{v}_k \mathbf{v}_k^t$. Iterating with $\mathbf{W} - \mathbf{Wv}_1\mathbf{v}_1^t$ instead of $\mathbf{W}$, we obtain that the second eigenfunction, $\mathbf{v}_2$, maximizes the residual variance after accounting for $\mathbf{v}_1$. The process is then iterated.

PCA and SVD are closely connected, as $\mathbf{W}^t\mathbf{W} = \mathbf{V}\mathbf{\Sigma}^t\mathbf{\Sigma}\mathbf{V}^t$. Thus, if d_k^2 are ordered such that $d_1^2 \geq \ldots \geq d_K^2 \geq 0$, the kth right singular vector of $\mathbf{W}$ is equal to the kth eigenvector of $\mathbf{W}^t\mathbf{W}$ and corresponds to the kth eigenvalue $\lambda_k = d_k^2$. Similarly, $\mathbf{WW}^t = \mathbf{U}\mathbf{\Sigma}\mathbf{\Sigma}^t\mathbf{U}$, indicating that the kth left singular vector of $\mathbf{W}$ is equal to the kth eigenvector of $\mathbf{WW}^t$ and corresponds to the kth eigenvalue $\lambda_k = d_k^2$.

SVD and PCA have been developed for multivariate data and can be applied to functional data. There are, however, some specific considerations that apply to FDA: (1) the data $W_i(s)$ are functions of s and are expressed in the same units for all s; (2) the mean function, $\overline{W}(s)$, and the main directions of variation in the functional space, $\mathbf{v}_k = \{v_k(s_j), j = 1, \ldots, p\}$, are functions of $s \in S$; (3) these functions inherit and abide by the rules induced by the organization of the space in S (e.g., they do not change too much for small variations in s); (4) the correlation structure between $W_i(s)$ and $W_i(s')$ may depend on (s, s'); and (5) the data may be observed with noise, which may substantially affect the calculation and interpretation of $\{v_k(s_j), j = 1, \ldots, p\}$. For these reasons, FDA often uses smoothing assumptions on $W_i(\cdot)$, $\overline{W}(\cdot)$ and $v_k(\cdot)$. These smoothing assumptions provide a different flavor to PCA and SVD and give rise to functional PCA (FPCA) and SVD (FSVD). While FPCA is better known in FDA, FSVD is a powerful technique that is indispensable for higher dimensional (large p) applications. A more in-depth look at smoothing in FDA is provided in Section 2.3.

2.1.3   SVD and PCA for High-Dimensional FDA

When data are high-dimensional (large p) the $n \times p$ dimensional matrix $\mathbf{W}$ cannot be loaded into the memory and SVD cannot be performed. Things are more difficult for PCA, which uses a $p \times p$ dimensional matrix $\mathbf{W}^t\mathbf{W}$. In this situation, feasible computational alternatives are needed.

Consider the case when p is very large but n is small to moderate. It can be shown that $\mathbf{WW}^t = \sum_{j=1}^{p} \mathbf{w}_j \mathbf{w}_j^t$, where $\mathbf{w}_j$ is the jth column of matrix $\mathbf{W}$. The advantage of this formulation is that it can be computed sequentially. For example, if $\mathbf{C}_k = \sum_{j=1}^{k} \mathbf{w}_j \mathbf{w}_j^t$, $\mathbf{C}_1 = \mathbf{w}_1\mathbf{w}_1^t$, $\mathbf{C}_{k+1} = \mathbf{C}_k + \mathbf{w}_{k+1}\mathbf{w}_{k+1}^t$, and $\mathbf{C}_p = \mathbf{WW}^t$. It takes $O(n^2)$ operations to calculate $\mathbf{C}_1$ because it requires the multiplication of an $n \times 1$ by a $1 \times n$ dimensional matrix. At every step, $k + 1$, only the $n \times n$ dimensional matrix $\mathbf{C}_k$ and the $n \times 1$ vector $\mathbf{w}_{k+1}$ need to be loaded in the memory. This avoids loading the complete data matrix. Thus, the matrix $\mathbf{WW}^t$ can be calculated in $O(n^2p)$ operations without ever loading the complete matrix, $\mathbf{W}$. The PCA decomposition of $\mathbf{WW}^t = \mathbf{U}\mathbf{\Sigma}\mathbf{\Sigma}^t\mathbf{U}$ yields the matrices $\mathbf{U}$ and $\mathbf{\Sigma}$. The matrix $\mathbf{V}$ can then be obtained as $\mathbf{V} = \mathbf{W}^t\mathbf{U}\mathbf{\Sigma}^{-1}$. Thus, each column of $\mathbf{V}$ is obtained by multiplying $\mathbf{W}^t$ with the corresponding column of $\mathbf{U}\boldsymbol{\Sigma}^{-1}$. This requires $O(n^2 p)$ operations. As, in general, we are only interested in the first K_0 columns of $\mathbf{V}$, the total number of operations is of the order $O(n^2 p K_0)$. Moreover, the operations do not require loading the entire data set in the computer memory. Indeed, $\mathbf{W}^t\mathbf{U}\boldsymbol{\Sigma}^{-1}$ can be done by loading one $1 \times n$ dimensional row of $\mathbf{W}^t$ at a time.

The essential idea of this computational trick is to replace the diagonalization of the large $p \times p$ dimensional matrix $\mathbf{W}^t\mathbf{W}$ with the diagonalization of the much smaller $n \times n$ dimensional matrix $\mathbf{W}\mathbf{W}^t$. When n is also large, this trick does not work. A simple solution to address this problem is to sub-sample the rows of the matrix $\mathbf{W}$ to a tractable sample size, say 2000. Sub-sampling can be repeated and right singular vectors can be averaged across sub-samples. Other solutions include incremental, or streaming, approaches [133, 203, 219, 285] and the power method [67, 158].

The incremental, or streaming, approaches start with a number of rows of $\mathbf{W}$ that can be handled computationally. Then covariance operators, eigenvectors, and eigenvalues are updated as new rows are added to the matrix $\mathbf{W}$. The power method starts with the $n \times n$ dimensional matrix $\mathbf{A} = \mathbf{W}\mathbf{W}^t$ and an $n \times 1$ dimensional random normal vector $\mathbf{u}_0$, which is normalized $\mathbf{u}_0 \leftarrow \mathbf{u}_0/||\mathbf{u}_0||$. Here $||\mathbf{a}|| = (\mathbf{a}^t\mathbf{a})^{1/2}$ is the norm induced by the inner product in $\mathbf{R}^n$. The power method consists of calculating the updates $\mathbf{u}_{r+1} \leftarrow \mathbf{A}\mathbf{u}_r$ and $\mathbf{u}_{r+1} \leftarrow \mathbf{u}_{r+1}/||\mathbf{u}_{r+1}||$. Under mild conditions, this approach yields the first eigenfunction, $\mathbf{v}_1$, which can be subtracted and the method can be iterated to obtain the subsequent eigenfunctions. The computational trick here is that diagonalization of matrices is replaced by matrix multiplications, which are much more computationally efficient.

We have found that sampling is a very powerful, easy-to-use method and we recommend it as a first line approach in cases when both n and p are very large.

2.1.4 SVD for US Excess Mortality

We show how SVD can be used to visualize and analyze the cumulative all-cause excess mortality data in 50 states and 2 territories (District of Columbia and Puerto Rico). Figure 2.1 displays these functions for each of the first 52 weeks of 2020. For each state or territory, i, the data are $W_i(s_j)$, where $s_j = j \in \{1, \ldots, p = 52\}$. The mean $\overline{W}(s_j)$ is obtained by averaging observations across states (i) for every week of 2020 $(s_j = j)$. The R implementation is

```
#Calculate the mean of Wr, the un-centered data matrix
mW <- colMeans(Wr)
#Construct a matrix with the mean repeated on each row
mW_mat <- matrix(rep(mW, each = nrow(Wr)), ncol = ncol(Wr))
#Center the data
W <- Wr - mW_mat
```

Here mW is the R notation for the mean vector that contains $\overline{W}(s_j)$, $j = 1, \ldots, p$. We have not divided by $\sqrt{np}$, as results are identical and it is more intuitive to work on the original scale of the data. Figure 2.1 displays the cumulative excess mortality per one million people in each state of the US and two territories in 2020 (light gray lines). This is the same data as in Figure 1.5 without emphasizing the mortality patterns for specific states. Instead, the dark red line is the average of these curves and corresponds to the mW variable. Figure 2.2 displays the same data as Figure 2.1 after centering the data (removing the mean at every time point). These data are stored as rows in the matrix W (in R notation) and have been denoted as $\mathbf{W}$ (in statistical notation). Five states are emphasized to provide examples of trajectories.

The centered data matrix $\mathbf{W}$ (W in R) is decomposed using the SVD. The left singular vectors, $\mathbf{U}$, are stored as columns in the matrix U, the singular values, d, are stored in the

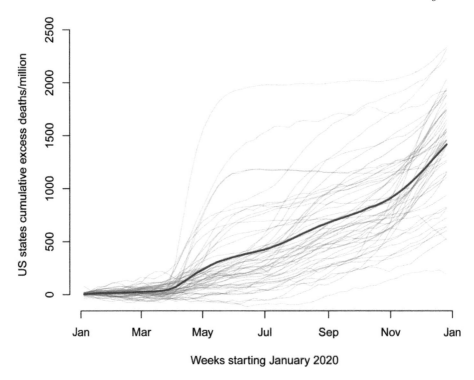

FIGURE 2.1: Each line represents the cumulative excess mortality for each state and two territories in the US. The mean cumulative excess mortality in the US per one million residents is shown as a dark red line.

vector d, and the right singular vectors, **V**, are stored as columns in the matrix V.

```
#Calculate the SVD of W
SVD_of_W <- svd(W)
#Left singular vectors stored by columns
U <- SVD_of_W$u
#Singular values
d <- SVD_of_W$d
#Right singular vectors stored by columns
V <- SVD_of_W$v
```

The individual and cumulative variance explained can be calculated from the vector of singular values, d. In R this is implemented as

```
#Calculate the eigenvalues
lambda <- SVD_of_W$d^2
#Individual proportion of variation
propor_var <- round(100 * lambda / sum(lambda), digits = 1)
#Cumulative proportion of variation
cumsum_var <- cumsum(propor_var)
```

Table 2.1 presents the individual and cumulative percent variance explained by the first five right singular vectors. The first two right singular vectors explain 84% and 11.9% of the variance, respectively, for a total of 95.9%. The first five right singular vectors explain

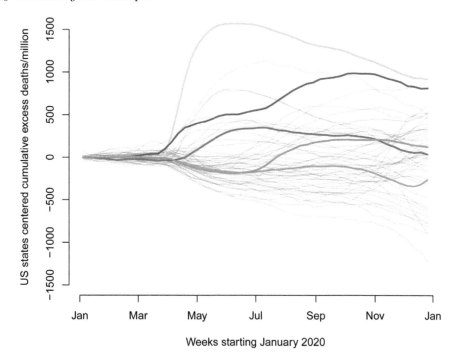

FIGURE 2.2: Each line represents the centered cumulative excess mortality for each state in the US. Centered means that the average at every time point is equal to zero. Five states are emphasized: New Jersey (green), Louisiana (red), Maryland(blue), Texas (salmon), and California (plum).

a cumulative 99.7%, indicating that dimension reduction is quite effective in this particular example. Recall that the right singular vectors are the functional principal components.

The next step is to visualize the two right singular vectors, which together explain 95.9% of the variability. These are the vectors V[,1] and V[,2] in R notation and $\mathbf{v}_1$ and $\mathbf{v}_2$ in statistical notation. Figure 2.3 displays the first (light coral) and second (dark coral) right singular vectors. The interpretation of the first right singular vector is that the mortality data for a state that has a positive coefficient (score) tends to (1) be closer to the US mean between January and April; (2) have a sharp increase above the US mean between April and June; and (3) be larger with a constant difference from the US mean between July and December. The mortality data for a state that has a positive coefficient on the second right singular vector tends to (1) have an even sharper increase between April and June

TABLE 2.1
All-cause cumulative excess mortality in 50 US states plus Puerto Rico and District of Columbia. Individual and cumulative percent variance explained by the first five right singular vectors (principal components).

Variance	**Right singular vectors**				
	1	**2**	**3**	**4**	**5**
Individual (%)	84.0%	11.9%	2.9%	0.6%	0.3%
Cumulative (%)	84.0%	95.9%	98.8%	99.4%	99.7%

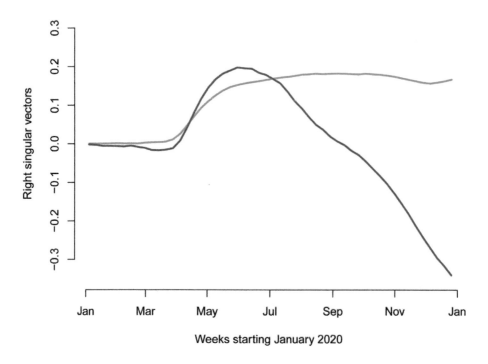

FIGURE 2.3: First two right singular vectors (principal components) for all-cause weekly excess US mortality data in 2020. First right singular vector: light coral. Second singular vector: dark coral.

relative to the US average; and (2) exhibit a decreased difference from the US mean as time progresses from July to December. Of course, things are more complex, as the mean and right singular vectors can compensate for one another in specific times of the year.

Individual state mortality data can be reconstructed for all states simultaneously. A $K_0 = 2$ rank reconstruction of the data can be obtained as

```
#Set the reconstruction rank
K0 <- 2
#Reconstruct the centered data using rank K0 approximation
rec <- SVD_of_W$u[,1:K0] %*% diag(SVDofW$d[1:K0]) %*% t(V[,1:K0])
#Add the mean to the rank K0 approximation of W
WK0 <- mW_mat + rec
```

The matrices W and WK0 contain the original and reconstructed data, where each state is recorded by rows. Figure 2.4 displays the original (solid lines) and reconstructed data (dashed lines of matching color) for five states: New Jersey (green), Louisiana (red), Maryland (blue), Texas (salmon), and California (plum). Even though the reconstructions are not perfect, they do capture the main features of the data for each of the five states. Better approximations can be obtained by increasing K_0, though at the expense of using additional right singular vectors.

Consider, for example, the mortality data from New Jersey. The rank $K_0 = 2$ reconstruction of the data is

$$W_{\mathrm{NJ}}(s) = \overline{W}_{\mathrm{US}}(s) + 0.49 v_1(s) + 0.25 v_2(s) \, ,$$

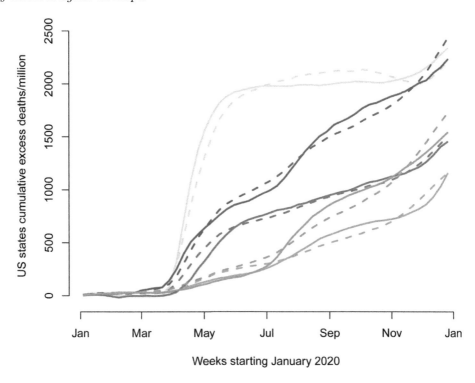

FIGURE 2.4: All-cause excess mortality (solid lines) and predictions based on rank 2 SVD (dashed lines) for five states in the US: New Jersey (green), Louisiana (red), Maryland (blue), Texas (salmon), and California (plum).

where the coefficients 0.49 and 0.25 correspond to u_{i1} and u_{i2}, the $(i, 1)$ and $(i, 2)$ entries of the matrix $\mathbf{U}$ (U in R), where i corresponds to New Jersey. These values can be calculated in R as

```
U[states=="New Jersey", 1:2]
```

where `states` is the vector containing the names of US states and territories. We have used the notation $\overline{W}_{\text{US}}(s)$ instead of $\overline{W}(s)$ and $W_{\text{NJ}}(s)$ instead of $W_i(s)$ to improve the precision of notation. Both coefficients for $v_1(\cdot)$ and $v_2(\cdot)$ are positive, indicating that for New Jersey there was a strong increase in mortality between April and June, a much slower increase between June and November and a further larger increase in December. Even though neither of the two components contained information about the increase in mortality in December, the effect was accounted for by the mean; see, for the example the increase in the November December period in the mean in Figure 2.1.

All the coefficients, also known as scores, are stored in the matrix $\mathbf{U}$. It is customary to display these scores using scatter plots. For example,

```
plot(U[,1], U[,2])
```

produces a plot similar to the one shown in Figure 2.5. Every point in this graph represents a state and the same five states were emphasized: New Jersey (green), Louisiana (red), Maryland (blue), Texas (salmon), and California (plum). Note that New Jersey is the point

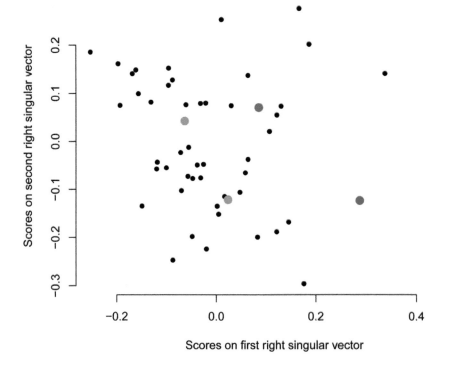

FIGURE 2.5: Scores on the first versus second right singular vectors for all-cause weekly excess mortality in the US. Each dot is a state, Puerto Rico, or Washington DC. Five states are emphasized: New Jersey (green), Louisiana (red), Maryland (blue), Texas (salmon), and California (plum).

with the largest score on the first right singular vector and the third largest score on the second right singular vector. Louisiana has the third largest score on the first right singular vector, which is consistent with being among the states with highest all-cause mortality. In contrast to New Jersey, the score for Louisiana on the second right singular vector is negative indicating that its cumulative mortality data continues to increase away from the US mean between May and November; see Figure 2.2.

2.2 Gaussian Processes

While all the data we observe will be sampled at discrete time points, observed functional data is thought of realizations of an underlying continuous process. Here we provide some theoretical concepts that will help with the interpretation of the analytic methods. A Gaussian Process (GP) is a collection of random variables $\{W(s), s \in S\}$ where every finite collection of random variables $\{W(s_1), \ldots, W(s_p)\}$, $s_j \in S$ for every $j = 1, \ldots, p$ and every p is a multivariate Gaussian distribution. For convenience, we consider $S = [0, 1]$ and interpret it as time, but Gaussian Processes can be defined over space, as well. A Gaussian Process is completely characterized by its mean $\mu(s)$ and covariance operator $K_W : S \times S \to \mathbb{R}$, where $K_W(s_1, s_2) = \text{Cov}\{W(s_1), W(s_2)\}$.

Assume now that the mean of the process is 0. By Mercer's theorem [199] there exists a set of eigenvalues and eigenfunctions λ_k, $\phi_k(s)$, where $\lambda_k \geq 0$, $\phi_k : S \to \mathbb{R}$ form an orthonormal basis in $L^2([0,1])$, $\int K_W(s,t)\phi_k(t)dt = \lambda_s\phi_k(s)$ for every $s \in S$ and $k = 1, 2, \ldots$, and

$$K_W(s_1, s_2) = \sum_{k=1}^{\infty} \lambda_k \phi_k(s_1)\phi_k(s_2) .$$

The Kosambi-Karhunen-Loève (KKL) [143, 157, 184] theorem provides the explicit decomposition of the process $W(s)$. Because $\phi_k(t)$ form an orthonormal basis, the Gaussian Process can be expanded as

$$W(s) = \sum_{k=1}^{\infty} \xi_k \phi_k(s) ,$$

where $\xi_k = \int_0^1 W(s)\phi_k(s)dt$, which does not depend on s. It is easy to show that the $E(\xi_k) = 0$ as

$$E(\xi_k) = E\{\int_0^1 W(s)\phi_k(s)ds\} = \int_0^1 E\{W(s)\}\phi_k(s)ds = 0 .$$

We can also show that the $\text{Cov}(\xi_k, \xi_l) = E(\xi_k\xi_l) = 0$ for $k \neq l$ and $\text{Var}(\xi_k) = \lambda_k$. The proof is shown below

$$\begin{aligned}
E(\xi_k\xi_l) &= E\left\{\int_0^1 \int_0^1 W(s)W(t)\phi_k(s)\phi_l(t)dtds\right\} \\
&= \int_0^1 \int_0^1 E\{W(s)W(t)\}\phi_k(t)\phi_l(s)dtds \\
&= \int_0^1 \left\{\int_0^1 K_W(s,t)\phi_k(t)dt\right\}\phi_l(s)ds \\
&= \lambda_k \int_0^1 \phi_k(s)\phi_l(s)ds \\
&= \lambda_k \delta_{kl} ,
\end{aligned} \qquad (2.8)$$

where $\delta_{kl} = 0$ if $k \neq l$ and 1 otherwise. The second equality holds because of the change of order of integrals (expectations), the third equality holds because of the definition of $K_W(s,t)$, the fourth equality holds because $\phi_k(s)$ is the eigenfunction of $K_W(\cdot, \cdot)$ corresponding to the eigenvalue λ_k, and the fifth equality holds because of the orthonormality of the $\phi_k(s)$ functions. These results hold for any $L^2[0,1]$ integrable process and does not require Gaussianity of the scores.

However, if the process is Gaussian, it can be shown that any finite collection $\{\xi_{k_1}, \ldots, \xi_{k_l}\}$ is jointly Gaussian. Because the individual entries are uncorrelated mean-zero, the scores are independent Gaussian random variables. One could reasonably ask, why should one care about all these properties and whether this theory has any practical implications. Below we identify some of the practical implications.

The expression "Gaussian Process" is quite intimidating, the definition is relatively technical, and it is not clear from the definition that such objects even exist. However, these results show how to generate Gaussian Processes relatively easily. Indeed, the only ingredients we need are a set of orthonormal functions $\phi_k(\cdot)$ in $L^2[0,1]$ and a set of positive numbers $\lambda_1 \geq \lambda_2 \geq \ldots$. For example, if $\phi_1(s) = \sqrt{2}\sin(2\pi s)$, $\phi_2(s) = \sqrt{2}\cos(2\pi s)$, $\lambda_1 = 4$, $\lambda_2 = 1$,

$\lambda_k = 0$, for $k \geq 3$, realizations of a Gaussian Process can simply be generated by simulating independently $\xi_1 \sim N(0, \lambda_1)$, $\xi_2 \sim N(0, \lambda_2)$ and calculating $W(s) = \xi_1 \phi_1(t) + \xi_2 \phi_2(t)$. Of course, more generally, if $\psi_1(s), \ldots, \psi_K(s)$ is a set of any K function that are $L^2[0, 1]$ integrable and $\boldsymbol{\xi} = (\xi_1, \ldots, \xi_K) \sim N(\mathbf{0}, \boldsymbol{\Sigma})$ then $W(s) = \sum_{k=1}^{K} \xi_k \psi_k(s)$ is a Gaussian Process with mean zero and covariance $K_W(s, t) = \text{Cov}\{W(s), W(t)\} = \boldsymbol{\psi}^t(s) \boldsymbol{\Sigma} \boldsymbol{\psi}(t)$, where $\boldsymbol{\psi}(s) = \{\psi_1(s), \ldots, \psi_K(s)\}^t$. In this case one would need to obtain the spectral decomposition of this covariance operator to obtain its KKL representation.

Another major implication is that we can easily obtain processes that are not Gaussian, but are generated by Gaussian Processes. For example, we can define $Y(s) \sim$ Bernoulli$\{p(s)\}$, where logit$\{p(s)\} = W(s)$, which is a binary process generated by the underlying Gaussian Process $W(s)$. Similarly, we can define $Y(s) \sim$ Poisson$\{\lambda(s)\}$, where $\log\{\lambda(s)\} = W(s)$, which is a continuous process with Poisson observations. There are many other possibilities, indicating the extraordinary diversity of processes that can be generated using Gaussian Processes.

Gaussian Processes have been used in time series and spatial analysis for a long time and are relatively well understood. Their application in functional data analysis tends to be different because in this case we often have repeated observations, that is, multiple samples of the process $W(s)$. These samples are denoted by $W_i(s)$, where $i = 1, \ldots, n$ are the observation units. Thus, there is no need to require stationarity or isotropy, as for every pair of sampling points (s_1, s_2) there are n pairs of observations $W_i(s_1, s_2)$ that can be used to estimate the covariance $K_W(s_1, s_2)$. Of course, a different set of questions can be raised about whether $K_W(s_1, s_2)$ is the same for every pair of observations, but this is the curse and blessing of having data on multiple observation units: being able to answer questions that would not even make sense with only one observation unit.

2.3 Semiparametric Smoothing

Nonparametric smoothing has witnessed substantial advances over the past two decades, including the development of (1) improved, stable software; (2) integrated models that include parametric, nonparametric and mixed effects (semiparametric models); and (3) a principled statistical inferential framework based on mixed effects models. We will show that these new ideas can be adapted to FDA, which will substantially simplify modeling, inference and software. By highlighting this connection, we will show that existing, user-friendly software can be adapted to complex functional data. This will provide easy to use, reproducible, inferential software.

To accomplish that, we first discuss the basic ideas of smoothing. Consider the case when the data structure consists of the pairs of observations (y_i, x_i) for $i = 1, \ldots, n$ and we are interested in regressing y_i on x_i. We are using this notation instead of $\{w(s_j), s_j\}$ notation to emphasize that smoothing can be conducted in any type of regression. In functional data, a direct application of smoothing would be to smooth observed functions $w_i(s_j)$ along the functional domain $s_j \in S$ separately for every study participant $i = 1, \ldots, n$. Note that in this case notation is changed slightly (indexing changes from i to j and the index i in $w_i(s_j)$ indicates a repeated observation of the function). However, smoothing will have many more applications in functional data, but for now we stay with the (y_i, x_i) notation, which is commonly used in smoothing.

More precisely, we would like to fit a model of the type

$$y_i = f(x_i) + \epsilon_i \,, \tag{2.9}$$

where ϵ_i are independent identically distributed $N(0, \sigma_\epsilon^2)$ random variables. We denote by $\mathbf{y} = (y_1, \ldots, y_n)^t$ and by $\mathbf{f} = \{f(x_1), \ldots, f(x_n)\}$. Here $f(\cdot)$ has either a specified parametric form, such as the linear parametric function $f(x_i) = \beta_0 + \beta_1 x_i$, or an unspecified nonparametric form with specific restrictions. Note that without any restrictions on $f(\cdot)$ the model is not identifiable and therefore unusable. Many different restrictions have been proposed for $f(\cdot)$ and almost all of them assume some degree of smoothness and/or continuity of the function. Despite the large number of approaches, all practical solutions have the following form $\widehat{\mathbf{f}} = \mathbf{S}\mathbf{y}$, where $\mathbf{y}$ is the $n \times 1$ dimensional vector with the ith entry equal to y_i and $\mathbf{S}$ is a $n \times n$ dimensional symmetric smoother matrix. The residual sum of squares for approximating $\mathbf{y}$ by $\widehat{\mathbf{f}} = \mathbf{S}\mathbf{y}$ is

$$\text{RSS} = ||\mathbf{y} - \widehat{\mathbf{f}}||^2 = ||\mathbf{y} - \mathbf{S}\mathbf{y}||^2 \, ,$$

where $||\mathbf{a}||^2$ is the sum of squares of the entries of the vector $\mathbf{a}$. The minimum RSS is zero and is obtained when $\widehat{\mathbf{f}} = \mathbf{y}$, when no restrictions are imposed on the function $f(\cdot)$. However, this is less interesting and smoothing is concerned with minimizing a version of the RSS when specific restrictions are imposed on $f(\cdot)$.

2.3.1   Regression Splines

Regression splines are a popular way to induce restrictions on $f(\cdot)$. We introduce ideas for univariate splines and then show that they can be generalized to bivariate and multivariate smoothing.

2.3.1.1   Univariate Regression Splines

Consider, for example, the restriction that $f(x_i) = \beta_1 + \beta_2 x_i$, that is, that $f(\cdot)$ is linear. With this restriction, equation (2.9) is a standard linear regression. The best linear unbiased predictor (BLUP) of $\mathbf{y}$ is $\widehat{\mathbf{f}} = \mathbf{S}\mathbf{y}$, where $\mathbf{S} = \mathbf{X}(\mathbf{X}^t\mathbf{X})^{-1}\mathbf{X}^t$ and

$$\mathbf{X} = \begin{bmatrix} 1 & x_1 \\ 1 & x_2 \\ \vdots & \vdots \\ 1 & x_p \end{bmatrix} = \begin{bmatrix} B_1(x_1) & B_2(x_1) \\ B_1(x_2) & B_2(x_2) \\ \vdots & \vdots \\ B_1(x_n) & B_2(x_n) \end{bmatrix} .$$

Here $B_1(x) = 1$ and $B_2(x) = x$ are basis functions. More flexible models are obtained by adding polynomial and/or spline terms. For example, a quadratic truncated polynomial regression spline has the form

$$f(x_i) = \beta_1 + \beta_2 x_i + \beta_3 x_i^2 + \sum_{k=1}^{K} \beta_{k+3}(x_i - \kappa_k)_+^2 \, ,$$

where $\kappa_1, \ldots, \kappa_K$ are knots and a_+^2 is equal to a^2 if $a > 0$ and 0 otherwise. For didactic purposes, we used the quadratic truncated polynomial regression spline, though any spline basis can be used. The smoother matrix has the same form, $\mathbf{S} = \mathbf{X}(\mathbf{X}^t\mathbf{X})^{-1}\mathbf{X}^t$, though $\mathbf{X}$ has more columns

$$\mathbf{X} = \begin{bmatrix} 1 & x_1 & x_1^2 & (x_1 - \kappa_1)_+^2 & \cdots & (x_1 - \kappa_K)_+^2 \\ 1 & x_2 & x_2^2 & (x_2 - \kappa_1)_+^2 & \cdots & (x_2 - \kappa_K)_+^2 \\ \vdots & \vdots & \vdots & \vdots & \vdots & \vdots \\ 1 & x_n & x_n^2 & (x_n - \kappa_1)_+^2 & \cdots & (x_n - \kappa_K)_+^2 \end{bmatrix} .$$

It is useful to write this matrix in the more general basis format

$$\mathbf{X} = \begin{bmatrix} B_1(x_1) & B_2(x_1) & B_3(x_1) & B_4(x_1) & \ldots & B_{K+3}(x_1) \\ B_1(x_2) & B_2(x_2) & B_3(x_2) & B_4(x_2) & \ldots & B_{K+3}(x_2) \\ \vdots & \vdots & \vdots & \vdots & \vdots & \vdots \\ B_1(x_p) & B_2(x_p) & B_3(x_p) & B_4(x_p) & \ldots & B_{K+3}(x_p) \end{bmatrix},$$

where $B_1(s) = 1$, $B_2(s) = s$, $B_3(s) = s^2$, and $B_{k+3}(s) = (s - \kappa_k)^2_+$, for $k = 1, \ldots, K$. This expression is general and emphasizes that other popular spline bases could be used, including B-splines, natural cubic splines, and thin-plate splines. Non-spline bases could also be used including Fourier, wavelets, or exponential functions. We prefer splines because they work well in many applications, though methods can be adapted to other bases. For more details on splines see, for example, [60, 258, 319].

Thus, the quadratic truncated polynomial regression spline model is

$$y_i = \sum_{k=1}^{K+3} \beta_k B_k(x_i) + \epsilon_i .$$

If $\boldsymbol{\beta} = (\beta_1, \ldots, \beta_{K+3})^t$ and $\boldsymbol{\epsilon} = (\epsilon_1, \ldots, \epsilon_n)$, the model has the following matrix format $\mathbf{y} = \mathbf{X}\boldsymbol{\beta} + \boldsymbol{\epsilon}$. The BLUP for $\mathbf{y}$ is $\widehat{\mathbf{f}} = \mathbf{Sy} = \mathbf{X}\widehat{\boldsymbol{\beta}}$, where $\mathbf{S} = \mathbf{X}(\mathbf{X}^t\mathbf{X})^{-1}\mathbf{X}$ and $\widehat{\boldsymbol{\beta}} = (\mathbf{X}^t\mathbf{X})^{-1}\mathbf{X}^t\mathbf{y}$ is the solution to the minimizing problem

$$\min_{\boldsymbol{\beta}} ||\mathbf{y} - \mathbf{X}\boldsymbol{\beta}||^2 . \tag{2.10}$$

Splines fit using minimization of the residual sum of squares (2.10) are referred to as regression splines. The type of basis (e.g., quadratic truncated polynomial) is used to define the type of spline, whereas the term "regression spline" refers to the method used for fitting. In Section 2.3.2 we will describe penalized splines, which use the same basis, but a different, penalized minimization criterion.

2.3.1.2 Regression Splines with Multiple Covariates

So far we have seen that spline regression with one predictor can be seen as a standard linear regression with some special regressors. We expand on this strategy and consider the case when the data structure is $(y_i, z_i, x_{i1}, x_{i2})$. For example, y_i could be the total amount of physical activity during the day, z_i could be sex, x_{i1} could be age, and x_{i2} could be body mass index (BMI) for individual i. Suppose that we have reason to believe that the association between y_i and z_i is linear, but we would like to allow for some flexibility in the associations between y_i and x_{i1} as well as y_i and x_{i2}. To do that we consider a model of the type

$$y_i = \gamma_0 + \gamma_1 z_i + \sum_{k=1}^{K_1} \beta_{k,1} B_{k,1}(x_{i1}) + \sum_{k=1}^{K_2} \beta_{k,2} B_{k,2}(x_{i2}) + \epsilon_i .$$

If we denote by $\boldsymbol{\beta} = (\gamma_0, \gamma_1, \beta_{1,1}, \ldots, \beta_{K_2,2})^t$ and the design matrix by

$$\mathbf{X} = \begin{bmatrix} 1 & z_1 & B_{1,1}(x_{11}) & \ldots & B_{K_1,1}(x_{11}) & B_{1,2}(x_{12}) & \ldots & B_{K_2,2}(x_{12}) \\ 1 & z_2 & B_{1,1}(x_{21}) & \ldots & B_{K_1,1}(x_{21}) & B_{1,2}(x_{22}) & \ldots & B_{K_2,2}(x_{22}) \\ \vdots & \vdots & \vdots & \vdots & \vdots & \vdots & \vdots & \vdots \\ 1 & z_n & B_{1,1}(x_{n1}) & \ldots & B_{K_1,1}(x_{n1}) & B_{1,2}(x_{n2}) & \ldots & B_{K_2,2}(x_{n2}) \end{bmatrix},$$

then we are back to the familiar linear regression. Thus, regression splines with multiple covariates can be represented exactly as the linear regression model $\mathbf{y} = \mathbf{X}\boldsymbol{\beta} + \boldsymbol{\epsilon}$ with the standard assumptions. The BLUP for $\mathbf{y}$ is $\hat{\mathbf{f}} = \mathbf{X}\hat{\boldsymbol{\beta}}$, where $\mathbf{X}$ is an $n \times (K_1 + K_2 + 2)$ dimensional matrix and $\hat{\boldsymbol{\beta}}$ minimizes the residual sum of squares criterion (2.10). The best unbiased estimator of $\boldsymbol{\beta}$ is $\hat{\boldsymbol{\beta}} = (\mathbf{X}^t\mathbf{X})^{-1}\mathbf{X}^t\mathbf{y}$ which requires the inversion of the $(K_1 + K_2 + 2) \times (K_1 + K_2 + 2)$ dimensional matrix $\mathbf{X}^t\mathbf{X}$. The rank of this matrix is equal to $\text{rank}(\mathbf{X}) \leq \min(n, K_1, K_2 + 2)$. Therefore the matrix $\mathbf{X}^t\mathbf{X}$ is not invertible when $n < K_1 + K_2 + 2$, that is when the number of parameters exceeds the number of observations. In fact, it is a good rule to ensure that $n \geq 5(K_1 + K_2 + 2)$, meaning that there are at least 5 observations for estimation of each parameter. However, in many applications the number of parameters far exceeds the number of observations or they are too many to provide reasonable estimation or inference. One solution around the problem is to consider shrinkage estimators, which is equivalent to penalized estimation as well as with imposing a prior on the model parameters. All of these seemingly unrelated strategies lead to regression models where some parameters are treated as random (random effects), transforming a vast majority of standard regression models into linear mixed effects (LME) regression models. We will discuss this general technique and transformations in Section 2.3.3.

2.3.1.3 Multivariate Regression Splines

Univariate regression splines smoothing extends easily to multivariate regression splines. We describe the extension to bivariate smoothing, but the same ideas can be applied to multivariate smoothing. Consider the case when the observations are (y_i, x_{1i}, x_{2i}) for $i = 1, \ldots, n$; we do not include z_i to reduce the notational burden, but this could be easily added, just as in Section 2.3.1.2. In the bivariate case the smoothing model becomes

$$y_i = f(x_{i1}, x_{i2}) + \epsilon_i .$$

The observed data consists of the $n \times 1$ dimensional vector $\mathbf{y} = \{y_1, \ldots, y_n\}^t$ and the corresponding covariates, $\mathbf{x}_1 = (x_{11}, x_{12}), \ldots, \mathbf{x}_n = (x_{n1}, x_{n2})$. The problem is to estimate the bivariate function $f(\cdot, \cdot)$. Bivariate spline models are of the type

$$f(\mathbf{x}) = \sum_{k=1}^{K} \beta_k B_k(\mathbf{x}) ,$$

where $B_k(\cdot)$ is a basis in $\mathbb{R}^2$. Examples of such bases include tensor products of univariate splines and thin-plate splines [258]. The bases for tensor products of univariate splines are of the form $B_{k_1,k_2}(x_1, x_2) = B_{k_1,1}(x_1)B_{k_2,2}(x_2)$ for $k_1 = 1, \ldots, K_1$ and $k_2 = 1, \ldots, K_2$, where K_1 and K_2 are the number of bases in the first and second dimension, respectively. In this notation the total number of basis functions is $K = K_1 K_2$ and

$$f(x_{i1}, x_{i2}) = \sum_{k_1=1}^{K_1} \sum_{k_2=1}^{K_2} \beta_{k_1 k_2} B_{k_1,1}(x_{i1}) B_{k_2,2}(x_{i2}) .$$

If we denote by $\boldsymbol{\beta} = (\beta_{11}, \beta_{12}, \ldots, \beta_{K_1 K_2})^t$ and the design matrix by

$$\mathbf{X} = \begin{bmatrix} B_{1,1}(x_{11})B_{1,2}(x_{12}) & B_{1,1}(x_{11})B_{2,2}(x_{12}) & \ldots & B_{K_1,1}(x_{11})B_{K_2,2}(x_{12}) \\ B_{1,1}(x_{21})B_{1,2}(x_{22}) & B_{1,1}(x_{21})B_{2,2}(x_{22}) & \ldots & B_{K_1,1}(x_{21})B_{K_2,2}(x_{22}) \\ \vdots & \vdots & \vdots & \vdots \\ B_{1,1}(x_{n1})B_{1,2}(x_{n2}) & B_{1,1}(x_{n1})B_{2,2}(x_{n2}) & \ldots & B_{K_1,1}(x_{n1})B_{K_2,2}(x_{n2}) \end{bmatrix} ,$$

then we are back to the familiar linear regression.

Thin plate spline bases are constructed a bit differently and require a set of points (knots) in space, say $\boldsymbol{\kappa}_1, \ldots, \boldsymbol{\kappa}_K \in \mathbb{R}^2$ and the function $\varphi : [0, \infty] \to \mathbb{R}$, $\varphi(r) = r^2 \log(r)$. The thin plate spline basis is $B_1(\mathbf{x}) = 1$, $B_2(\mathbf{x}) = x_1$, $B_3(\mathbf{x}) = x_2$, and $B_{k+3}(\mathbf{x}) = \varphi(\|\mathbf{x} - \kappa_k\|)$, for $k = 1, \ldots, K$. The corresponding design matrix is

$$\mathbf{X} = \begin{bmatrix} B_1(\mathbf{x}_1) & B_2(\mathbf{x}_1) & \ldots & B_{K+3}(\mathbf{x}_1) \\ B_1(\mathbf{x}_2) & B_2(\mathbf{x}_2) & \ldots & B_{K+3}(\mathbf{x}_2) \\ \vdots & \vdots & \vdots & \vdots \\ B_1(\mathbf{x}_n) & B_2(\mathbf{x}_n) & \ldots & B_{K+3}(\mathbf{x}_n) \end{bmatrix},$$

and the model parameter is $\boldsymbol{\beta} = (\beta_1, \ldots, \beta_{K+3})$. Thin plate splines are a type of spline smoother with a radial basis. Here we acknowledge the more general class of radial smoothers, but we focus on thin plate splines. For thin plate splines we used $K + 3$ instead of K to emphasize the special nature of the intercept, and first and second coordinate bases.

In summary, just as in the univariate case, the multivariate regression spline model has the following matrix format $\mathbf{y} = \mathbf{X}\boldsymbol{\beta} + \boldsymbol{\epsilon}$. The best predictor for $\mathbf{y}$ is $\widehat{\mathbf{f}} = \mathbf{X}\widehat{\boldsymbol{\beta}}$, where $\widehat{\boldsymbol{\beta}}$ minimizes the residual sum of squares criterion (2.10). The difference between tensor products of splines and thin plate splines is that the knots for tensor products are arranged in a rectangular shape. This makes them better suited for fitting rectangular or close to rectangular surfaces. Thin plate splines can better adapt to irregular surfaces, as knots can be placed where observations are. For standard regression to work one needs the matrix $\mathbf{X}^t\mathbf{X}$ to be invertible. A minimum requirement for that is for the number of observations, n, to exceed the number of columns in matrix $\mathbf{X}$, as was discussed in Section 2.3.1.2. For example, the number of columns in $\mathbf{X}$ using a tensor product of splines is $K_1 K_2$, which is much larger than $K_1 + K_2 + 2$ in the case discussed in Section 2.3.1.2. In general, the number of columns in the design matrix grows much faster when we consider smoothing in higher dimensions. While imperfect, it is a good rule to have at least $n > 5K_1 K_2$ observations when running a tensor product regression. When the number of parameters exceeds the number of observations our strategy will be to use penalized approaches, which impose smoothing on parameters. The next section provides the necessary details. But, for now, remember that every regression spline model is a standard regression. It can be easily extended to non-Gaussian outcomes by simply changing the distribution of the error, ϵ_i. From an implementation perspective, one simply replaces the `lm` function with the `glm` function in R.

2.3.2 Penalized Splines

Penalized splines are a class of nonparametric models that balance signal complexity and computational efficiency. Penalized splines is the general term for spline fitting with a small to moderate number of fixed bases and spline parameters subject to penalties [306]. Here we follow the principle of using a moderate number of bases coupled with a roughness penalty, as described in [116, 258, 322]. Early work in this area is due to [71, 221, 150, 54, 304], while their connection to mixed effects models [26, 258, 322] extended their use in the context of semiparametric models.

Consider first the case of scatterplot smoothing, which is the case when we have pairs of observations (y_i, x_i), $i = 1, \ldots, n$, where both y_i and x_i are scalars and we are interested in models of the type $y_i = f(x_i) + \epsilon_i$, where $\epsilon_i \sim N(0, \sigma_\epsilon^2)$ are independent. In Section 2.3 we have shown that, irrespective of the spline basis chosen, the model can be written in matrix format as

$$\mathbf{y} = \mathbf{X}\boldsymbol{\beta} + \boldsymbol{\epsilon},$$

where $\mathbf{y} = (y_1, \ldots, y_n)^t$, the columns of the matrix $\mathbf{X}$ correspond to the spline basis, $\boldsymbol{\beta}$ are the parameters of the spline, and $\boldsymbol{\epsilon} = (\epsilon_1, \ldots, \epsilon_n)^t$. As discussed in Section 2.3, estimating the model is equivalent to minimizing the sum of squares of residuals criterion described in equation (2.10).

The problem with this approach is that it is hard to know how many basis functions are enough to capture the complexity of the underlying mean function. One option is to estimate the number of knots, but we have found this idea quite impractical especially in the context when we might have additional covariates, nonparametric components, and/or non-Gaussian data. An idea that works much better is to consider a rich spline basis and add a quadratic penalty on the spline coefficients in equation (2.10). This is equivalent to minimizing the following penalized sum of squares criterion

$$\min_{\beta} ||\mathbf{y} - \mathbf{X}\boldsymbol{\beta}||^2 + \lambda \boldsymbol{\beta}^t \mathbf{D} \boldsymbol{\beta} \, , \tag{2.11}$$

where $\lambda \geq 0$ is a scalar and $\mathbf{D}$ is a matrix that provides the penalty structure for a specific choice of spline basis. For example, the penalized quadratic truncated polynomial splines use the penalty matrix

$$\mathbf{D} = \begin{bmatrix} \mathbf{0}_{3 \times 3} & \mathbf{0}_{3 \times K} \\ \mathbf{0}_{K \times 3} & \mathbf{I}_K \end{bmatrix} \, ,$$

where $\mathbf{0}_{a \times b}$ is a matrix of zero entries with a rows and b columns and $\mathbf{I}_K$ is the identity matrix of dimension K. In this example, the penalty is equal to $\lambda \sum_{k=4}^{K+3} \beta_k^2$ and leaves the parameters β_1, β_2, β_3 unpenalized.

For every fixed λ, by setting the derivative with respect to $\boldsymbol{\beta}$ equal to zero in expression (2.11), it can be shown that the minimum is achieved at $\widehat{\boldsymbol{\beta}}_\lambda = (\mathbf{X}^t \mathbf{X} + \lambda \mathbf{D})^{-1} \mathbf{X}^t \mathbf{y}$. Thus, the predictor of $\mathbf{y}$ is $\widehat{\mathbf{f}}_\lambda = \mathbf{S}_\lambda \mathbf{y}$, where

$$\mathbf{S}_\lambda = \mathbf{X}(\mathbf{X}^t \mathbf{X} + \lambda \mathbf{D})^{-1} \mathbf{X}^t \, . \tag{2.12}$$

The scalar parameter λ controls the amount of smoothing, which varies from the saturated parametric model when $\lambda = 0$ (no penalty) to a parsimonious parametric model when $\lambda = \infty$ (e.g., the quadratic regression model in the penalized truncated polynomial case.) The trace of the $\mathbf{S}_\lambda$ matrix is referred to as the number of degrees of freedom of the smoother.

With this approach the smoothing problem is reduced to estimating λ. Some of the most popular approaches for estimating λ are cross-validation (CV) [160, 209], generalized cross validation (GCV) [54], Akaike's Information Criterion (AIC) [3] and restricted maximum likelihood (REML) [121, 228]. We describe the first three approaches here and the REML approach in Section 2.3.3.

Cross-validation estimates λ as the value that minimizes $\mathrm{CV}(\lambda) = \frac{1}{n} \sum \{y_i - \widehat{f}_{\lambda, -i}(x_i)\}^2$, where $\widehat{f}_{\lambda, -i}(x_i)$ is the estimator of y_i based on the entire data, except (y_i, x_i). It can be shown that this formula can be simplified to

$$\mathrm{CV}(\lambda) = \frac{1}{n} \sum_{i=1}^{n} \left\{ \frac{y_i - \widehat{f}_\lambda(x_i)}{1 - S_{\lambda, ii}} \right\}^2 \, ,$$

where $\widehat{f}_\lambda(x_i)$ is the estimator of y_i based on the entire data and $S_{\lambda, ii}$ is the ith diagonal entry of the smoother matrix $\mathbf{S}_\lambda$. The advantage of this formula is that it requires only one regression for each λ using the entire data set. The original formulation would require n regressions, each with a different data set. Generalized cross validation (GCV) further simplifies the $\mathrm{CV}(\lambda)$ formula by replacing $S_{\lambda, ii}$ by $\mathrm{tr}(\mathbf{S}_\lambda)/n$. With this replacement,

$$\mathrm{GCV}(\lambda) = \frac{1}{n\{1 - \mathrm{tr}(\mathbf{S}_\lambda)/n\}^2} \sum \{y_i - \widehat{f}_\lambda(x_i)\}^2 \, .$$

The AIC criterion minimizes

$$\mathrm{AIC}(\lambda) = \log \sum_{i=1}^{n}\{y_i - \widehat{f}_\lambda(x_i)\}^2 + 2\frac{\mathrm{tr}(\mathbf{S}_\lambda)}{n} \ .$$

There is a close connection between $\mathrm{GCV}(\lambda)$ and $\mathrm{AIC}(\lambda)$. Indeed,

$$\log\{\mathrm{GCV}(\lambda)\} = \log \sum_{i=1}^{n}\{y_i - \widehat{f}_\lambda(x_i)\}^2 - \frac{2}{n}\log\left\{1 - \frac{\mathrm{tr}(\mathbf{S}_\lambda)}{n}\right\}^n - \log(n) \ .$$

Here, the last term, $\log(n)$, can be ignored, as this is a constant. For large n, $\log(1-x/n)^n \approx -x$, indicating that the second term in the $\log\{\mathrm{GCV}(\lambda)\}$ formula can be approximated by $2\mathrm{tr}(\mathbf{S}_\lambda)/n$.

All these methods apply without any change to multivariate thin plate splines. For tensor products of splines, things are a bit different. A standard, though not unique, strategy is to parameterize $\beta_{k_1,k_2} = b_{k_1}c_{k_2}$, which transforms the function to

$$f(x_{i1}, x_{i2}) = \{\sum_{k_1=1}^{K_1} b_{k_1} B_{k_1,1}(x_{i1})\}\{\sum_{k_2=1}^{K_2} c_{k_2} B_{k_2,2}(x_{i2})\} \ .$$

In this case, the parameter vector is $K_1 + K_2$ dimensional $\beta = (b_1,\dots,b_{K_1},c_1,\dots,c_{K_2})$ and one penalty could be $\beta^t \mathbf{D}\beta$, where

$$\mathbf{D} = \begin{bmatrix} \lambda_1\mathbf{D}_1 & \mathbf{0}_{K_1\times K_2} \\ \mathbf{0}_{K_2\times K_1} & \lambda_2\mathbf{D}_2 \end{bmatrix} \ .$$

The penalties $\lambda_1\mathbf{D}_1$ and $\lambda_2\mathbf{D}_2$ are the univariate "row" and "column" penalties, respectively. The advantage of this approach is that it combines two univariate smoothers into a bivariate smoother. The disadvantage is that it requires two smoothing parameters, which increases computational complexity.

As we have seen in Sections 2.3.1.2, we could have multiple covariates that we want to model nonparametrically. The data structure is $(y_i, z_i, x_{i1},\dots,x_{iq})$ and we would like to fit a model of the type

$$y_i = \gamma_0 + \gamma_1 z_i + \sum_{q=1}^{Q} f_q(x_{iq}) + \epsilon_i \ ,$$

where $f_q(\cdot)$ are functions that are modeled as splines. As we have discussed, this spline model can be written in matrix format as

$$\mathbf{y} = \mathbf{Z}\gamma + \sum_{q=1}^{Q}\mathbf{X}_q\beta_q + \epsilon \ ,$$

where the ith row of $\mathbf{Z}$ is $(1, z_i)$, $\gamma = (\gamma_0,\gamma_1)^t$, and $\mathbf{X}_q$ and β_q are the spline design matrix and coefficients for the function $f_q(\cdot)$, respectively. Just as in the case of univariate penalized spline smoothing, one can control the roughness of the functions $f_q(\cdot)$ by minimizing the penalized criterion

$$||\mathbf{y} - \mathbf{Z}\gamma - \sum_{q=1}^{Q}\mathbf{X}_q\beta_q||^2 + \sum_{q=1}^{Q}\lambda_q\beta_q^t\mathbf{D}_q\beta_q \ ,$$

where $\lambda_q \geq 0$ is a scalar and $\mathbf{D}_q$ is the matrix that provides the penalty structure for function $f_q(\cdot)$. Note that the parameters γ are not penalized, but for consistency, one could add the "penalty" $\lambda_0\boldsymbol{\gamma}^t\mathbf{D}_0\boldsymbol{\gamma}$, where $\mathbf{D}_0$ is a 2×2 dimensional matrix of zeros. So, the criterion can be rewritten as

$$||\mathbf{y} - \mathbf{Z}\boldsymbol{\gamma} - \sum_{q=1}^{Q} \mathbf{X}_q\boldsymbol{\beta}_q||^2 + \lambda_0\boldsymbol{\gamma}^t\mathbf{D}_0\boldsymbol{\gamma} + \sum_{q=1}^{Q} \lambda_q\boldsymbol{\beta}_q^t\mathbf{D}_q\boldsymbol{\beta}_q . \tag{2.13}$$

Of course, the parameter λ_0 is not identifiable, but the expression is useful for understanding the effects of penalization or lack thereof on the optimization criterion. The model expression can be made even more compact if we denote $\mathbf{X} = [\mathbf{Z}|\mathbf{X}_1|\dots|\mathbf{X}_q]$, the matrix obtained by column binding these matrices, by $\boldsymbol{\beta} = \{\boldsymbol{\gamma}^t, \boldsymbol{\beta}_1^t, \dots, \boldsymbol{\beta}_Q^t\}^t$ and by

$$\mathbf{D}_{\boldsymbol{\lambda}} = \begin{bmatrix} \lambda_0\mathbf{D}_0 & \mathbf{0} & \mathbf{0} & \dots & \mathbf{0} \\ \mathbf{0} & \lambda_1\mathbf{D}_1 & \mathbf{0} & \dots & \mathbf{0} \\ \dots & \dots & \dots & \dots & \dots \\ \mathbf{0} & \dots & \dots & \mathbf{0} & \lambda_Q\mathbf{D}_Q \end{bmatrix} ,$$

where $\mathbf{0}$ is a generic matrix of zeros with the dimensions conforming to each entry. With this notation the penalized criterion can be rewritten as

$$||\mathbf{y} - \mathbf{X}\boldsymbol{\beta}||^2 + \boldsymbol{\beta}^t\mathbf{D}_{\boldsymbol{\lambda}}\boldsymbol{\beta} . \tag{2.14}$$

The only difference is that the design matrix depends now on more than one smoothing parameter $\boldsymbol{\lambda} = (\lambda_0, \lambda_1, \dots, \lambda_Q)$, but all criteria for estimating the smoothing parameters in the case of one smoothing parameter can be used to estimate the vector of smoothing parameters.

The most elegant part of this is that we can seamlessly combine penalized and non-penalized parameters in the same notation. It also becomes clear that parametric and nonparametric components can be integrated and considered together. Adding bivariate or multivariate smoothing follows the exact same principles discussed here.

2.3.3 Smoothing as Mixed Effects Modeling

Here we revisit the penalized spline criterion (2.13) and its more compact form (2.14) and make a few modifications. We divide the expression by $-1/2\sigma_\epsilon^2$ and reparametrize $\lambda_q = \sigma_\epsilon^2/\sigma_q^2$ for $q = 0, \dots, Q$, where $\sigma_q^2 \geq 0$ are positive parameters. With this notation criterion (2.13) becomes

$$-\frac{||\mathbf{y} - \mathbf{Z}\boldsymbol{\gamma} - \sum_{q=1}^{Q} \mathbf{X}_q\boldsymbol{\beta}_q||^2}{2\sigma_\epsilon^2} - \frac{\boldsymbol{\gamma}^t\mathbf{D}_0\boldsymbol{\gamma}}{2\sigma_0^2} - \sum_{q=1}^{Q} \frac{\boldsymbol{\beta}_q^t\mathbf{D}_q\boldsymbol{\beta}_q}{2\sigma_q^2} . \tag{2.15}$$

Thus, for a fixed σ_ϵ^2 and σ_θ^2, the solution that maximizes (2.15) is the best linear unbiased predictor in the model

$$\begin{cases} [\mathbf{y}|\boldsymbol{\beta}, \sigma_\epsilon^2] = N(\mathbf{Z}\boldsymbol{\gamma} + \sum_{q=1}^{Q} \mathbf{X}_q\boldsymbol{\beta}_q, \sigma_\epsilon^2\mathbf{I}_p) ; \\ [\boldsymbol{\beta}_q|\sigma_q^2] = \dfrac{\det(\mathbf{D}_q)^{1/2}}{(2\pi)^{K_q/2}\sigma_q} \exp\left(-\dfrac{\boldsymbol{\beta}_q^t\mathbf{D}_q\boldsymbol{\beta}_q}{2\sigma_q^2}\right) , \text{ for } q = 1, \dots, Q , \end{cases} \tag{2.16}$$

where the notation $[\mathbf{y}|\mathbf{x}]$ denotes the conditional probability density function (pdf) of $\mathbf{y}$ given $\mathbf{x}$ and K_q is the dimension of the square penalty matrix $\mathbf{D}_q$. This indicates that $\boldsymbol{\beta}_q$ can be viewed as random effects in a specific mixed effects model, where σ_ϵ^2, σ_q^2 and, implicitly, $\lambda_q = \sigma_\epsilon^2/\sigma_q^2$, for $q = 1, \dots, Q$ can be estimated using the usual mixed effects model estimation. Using maximum likelihood (ML) or restricted maximum likelihood (REML) in the mixed effects model (2.16) is equivalent to the ML and REML criteria for estimating the smoothing parameters in the penalized spline model.

For some penalties the matrix $\mathbf{D}_q$ is not full rank and hence, not invertible. This is a problem only because we wanted to keep notation general. In practice, this means that the penalty (or, equivalently, the shrinkage prior) applies only to a subset of the parameters $\boldsymbol{\beta}_q$. For example, in the case of quadratic truncated penalized splines $\boldsymbol{\beta}_q = (\beta_{q,1}, \beta_{q,2}, \ldots, \beta_{q,K_q+3})$ $\boldsymbol{\beta}_q^t \mathbf{D}_q \boldsymbol{\beta}_q = \sum_{k=1}^{K} \beta_{q,k+3}^2$, which is equivalent to assuming $\beta_{q,k+3} \sim N(0, \sigma_q^2)$ for $k = 1, \ldots, K_q$. Thus, in this case the last K_q entries of the $\boldsymbol{\beta}_q$ vector are treated as random effects, while the first three are treated as fixed effects.

The general form of model (2.16) suggests a large number of possible extensions. First, the conditional distribution $[\mathbf{y}|\boldsymbol{\beta}]$ does not need to be Gaussian. Indeed, the same exact idea applies to any other family of likelihoods including Bernoulli, Poisson, and Gamma. Second, the conditional distribution of $\boldsymbol{\beta}$ does not need to be normal (which is equivalent with a quadratic, or L^2, penalty) and could be a multivariate t or double exponential (e.g., L^1 penalty). In this book we will work only with quadratic penalties, which are equivalent to using normal priors on the random effects. Third, the conditional likelihood $[\mathbf{y}|\boldsymbol{\beta}]$ can be enriched with additional fixed and random effects that represent the known data structure. Adding penalties is equivalent to inducing structure on random effects.

Understanding model (2.16) is a crucial step in understanding smoothing in particular and statistics in general. Fundamentally, the result says that most regression models, traditional and modern, standard or penalized, are mixed effects models. We will show that functional regression models can also be viewed as mixed effects models. This will provide the inferential and computational infrastructure for many approaches described in this book.

For presentation simplicity we focused on one covariate and one smoothing parameter. However, ideas generalize directly to the case when we have multiple predictors and a combination of parametric and nonparametric components. For a detailed treatment of these concepts, see [258].

2.3.4 Penalized Spline Smoothing in NHANES

We now illustrate how smoothing can be used in practice and provide the associated software. In Section 1.2.1, we introduced the NHANES physical activity data measured by accelerometers as a function of time of the day. We provide two smoothing examples: scatter plot smoothing, and nonparametric regression with standard covariates and multiple predictors with smooth effects.

2.3.4.1 Mean PA among Deceased and Alive Individuals

Mortality information was available for 8,713 of the 12,610 study participants in NHANES. The left panel in Figure 1.2 displays the smooth means for physical activity data for individuals who died (blue line) and survived (red line). There were 832 deceased individuals and 7,881 who were still alive according to the mortality release file that included events up to December 31, 2019.

Here we show how these smooth estimators are obtained. First, within each group and at each minute of the day, s_j, $j = 1, \ldots, 1440$, the raw average $W(s_j)$ is calculated. These raw averages are displayed in Figure 2.6 as black dots, where the two groups are clearly distinguishable. The raw means exhibit small fluctuations around smooth trends, with larger fluctuations for the groups of individuals who died. This is likely due to the difference in sample size between the two groups: 832 deceased and 7,881 alive individuals, respectively.

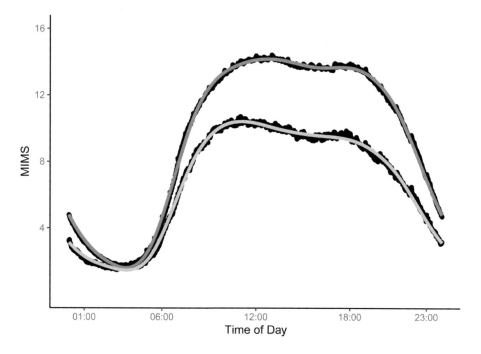

FIGURE 2.6: Raw averages (black dots) and corresponding smooth averages of physical activity data at every minute of the day in the NHANES study. Data are separate in two groups: deceased (blue line) and alive (red line) individuals as of December 31, 2019. The smooth averages were obtained by smoothing the raw averages and not the original data. The x-axis is expressed as time from midnight to midnight and the y-axis is expressed in MIMS.

For each group, we fit a model of the type

$$W_{\text{group}}(s_j) = f_{\text{group}}(s_j) + \epsilon_j \, ,$$

where $f_{\text{group}}(s_j)$ is modeled using a penalized cyclic cubic regression spline with 30 knots placed at the quantiles of the minutes. Because minutes are evenly spaced, the knots are evenly spaced. The function `gam` in the `mgcv` package can be used to provide both fits. As an example, consider the case when the average activity data for individuals who are alive is contained in the variable `nhanes_mean_alive` at the times contained in variable `minute`. Both vectors are of length 1440, the number of minutes in a day. The code below shows how to fit the model and extract the smooth estimators. Notice that here we use REML to select the scalar parameter λ by specifying `method = "REML"`.

```
#Fit penalized cyclical splines with 30 knots using the REML criterion
MIMS_sm_alive <- gam(nhanes_mean_alive ~ s(minute, bs = "cc", k = 30),
                     method = "REML")
#Obtain the smooth estimator
pred_sm_alive <- MIMS_sm_alive$fitted.values
```

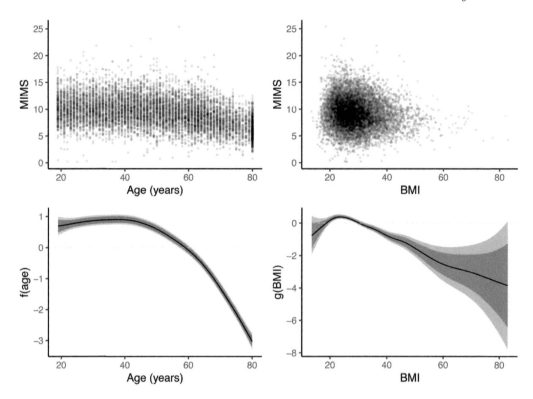

FIGURE 2.7: Upper panels: scatter plots of age and BMI versus average MIMS across days and within days. Each dot corresponds to a study participant. Lower panels: estimated smooth associations (black solid lines) between age (left) and BMI (right) and average MIMS. The 95% pointwise confidence intervals are shown in dark gray, while the 95% correlation and multiplicity adjusted confidence intervals are shown in light gray.

2.3.4.2 Regression of Mean PA

Consider now a slightly more complex model, where the outcome is the average MIMS for every study participant across all days and all minutes of the day. The predictors are age, BMI, gender, and poverty-income ratio (PIR), where the effect of age and BMI is allowed to vary smoothly. PIR is the total family income divided by the poverty threshold in the US. Here we focus on the 7,805 study participants who were older than 18 years at the time of wearing the accelerometer, and had no missing data on BMI, gender, PIR. The upper panels in Figure 2.7 show the scatter plots of average MIMS versus age (upper-left) and BMI (upper-right). Each dot corresponds to a study participant.

We fit the following model

$$\text{MIMS}_i = f(\text{age}_i) + g(\text{BMI}_i) + \text{gender}_i \beta_{\text{gender}} + \text{PIR}_i \beta_{\text{PIR}} + \epsilon_i \,,$$

where the $f(\cdot)$ and $g(\cdot)$ are modeled using penalized splines. If the NHANES data set is stored in the data frame **nhaned_df**, the R code for fitting this model is provided below.

```
#Fit a penalized spline regression with two nonparametric predictors
fit_mims <- gam(MIMS ~ s(age, k = 30) + s(BMI, k = 30) + gender + PIR,
                method = "REML", data = nhanes_df)
```

The syntax is similar to the structure used in the standard `glm` functions, though the nonparametric components are now specified via `s(age, k = 30)` and `s(BMI, k = 30)`. These indicate that both functions are modeled using thin-plate regression splines with 30 knots placed at the quantiles of `age` and `BMI` variables, respectively. There are many ways to extract estimates from the fitted object `fit_mims`. One way is to leverage the `plot.gam` function, which extracts the mean and standard error for each predictor. The code to obtain estimates for age effect is shown below.

```
#Extract the mean and standard error for age effect
plot_fit<- plot(fit_mims, select = 0)
age_xfit<- plot_fit[[1]]$x
age_yfit<- plot_fit[[1]]$fit
age_sefit <- plot_fit[[1]]$se
```

Here we specify `select = 0` in the `plot.gam` function, so that it will not actually return the plot. The estimated `s(age, k = 30)` function at the functional arguments `plot_fit[[1]]$x` is stored as `plot_fit[[1]]$fit`. The 95% pointwise confidence interval can be constructed by adding and subtracting `plot_fit[[1]]$se`, which contains the standard error of the mean already multiplied by 2. For BMI, the same procedure is applied but `plot_fit[[1]]` is replaced by `plot_fit[[2]]`.

The estimated effects of age and BMI on MIMS are shown in the lower panels of Figure 2.7. The estimated smooth associations are shown using black solid lines, with dark gray region representing 95% pointwise confidence intervals. In addition, the light gray region represents 95% correlation and multiplicity adjusted (CMA) confidence intervals; for more details see Section 2.4. Results indicate that average physical activity increase slightly between age 20 and 40 with a strong decline after age 60. Physical activity is also higher for BMI between 22 and 27 with lower levels both for lower and higher BMI levels. If the outcome is binomial or Poisson, we can simply add `family = binomial()` or `family = poisson()` to the `gam` function.

2.4 Correlation and Multiplicity Adjusted (CMA) Confidence Intervals and Testing

So far, we have focused on the case when we observe data (y_i, x_i), $i = 1, \ldots, n$ and are interested in fitting a model of the type

$$
\begin{aligned}
y_i &= f(x_i) + \epsilon_i \\
&= \mathbf{X}_i^t \boldsymbol{\beta} + \epsilon_i \,,
\end{aligned}
\tag{2.17}
$$

for $i = 1, \ldots, n$, where $\mathbf{X}_i^t = \{B_1(x_i), \ldots, B_K(x_i)\}$ is the vector of basis functions evaluated at x_i and ϵ_i are independent random variables that are typically assumed to have a $N(0, \sigma_\epsilon^2)$ distribution. If the model is enriched with additional covariates, nonlinear functions of other covariates, or multivariate functions, the row vector $\mathbf{X}_i^t$ becomes more complex, but the structure of the problem remains unchanged.

Suppose that we are interested in constructing confidence intervals for the $n \times 1$ dimensional vector $\mathbf{f} = \{f(x_1), \ldots, f(x_n)\}^t$, where, for notation simplicity, we considered the same grid of points $x_1, \ldots, x_n$ as that of the observed data. This need not be the case, and the same exact procedure would apply to any grid of points.

As we have shown in Section 2.3, model (2.17) can be fit using either unpenalized or penalized likelihoods and they provide a point estimator $\widehat{\boldsymbol{\beta}}$ and an estimator of the $K \times K$ dimensional variance-covariance matrix $\widehat{\mathbf{V}}_{\boldsymbol{\beta}} = \widehat{\mathrm{Var}}(\widehat{\boldsymbol{\beta}})$. The point estimator of $\widehat{f}(x_i)$ is thus $\mathbf{X}_i^t \widehat{\boldsymbol{\beta}}$ and the point estimator $\widehat{\mathbf{f}}$ of $\mathbf{f}$ can be written as

$$\widehat{\mathbf{f}} = \mathbf{X}\widehat{\boldsymbol{\beta}} , \tag{2.18}$$

where $\mathbf{X}$ is the $n \times K$ dimensional matrix with $\mathbf{X}_i^t$ as row i. For example, in the simplest, unpenalized regression case, $\widehat{\mathbf{V}}_{\boldsymbol{\beta}} = \sigma_\epsilon^2 (\mathbf{X}^t \mathbf{X})^{-1}$ and $\widehat{\mathbf{V}}_f = \widehat{\mathrm{Var}}(\widehat{\mathbf{f}}) = \sigma_\epsilon^2 \mathbf{X}(\mathbf{X}^t \mathbf{X})^{-1}\mathbf{X}^t$. For the case of penalized regression, mixed effects models are used to produce similar variance estimators for $\widehat{\boldsymbol{\beta}}$ and $\widehat{\mathbf{f}}$.

Under the assumption of joint normality of $\widehat{\boldsymbol{\beta}}$, $100(1-\alpha)\%$ confidence intervals that are not adjusted for correlation or multiplicity can be obtained for $\boldsymbol{\beta}$ as

$$\widehat{\boldsymbol{\beta}} \pm z_{1-\alpha/2}\mathrm{diag}(\widehat{\mathbf{V}}_{\boldsymbol{\beta}}) ,$$

and for $\mathbf{f}$ as

$$\widehat{\mathbf{f}} \pm z_{1-\alpha/2}\mathrm{diag}(\widehat{\mathbf{V}}_{\mathbf{f}}) .$$

Here $z_{1-\alpha/2}$ is the $1 - \alpha/2$ quantile of a $N(0,1)$ distribution and $\mathrm{diag}(\mathbf{A})$ is the diagonal vector of the symmetric matrix $\mathbf{A}$.

These confidence intervals for $\mathbf{f}$ are not unadjusted for correlation and multiplicity. Many books and manuscripts refer to these intervals as "pointwise" confidence intervals to differentiate them from the "joint" confidence intervals that take into account the correlation and multiplicity of the tests. Here we prefer the terms "unadjusted" instead of "pointwise" and "correlation and multiplicity adjusted (CMA)" instead of "joint" confidence intervals. Note that the CMA confidence intervals are calculated at every point and it would be difficult to refer to them as "joint," but at every point.

Some drawbacks of the unadjusted confidence intervals are that they (1) do not account for the correlation among tests, which can be quite large given the inherent correlation of functional data; (2) do not address the problem of testing multiplicity (whether or not the confidence interval crosses zero at every time point along a function); and (3) cannot be used directly to conduct tests of significance. To address these problems, we describe three complementary procedures that can be used to construct α-level correlation and multiplicity adjusted (CMA) confidence intervals that account for correlation. We consider the case of univariate smoothing, but the same ideas apply more generally to semiparametric and multivariate smoothing.

2.4.1   CMA Confidence Intervals Based on Multivariate Normality

Under the assumption that the bias of $\widehat{\mathbf{f}}$ is negligible and that $\widehat{\boldsymbol{\beta}}$ has a multivariate normal distribution, it follows that $\widehat{\mathbf{f}}$ has the following approximate multivariate normal distribution

$$\widehat{\mathbf{f}} \approx N(\mathbf{f}, \widehat{\mathbf{V}}_f) , \tag{2.19}$$

where $\widehat{\mathbf{V}}_f = \mathbf{X}\widehat{\mathbf{V}}_{\boldsymbol{\beta}}\mathbf{X}^t$. Let $\mathbf{D}_f = \sqrt{\mathrm{diag}(\widehat{\mathbf{V}}_f)}$ be the $n \times 1$ dimensional vector obtained as the entry-wise square root of the diagonal elements of $\mathbf{V}_f$. This vector consists of the pointwise standard deviations of $\widehat{\mathbf{f}}$. It follows that

$$(\widehat{\mathbf{f}} - \mathbf{f})/\mathbf{D}_f \approx N(\mathbf{0}_n, \mathbf{C}_f) , \tag{2.20}$$

where $\mathbf{0}_n$ is the $n \times 1$ dimensional vector of zeros, $(\widehat{\mathbf{f}} - \mathbf{f})/\mathbf{D}_f$ is obtained by dividing entry-wise the $n \times 1$ dimensional vector $\widehat{\mathbf{f}} - \mathbf{f}$ by the $n \times 1$ dimensional vector $\mathbf{D}_f$, and

$\mathbf{C}_f = \mathbf{V}_f/(\mathbf{D}_f\mathbf{D}_f^t)$ is the correlation matrix obtained by dividing entry-wise the $n \times n$ dimensional matrix $\mathbf{V}_f$ by the $n \times n$ dimensional matrix $\mathbf{D}_f\mathbf{D}_f^t$. Thus, if we can find a value $q(\mathbf{C}_f, 1 - \alpha)$ such that

$$P\{q(\mathbf{C}_f, 1 - \alpha) \times \mathbf{e} \leq \mathbf{X} \leq q(\mathbf{C}_f, 1 - \alpha) \times \mathbf{e}\} = 1 - \alpha\,,$$

where $e = (1, \dots, 1)^t$ is the $n \times 1$ dimensional vector of ones, and $\mathbf{X} = (\widehat{\mathbf{f}} - \mathbf{f})/\mathbf{D}_f \approx N(\mathbf{0}_n, \mathbf{C}_f)$, we can obtain a CMA $(1 - \alpha)$ level confidence interval for $\mathbf{f}$ as

$$\widehat{\mathbf{f}} \pm q(\mathbf{C}_f, 1 - \alpha) \times \mathbf{D}_f\,.$$

Luckily, the function `qmvnorm` in the R package `mvtnorm` [96, 97] does exactly that. To see that, assume that $\widehat{\mathbf{f}}$ is stored in `fhat` and $\widehat{\mathbf{V}}_f$ is stored in `Vf`. The code below describes how to obtain the correlation matrix `Cf`, the $1 - \alpha$ quantiles $q(\mathbf{C}_f, 1 - \alpha)$ for the joint distribution, and the lower and upper bounds of the correlation and multiplicity adjusted (CMA) confidence interval.

```
#Code for calculating the joint CI for the mean of a multivariate Gaussian
#Calculate the standard error along the diagonal
Df <- sqrt(diag(Vf))
#Calculate the correlation matrix
Cf <- cov2cor(Vf)
#Obtain the critical value for the joint confidence interval
qCf_alpha <- qmvnorm(1 - alpha, corr = Cf, tail = "both.tails")
#Obtain the upper and lower bounds of the joint CI
uCI_joint <- fhat + qCf_alpha * Df
lCi_joint <- fhat - qCf_alpha * Df
```

To build intuition about the differences between unadjusted and CMA confidence intervals we will investigate the effect of number of tests and correlation between tests on the critical values used for building confidence intervals. We consider the case when n tests are conducted and the correlation matrix between tests has an exchangeable structure with correlation ρ between any two tests. Thus, the correlation matrix has the following structure

$$\mathbf{C}_f = \begin{bmatrix} 1 & \rho & \rho & \cdots & \rho \\ \rho & 1 & \rho & \cdots & \rho \\ \vdots & \vdots & \vdots & \vdots & \vdots \\ \rho & \rho & \rho & \cdots & 1 \end{bmatrix}\,.$$

Note that unadjusted two-sided confidence intervals would use a critical value of $z_{1-0.05/2} = 1.96$, which is the 0.975 quantile of a $N(0,1)$ distribution. Using a Bonferonni correction for n independent tests would replace this critical value with $z_{1-0.05/(2n)}$. The first column in Table 2.4.2 (corresponding to correlation $\rho = 0$) provides the Bonferonni correction for $n = 10$ (2.80), $n = 100$ (3.48), and $n = 200$ (3.66) tests. These critical values provide an upper bound on the correlation and multiplicity adjusted (CMA) critical values. The second, third, and fourth columns correspond to increasing correlations between tests $\rho = 0.25, 0.50, 0.75$, respectively. Note that for small correlations ($\rho = 0.25$) there is little difference between the Bonferonni correction and the CMA critical values. For example, when $n = 100$ and $\rho = 0.25$ the CMA critical value is 3.43 versus the Bonferonni correction 3.48, a mere 1.5% difference. However, when the correlation is moderate or large, the differences are more sizeable. For example, when $n = 100$ and $\rho = 0.75$, the CMA critical value is 3.01 compared to 3.48, which corresponds to a 13.5% difference.

TABLE 2.2
Critical values for correlation and
multiplicity adjusted $\alpha = 0.05$ level tests
with an exchangeable correlation structure.

# tests (n)	Correlation (ρ)			
	0.00	0.25	0.50	0.75
n=10	2.80	2.78	2.72	2.57
n=100	3.48	3.43	3.30	3.01
n=200	3.66	3.61	3.44	3.12

In functional data, as observations are closer together, the tests become more correlated, which in turn leads to larger differences between the Bonferonni and CMA critical values. Indeed, the Bonferonni correction uses $z_{1-0.05/(2n)}$, which slowly converges to infinity as $n \to \infty$. That means that if we do not account for correlation, the joint confidence intervals have infinite length and are quite useless. In contrast, as the number of observations are sampled more finely, the correlation between observations also increases, keeping the CMA critical values meaningful.

There is some lack of clarity in the literature in terms of exactly what needs to be correlated for the correlation to have an effect on the CMA critical values and confidence intervals. A closer look at equation (2.20) shows that the correlation matrix $\mathbf{C}_f$ is the correlation of the residuals $\widehat{\mathbf{f}} - \mathbf{f}$ and not of the true underlying function $\mathbf{f}$. Thus, correlation does not affect the joint confidence intervals when the observed data are correlated, but when the residuals are correlated. This statement requires reflection as it is neither intuitive nor universally understood.

Figure 2.7 displays the 95% CMA confidence intervals (shown in light gray) for the smooth estimators of the associations between average PA and age and BMI. These confidence intervals overlay the unadjusted 95% confidence intervals (shown in darker gray). In this example, we used the procedure described in this section.

2.4.2 CMA Confidence Intervals Based on Parameter Simulations

When the dimension of the vector $\mathbf{f}$ is very large, it could be impractical to use the approach described in Section 2.4.1. One way around this problem is to use simulations. Our goal is still the same, to obtain $q(\mathbf{C}_f, 1 - \alpha)$, which is the $1 - \alpha$ quantile of $|\widehat{\mathbf{f}} - \mathbf{f}|/\mathbf{D}_f$, where $(\widehat{\mathbf{f}} - \mathbf{f})/\mathbf{D}_f$ has a multivariate distribution $N(\mathbf{0}_n, \mathbf{C}_f)$. Thus, if $\mathbf{X}_1, \dots, \mathbf{X}_B$ are B samples of $n \times 1$ dimensional vectors simulated from the $N(\mathbf{0}_n, \mathbf{C}_f)$ then an excellent approximation of $q(\mathbf{C}_f, 1 - \alpha)$ is obtained by calculating the $100 \times (1 - \alpha)$ percentile of $\{d_1 = \max(|\mathbf{X}_1|), \dots, d_B = \max(|\mathbf{X}_B|)\}$, where the absolute value of the vector $\mathbf{X}$, $|\mathbf{X}|$, is calculated element wise and the maximum is calculated over all entries of the vector. The problem that remains is still the high dimensionality of these variables, as simulating very high-dimensional multivariate normals can be difficult. Instead, we take advantage of the fact that we can simulate $\boldsymbol{\beta}_1, \dots, \boldsymbol{\beta}_B$ from the K-dimensional ($K << n$) distribution $N(\widehat{\boldsymbol{\beta}}, \widehat{\mathbf{V}}_{\boldsymbol{\beta}})$. It follows that $\mathbf{X}_b = \mathbf{B}(\boldsymbol{\beta}_b - \widehat{\boldsymbol{\beta}})/\mathbf{D}_f$, are approximately distributed as $N(\mathbf{0}_n, \mathbf{C}_f)$, which is the distribution of $(\widehat{\mathbf{f}} - \mathbf{f})/\mathbf{D}_f$. Here the division of the $n \times 1$ dimensional vectors $\mathbf{B}(\boldsymbol{\beta}_b - \widehat{\boldsymbol{\beta}})$ and $\mathbf{D}_f$ is conducted element-wise.

Note that generating the $\boldsymbol{\beta}_b$ simulations is typically very fast because the dimension K is not too large. Calculating $\mathbf{B}(\boldsymbol{\beta}_b - \widehat{\boldsymbol{\beta}})$ requires K operations for the difference $\boldsymbol{\beta}_b - \widehat{\boldsymbol{\beta}}$ and $n(2K - 1)$ operations for the matrix product. Conducting the division by $\mathbf{D}_f$ requires an additional n operations for a total of $2Kn + K$ operations. This can be quite large when n is large, but it is linear in n and works in many cases when the exact approach in Section 2.4.1 cannot be used.

2.4.3 CMA Confidence Intervals Based on the Nonparametric Bootstrap of the Max Absolute Statistic

The joint confidence intervals obtained in Sections 2.4.1 and 2.4.2 are based on the assumption that $\widehat{\boldsymbol{\beta}}$ or $\widehat{\mathbf{f}}$ has a multivariate normal distribution. An alternative approach that does not require normality, but requires symmetry of the distribution around the mean, was proposed by [258]. A class of such distributions is the multivariate spherically symmetric distributions; see, for example, [89]. Consider the case when we have B realizations $\widehat{\mathbf{f}}^b$, $b = 1, \ldots, B$ from the distribution of the estimators of $\mathbf{f}$. These realizations could be obtained, for example, using a nonparametric bootstrap of residuals, though other methods could be used. Denote by

$$\bar{\mathbf{f}} = \frac{1}{B} \sum_{b=1}^{B} \widehat{\mathbf{f}}^b$$

the $n \times 1$ dimensional vector corresponding to the mean of $\widehat{\mathbf{f}}^b$, for $b = 1, \ldots, B$. We are interested in the diagonal of the $n \times n$ dimensional covariance matrix

$$\text{diag}(\widehat{\mathbf{V}}_f) = \text{diag}\left\{ \frac{1}{B} \sum_{b=1}^{B} (\widehat{\mathbf{f}}^b - \bar{\mathbf{f}})(\widehat{\mathbf{f}}^b - \bar{\mathbf{f}})^t \right\} = \frac{1}{B} \sum_{b=1}^{B} \text{diag}\left\{ (\widehat{\mathbf{f}}^b - \bar{\mathbf{f}})(\widehat{\mathbf{f}}^b - \bar{\mathbf{f}})^t \right\} .$$

To calculate this diagonal, we do not actually have to obtain the entire $n \times n$ dimensional matrix $\widehat{\mathbf{V}}_f$. Indeed, the $n \times 1$ dimensional vector $\text{diag}(\mathbf{V}_f)$ has the ith entry equal to

$$\frac{1}{B} \sum_{b=1}^{B} \{ \widehat{f}^b(x_i) - \bar{f}(x_i) \}^2 ,$$

where $\widehat{f}^b(x_i)$ is the ith entry of the vector $\widehat{\mathbf{f}}^b$ and $\bar{f}(x_i)$ is the ith entry of the vector $\bar{\mathbf{f}}$. Denote by $\widehat{\mathbf{D}}_f = \sqrt{\text{diag}(\widehat{\mathbf{V}}_f)}$ and consider the variables

$$d_b = \max_{i=1,\ldots,n} |\widehat{f}^b(x_i) - \bar{f}(x_i)| / \widehat{D}_{f,i} ,$$

where $\widehat{D}_{f,i}$ is the ith entry of the vector $\widehat{\mathbf{D}}_f$. If the the $100(1-\alpha)$ percentile of the distribution of $\{ d_b : b = 1, \ldots, B \}$ is denoted by $q(\mathbf{C}_f, 1 - \alpha)$, the $1 - \alpha$ CMA confidence interval for $\mathbf{f}$ can be obtained as

$$\bar{\mathbf{f}} \pm q(\mathbf{C}_f, 1 - \alpha) \times \widehat{\mathbf{D}}_f .$$

This approach does not use the normality assumption of the vectors $\widehat{\mathbf{f}}^b$, but requires a procedure that can produce these predictors. In this book, we focus primarily on nonparametric bootstrap of units that are assumed to be independent to obtain these estimators. Drawbacks of this approach include the need to fit the models multiple times, the reduced precision of estimating low-probability quantiles, and unknown performance when confidence intervals are asymmetric.

2.4.4 Pointwise and Global Correlation and Multiplicity Adjusted (CMA) p-values

Consider the simple case when we would like to test whether

$$H_0 : f(x_i) = 0 \text{ versus } H_A : f(x_i) \neq 0 .$$

The test statistics for this test is $\widehat{f}(x_i)/\widehat{D}_{f,i}$, which in large samples has a $N(0,1)$ distribution. The p-value for such a test is defined as the probability under the null hypothesis of observing a value as extreme or more extreme than the realized value of the test statistic. More precisely, if $X \sim N(0,1)$

$$
\begin{aligned}
\mathrm{p-value} \quad &= \quad P(X \geq |\widehat{f}(x_i)|/\widehat{D}_{f,i}) + P(X \leq -|\widehat{f}(x_i)|/\widehat{D}_{f,i}) \\
&= \quad 2P(X \geq |\widehat{f}(x_i)|/\widehat{D}_{f,i}) \\
&= \quad 2(1 - \Phi\{|\widehat{f}(x_i)|/\widehat{D}_{f,i}\}) ,
\end{aligned}
$$

where $\Phi(\cdot) : \mathbb{R} \to [0,1]$ is the cumulative distribution function of the $N(0,1)$ distribution. It can be easily verified that this p-value is the minimum value of α such that the $1 - \alpha$ level confidence interval $\widehat{f}(x_i) \pm z_{1-\alpha/2}\widehat{D}_{f,i}$ does not contain zero (null hypothesis is rejected at level α).

This is Statistics 101 and one could easily stop here. However, when we focus on a function we conduct not one, but n tests and these tests are correlated. In Sections 2.4.1, 2.4.2, and 2.4.3 we have shown that we can construct a correlation and multiplicity adjusted (CMA) confidence interval $\widehat{f}(x_i) \pm q(\mathbf{C}_f, 1-\alpha)\widehat{D}_{f,i}$. Just as in the case of the normal unadjusted pointwise confidence intervals, these confidence intervals can be calculated for any value of α. For every x_i we can find the largest value of α for which the confidence interval does not include zero. We denote this probability by $p_{\mathrm{pCMA}}(x_i)$ and refer to it as the pointwise correlation and multiplicity adjusted (pointwise CMA) p-value. As $q(\mathbf{C}_f, 1-\alpha) > z_{1-\alpha/2}$, the pointwise CMA confidence intervals are wider than the pointwise unadjusted confidence intervals. Therefore, the pointwise CMA p-values will be larger than pointwise unadjusted p-values and fewer observations will be deemed "statistically significantly different from zero." However, the tests will preserve the family-wise error rate (FWER) while accounting for test correlations.

Similarly, we define the global pointwise correlation and multiplicity adjusted (global CMA) p-value as the largest α level at which at least one confidence interval $\widehat{f}(x_i) \pm q(\mathbf{C}_f, 1-\alpha)\widehat{D}_{f,i}$, for $i = 1, \ldots, n$ does not contain zero. If we denote by $p_{\mathrm{gCMA}}(x_1, \ldots, x_n)$ this p-value, it can be shown that

$$p_{\mathrm{gCMA}}(x_1, \ldots, x_n) = \min\{p_{\mathrm{pCMA}}(x_1), \ldots, p_{\mathrm{pCMA}}(x_n)\}$$

because the null hypothesis is rejected if it is rejected at any point in the domain of $f(\cdot)$.

The advantage of using these p-values over the unadjusted p-values is that the tests preserve their nominal level. The pointwise tests are focused on testing whether a particular value of the function is zero, whereas the global tests focus on testing whether the entire function is zero simultaneously.

We will use the techniques described in this section throughout the book to obtain CMA confidence intervals for high-dimensional functional parameters and conduct inference, including tests of associations that account for the correlation of functional data and test multiplicity.

Throughout this section we have used the nomenclature correlation and multiplicity adjusted (CMA) to refer both to p-values and confidence intervals. This is the first time this nomenclature is used and we hope that it will be adopted. The main reason for this nomenclature is that it is precise and states explicitly that adjustments are for correlation and multiplicity.

2.4.5 The Origins of CMA Inference Ideas in this Book

Tracking the exact origins of ideas is difficult, imprecise, and may overlook important related contributions in the vast FDA and semiparametric regression literature. Here we point out how these ideas percolated into our own work and how they evolved to the level introduced in this section.

The idea of CMA confidence intervals based on multivariate normality was first highlighted by Andrew Leroux, who also identified the connection with the `mvtnorm::qmvnorm` function in R. Furthermore, Cui and Leroux [57] introduced the CMA confidence intervals based on parameter simulations for FoSR regression; see Section 4 of their paper. This provides a quick practical solution in scenarios when `mvtnorm::qmvnorm` does not work. Such scenarios include very large dimensional parameters and/or cases when the distribution is degenerate (correlation matrix is not full rank).

The primary inspiration for the CMA confidence intervals based on the nonparametric bootstrap of the maximum absolute statistic is from Section 6.5 in [258]; see equations (6.16) and (6.17) in their book. The method was proposed there for scatter-plot smoothing and was used for the first time in functional data analysis in [53] for testing the difference in the means of two correlated functional processes. This method was then used extensively in other functional data analysis contexts; see, for example, [57, 223, 270, 327]. This approach works well in practice as long as it does not take too long to conduct the bootstrap analyses. In general, the estimation of very small p-values requires a very large number of simulations. An advantage of this approach is that it provides a tool for checking whether the distribution of the functional parameter is reasonably Gaussian. This assumption is often made, but rarely checked in complex scenarios.

The idea of inverting $1-\alpha$ level CMA confidence intervals to obtain global CMA-adjusted p-values was first described by Ciprian M. Crainiceanu and was published for the first time in [270]. The idea was further extended to obtaining pointwise CMA-adjusted p-values and to using them for testing null hypotheses on functional parameters.

As of the writing of this book, not much is known about CMA-adjusted confidence intervals for cases when the distribution of the functional effect estimators is non-spherically symmetric (e.g., when the distribution is multivariate, but not unimodal). Such behavior of estimators was observed especially on the boundary of the space (close to the parametric envelope) and may be related to the extreme skewness of variance or smoothing parameter estimators at the limit; see, for example, [49, 50, 287].

Other techniques for CMA inference exist and we refer here to the adjusted p-values for functional hypotheses based on the permutation F-tests [253] and global envelope tests [332]. Having multiple options is useful especially when methods have not been tested extensively. We suggest trying multiple methods, and comparing and identifying potential discrepancies between results.

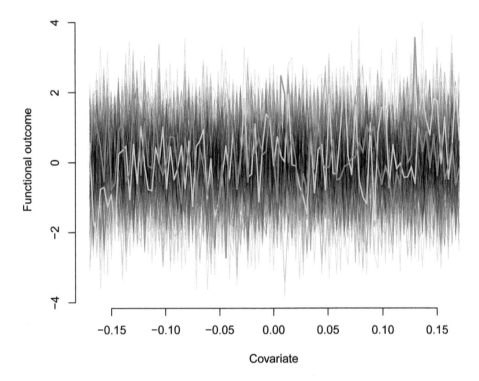

FIGURE 2.8: Simulated data (black curves) from the model (2.21), where the vectors $\mathbf{v}_1$ and $\mathbf{v}_2$ are orthonormal. Two individual trajectories are highlighted in blue and red.

2.5 Covariance Smoothing

An important component of FDA is to combine SVD and PCA with smoothing techniques, which results in functional SVD (FSVD) and functional PCA (FPCA). Using simulations, we explain the combined effect of using SVD and smoothing approaches and discuss the type of practical situations when smoothing is important.

Consider a simple data generating example

$$y_{ij} = b_{1i}v_1(j) + b_{2i}v_2(j) + \epsilon_{ij} , \qquad (2.21)$$

for $j = 1, \ldots, p+1$, where $b_{1i} \sim N(0, \sigma_1^2)$, $b_{2i} \sim N(0, \sigma_2^2)$, and $\epsilon_{ij} \sim N(0, \sigma_\epsilon^2)$ are mutually independent. Here $v_1(j)$ and $v_2(j)$ are the jth entries of the vectors $\mathbf{v}_1 = \mathbf{x}_1/||\mathbf{x}_1||$ and $\mathbf{v}_2 = \mathbf{x}_2/||\mathbf{x}_2||$, respectively, where $||\mathbf{x}||$ denotes the L_2 norm of the vector $\mathbf{x}$, $x_1(j) = 1$ and $x_2(j) = (j-1)/p - 1/2$, for $j = 1, \ldots, p+1$. This ensures that the vectors $\mathbf{v}_1$ and $\mathbf{v}_2$ are orthonormal. We use $p = 100$ and generate $n = 150$ samples with $\sigma_1^2 = 4$, $\sigma_2^2 = 1$, and $\sigma_\epsilon^2 = 1$. Figure 2.8 displays the 150 functions simulated (black curves) and highlights two trajectories in blue and red, respectively.

Model (2.21) is a particular case of the SVD, but it contains the additional term, $\epsilon_{ij} \sim N(0, \sigma_\epsilon^2)$. We investigate what happens if we apply an SVD decomposition to the data generated from a rank 2 SVD model with noise. Figure 2.9 displays the first two estimated right singular vectors. The first right singular vector is shown as light coral, while the true component ($\mathbf{v}_1$) is shown in dark blue. The second right singular vector is shown as dark

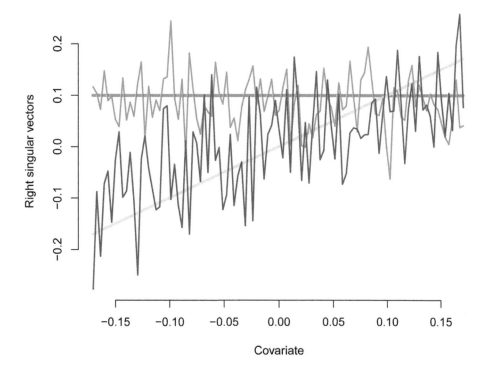

FIGURE 2.9: The first two estimated right singular vectors of the $n = 150$ functions simulated from model (2.21) using SVD. The first right singular vector is shown as light coral while the true component is shown in dark blue. The second right singular vector is shown as dark coral while the true component is shown as light blue.

coral, while the true component ($\mathbf{v}_1$) is shown as light blue. Both estimated right singular vectors exhibit substantial variability around the true components. This increases their complexity (roughness) and reduces their interpretability. In this context, applying a linear smoother either to the data or the right singular vectors addresses most of the observed problems.

Figure 2.10 displays the proportion of variance explained by the estimated right singular vectors. The slow decrease in variance explained is a tell-tale sign of noise contamination for observations. Similar behavior can be observed even if the sample size increases. As useful as SVD and PCA are, it is undeniable that they are difficult to explain to collaborators and be translated into actionable information. When they are contaminated by noise, the problem becomes much harder. Thus, the idea of smoothing the data, the covariance operator and/or the right singular vectors appears naturally in this context. In the next section we will discuss exactly how to do this smoothing and explain the intrinsic connection between these options.

2.5.1 Types of Covariance Smoothing

There are two major types of functional data: dense and sparse. Each type of data requires a specialized set of tools for smoothing. We provide methods for each type of data separately.

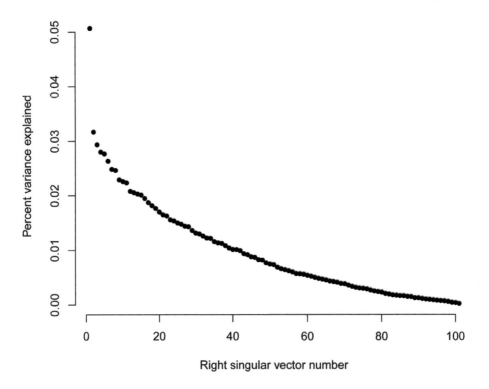

FIGURE 2.10: Percent variance explained by the right singular vectors when the true model is a rank 2 SVD with noise. Results are for the simulated data (black curves) from the model (2.21), where the vectors $\mathbf{v}_1$ and $\mathbf{v}_2$ are orthonormal.

2.5.1.1   Covariance Smoothing for Dense Functional Data

Recall that the ith row of the data matrix, $\mathbf{W}$, is the functional data stored in the $1 \times p$ dimensional vector $\mathbf{W}_i^t = \{W_i(s_j) : j = 1, \ldots, p\}$ for study participant i. The transposed sign is used because, by the convention in this book, person-specific data are recorded by row not by column. As discussed in Section 2.3, most smoothers are of the type $\mathbf{SW}_i$, or in its transposed form, $\mathbf{W}_i^t \mathbf{S}$. The smoother matrix $\mathbf{S}$ is typically a $p \times p$ symmetric semi-positive definite matrix. In matrix format, if the same smoother matrix, $\mathbf{S}$, is applied to each row of $\mathbf{W}$ the result can be written as $\mathbf{WS}$. Thus, smoothing each row of the data is equivalent with replacing the data matrix, $\mathbf{W}$ with

$$\mathbf{WS} = \mathbf{U\Sigma}(\mathbf{V}^t \mathbf{S}) , \qquad (2.22)$$

where $\mathbf{W} = \mathbf{U\Sigma V}^t$ is the SVD of $\mathbf{W}$. This indicates that each row of $\mathbf{V}^t$ (column of $\mathbf{V}$) is smoothed with the same smoother matrix as the rows of $\mathbf{W}$. This shows that (1) smoothing the data is equivalent to smoothing the right singular vectors, and (2) the smoothing matrix is the same for the data and right singular vectors. Thus, smoothing one or another is a stylistic rather than a consequential choice.

Another approach is to smooth the covariance operator $\mathbf{W}^t \mathbf{W}$. Replacing the data matrix $\mathbf{W}$ with $\mathbf{WS}$ results in the following class of "sandwich covariance estimators" [328, 331]

$$\widehat{\mathbf{K}}_S = \mathbf{S}^t \mathbf{W}^t \mathbf{WS} .$$

Irrespective of the smoother matrix, $\mathbf{S}$, these estimators are semi-definite positive and symmetric. That is, the estimators are guaranteed to be covariance matrices.

This approach works with any type of smoother, including parametric or penalized smoothers. Choosing the smoothing parameter is slightly different from the univariate context described in Section 2.3. The residuals after smoothing can be written in matrix format as $\mathbf{W} - \mathbf{WS}_\lambda$. Therefore, the sum of squares for error is $||\mathbf{W} - \mathbf{WS}_\lambda||_F^2$, where $||\mathbf{A}||_F^2$ is the sum of squares of the entries of the matrix $\mathbf{A}$ (square of the Frobenius norm). With this change, the smoothing parameter λ can be estimated using the same smoothing criteria introduced for univariate smoothing. For example, [331] uses the pooled GCV adapted for matrix operators, an approach that scales up to high-dimensional functional data. This is implemented in the `fpca.face` function in the `refund` package. The `fpca.sc` function in `refund` uses the tensor-product bivariate penalized splines. This approach does not scale up as well as the `fpca.face` approach, but provides an automatic smoothing alternative.

2.5.1.2 Covariance Smoothing for Sparse Functional Data

So far, we have discussed cases when data are observed at every time point and the number of observations is large. In many applications, this may not be the case. For example, in the CD4 example in Section 1.2.3 there are between 1 and 11 (median=5) observations per person at times that are specific to each study participant. This type of data structure requires substantial changes in the smoothing approaches. Indeed, consider the case when the observed data are

$$W_i(s_{ij}) = X_i(s_{ij}) + \epsilon_{ij} , \qquad (2.23)$$

where $j = 1, \ldots, p_i$. A subtle, but consequential difference from equation (2.9) is that the location of observations, s_{ij}, are indexed by i and j. This indicates that the locations and number of observations depend on the study participant. Another difference is that the number of observations for every study participant, p_i, is small, which may raise problems for smoothing of the data at the individual level without borrowing information from other study participants.

There are many methods for covariance smoothing for sparse data, including (1) bivariate kernel smoothing of products of residuals [283]; (2) reduced rank spline mixed effects models [134]; (3) tensor product of splines [255]; (4) local polynomials [334]; (4) reproducing kernel Hilbert spaces [28]; (5) geometric maximum likelihood estimation [129].

We cannot cover all these ideas, and we focus on the early, but powerful, idea introduced in [283]. Assume that $W_i(\cdot)$ and $X_i(\cdot)$ are continuous processes and denote by K_W and K_X their covariance operators, respectively. It can be shown that $K_W(s_1, s_2) = K_X(s_1, s_2)$ if $s_1 \neq s_2$ and $K_W(s, s) = K_X(s, s) + \sigma_\epsilon^2$. Therefore, if $\overline{W}(s)$ is an estimator of the marginal mean of $W_i(s)$ and $r_i(s) = W_i(s) - \overline{W}(s)$ are the residuals, $r_i(s_{ij_1})r_i(s_{ij_2})$ is an estimator of $K_W(s_{ij_1}, s_{ij_2})$ and of $K_X(s_{ij_1}, s_{ij_2})$ for $j_1 \neq j_2$. An estimator, $\widehat{K}_X$ of K_X can be obtained using a bivariate smoother for $\{(s_{ij_1}, s_{ij_2}), r_i(s_{ij_1})r_i(s_{ij_2})\}$ for all $i = 1, \ldots, n$ and all $j_1, j_2 = 1, \ldots, p_i$ with $j_1 \neq j_2$. Most bivariate smoothers will produce estimators on the diagonal $\widehat{K}_X(s, s)$. It follows that $r_i(s_{ij})r_i(s_{ij}) - \widehat{K}_X(s_{ij}, s_{ij})$ are estimators of σ_ϵ^2. Therefore, an estimator of σ_ϵ^2 is

$$\widehat{\sigma}_\epsilon^2 = \frac{1}{n} \sum_{i=1}^{n} \frac{1}{p_i} \sum_{j=1}^{p_i} \{r_i(s_{ij})r_i(s_{ij}) - \widehat{K}_X(s_{ij}, s_{ij})\} .$$

Some estimators replace $r_i(s_{ij})r_i(s_{ij})$ with $\widehat{K}_W(s_{ij}, s_{ij})$, where $\widehat{K}_W(\cdot, \cdot)$ is obtained by smoothing $\{(s_{ij_1}, s_{ij_2}), r_i(s_{ij_1})r_i(s_{ij_2})\}$ for all $i = 1, \ldots, n$ and all $j_1, j_2 = 1, \ldots, p_i$. In the original paper [283] kernel bivariate smoothers were proposed, though local polynomial regression and smoothing splines were also suggested as alternatives. Local polynomial smoothing was proposed by [334], where smoothing parameters were estimated by leave-one-subject-out cross-validation. To date, automatic implementation of this approach has proven to be computationally expensive. Bivariate penalized splines regression was proposed by [62], where the smoothing parameter is estimated using a number of popular criteria. The default choice in the `fpca.sc` function in the R package `refund` is the tensor-product of bivariate penalized splines, but other choices are available. Finally, a fast penalized splines approach using leave-one-subject-out cross-validation was proposed by [328]. The function `face.sparse` in the `face` package [329] in R provides this implementation. We will rely on this function for sparse FPCA because it is fast, stable and uses automatic smoothing.

2.5.2   Covariance Smoothing in NHANES

We first illustrate covariance smoothing for dense functional data using the NHANES accelerometry data introduced in Section 1.2.1. For each of the 12,610 study participants, the objectively measured physical activity data were collected as minute-level MIMS and were averaged across available days at each minute of the day. The observations are dense because we have the data at each minute of the 1,440 minutes of the day, and the grid is the same across 12,610 study participants.

The data are stored as a $12,610 \times 1,440$ matrix named `MIMS`. Each column of the matrix represents the data collected at one minute of a day from midnight to midnight. Once the data are in this format, the raw covariance can be obtained as follows.

```
#Calculate raw covariance for NHANES accelerometry data
cov_nhanes_raw <- cov(MIMS, use = "pairwise.complete.obs")
```

Here we specify `use = "pairwise.complete.obs"`, since the minute-level MIMS values were missing for some study participants at certain times of the day.

To perform the covariance smoothing for NHANES accelerometry data, we use the `fpca.face` function in the R package `refund`, which performs Functional Principal Component Analysis (FPCA) for large-scale high-dimensional functional data and implements the covariance smoothing internally. The technical details behind this FPCA software will be introduced in Chapter 3. For illustration purposes, here we focus only on extracting the necessary elements from the fitted object to calculate the smooth covariance. The first step is to perform FPCA on the data using the code shown below.

```
#FPCA using pre-specified smoothing parameter
fit_nhanes_smth_0.01 <- fpca.face(Y = MIMS, lambda = 0.01)
```

By default, the smoothing parameter is chosen using `optim` or a grid search. However, the smoothing parameter can also be manually specified using the `lambda` argument. In the code above, we use 0.01 as the smoothing parameter by specifying `lambda = 0.01`. The eigenvalues and eigenfunctions are stored as `evalues` and `efunctions` elements in the fitted object. Therefore, the smooth covariance can be calculated as follows.

```
#Calculate smooth covariance using the FPCA output
cov_nhanes_smth_0.01 <- tcrossprod(fit_nhanes_smth_0.01$efunctions %*%
                                   diag(sqrt(fit_nhanes_smth_0.01$evalues)))
```

Figure 2.11 displays the raw covariance function (upper-left panel) and the smooth estimator of the covariance functions using `fpca.face` function with different smoothing parameters, including when $\lambda = 0.01$ (upper-right), $\lambda = 1$ (bottom-left), and $\lambda = 100$ (bottom-right). Both the x-axis and y-axis represent the time of day from midnight to midnight. Higher covariance values are colored in red while lower covariance values are colored

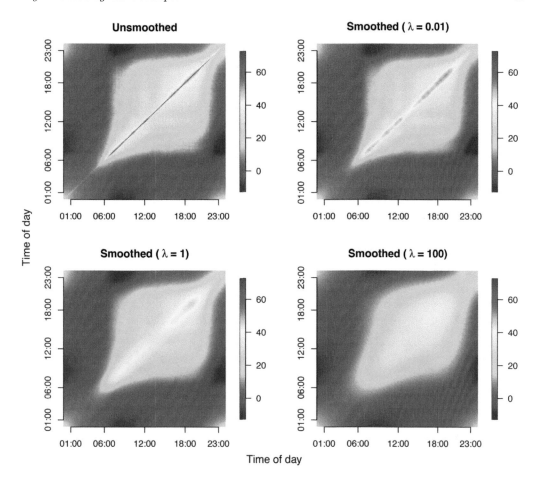

FIGURE 2.11: Estimated raw covariance (upper-left) and smooth covariance functions for the NHANES accelerometry data using the `fpca.face` function with different smoothing parameters shown in the title of each panel.

in blue. In the raw covariance, the highest values correspond to the morning period between 6 AM and 11 AM and the evening period between 5 PM and 9 PM. This result suggests that the physical activity intensity tends to have higher variability in the morning and evening. As the smoothing parameter increases, we observe a smoother covariance function across the domain and the highest covariance value becomes smaller. The color coding was kept identical in all panels to make the comparison among various covariance estimators possible.

```
#Calculate correlation from covariance
cor_nhanes_smth_0.01 <- cov2cor(cov_nhanes_smth_0.01)
```

To obtain the correlation function after covariance smoothing, one way is to use the `cov2cor` function in the `stats` package. Figure 2.12 displays the correlation plots corresponding to the covariance plots displayed in Figure 2.11. The top-left panel corresponds to un-smoothed correlation, indicating mostly positive estimators (shades or red) with stronger correlations around the main diagonal. This is due to the fact that physical activity tends to persist to be high or low for short periods of time. However, some light shades of blue can

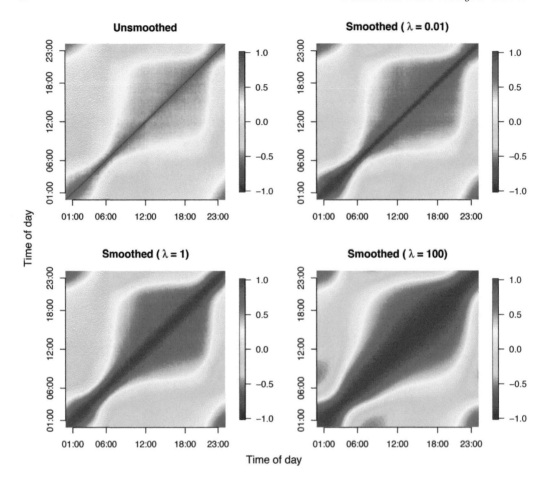

FIGURE 2.12: Estimated raw correlation (upper-left) and smooth correlation functions for the NHANES accelerometry data using the `fpca.face` function with different smoothing parameters shown on the title of each panel.

be observed farther away from the diagonal, which could indicate that individuals who tend to be more active during the day, may be less active during the night. As the amount of smoothing increases, correlations become stronger and at times that are farther away; note the changes toward darker shades of red in wider bands around the main diagonal as the smoothing parameter, λ, increases. The negative correlations off diagonal (shades of blue) become more localized and clearly defined.

2.5.3 Covariance Smoothing for CD4 Counts

We illustrate covariance smoothing for sparse functional data using the CD4 counts data described in Section 1.2.3. Recall that data are the log CD4 counts at a number of time points that are different for every study participant. As observations are sparse within the original data matrix, a more efficient way to store them is in the following format.

```
#Efficient data structure for CD4 data
head(data)
w           argvals       subj
6.306       -9            1
6.795       -3            1
6.488       3             1
6.623       -3            2
6.129       3             2
```

The first line provides the first log CD4 count for study participant 1, which was recorded on month 9 before sero-conversion. Recall that 0 indicates the time of sero-conversion, while positive times are months from sero-conversion, and negative times are months before sero-conversion. The first study participant has three observations at 9 and 3 months before sero-conversion and at 3 months after. The second study participant has their first observation 3 months before sero-conversion. In this format, the data is roughly 4 times smaller, as it is recorded in a 1877×3 dimensional matrix compared to the original format, which is a 366×61 dimensional matrix.

Once data are in this format, the powerful `face.sparse` function in the R package `face` [329] can be used. It uses automatic penalized splines smoothing and fits the data in less than 10 seconds on a standard laptop.

```
#Load the face package
library(face)
#Use face to conduct sparse FDA
fit_face <- face.sparse(data, argvals.new = c(-20:40))
```

The function produces estimators of the mean and covariance as well as predictors for each study participant at each of the time points in the vector `argvals.new`. The estimated smooth mean, covariance, and correlations can be extracted as described below.

```
#Extract the smooth mean
wbar <- fit_face$mu.new
#Extract the covariance function
Chat <- fit_face$Chat.new
#Extract the correlation function
Rhohat <- fit_face$Cor.new
```

We will visualize now the data $r_i(s_{ij_1})r_i(s_{ij_2})$ for every $i = 1, \ldots, n = 366$ and $j_1, j_2 = 1, \ldots, p_i$. To better understand how these data are constructed, we will focus on the first study participant, who had the observations 6.306, 6.795, and 6.488 at times -9, -3, and 3 months, respectively. Thus, $W_1(-9) = 6.306$, $W_1(-3) = 6.795$, and $W_1(3) = 6.488$, where $s_{11} = -9$, $s_{12} = -3$, $s_{13} = 3$, and $p_i = 3$. At these time points, the estimated smooth mean (stored in variable `wbar`) is $\overline{W}(-9) = 6.852$, $\overline{W}(-3) = 6.779$, $\overline{W}(3) = 6.605$, respectively. Thus, the residuals are $r_1(-9) = 6.306 - 6.852 = -0.546$, $r_1(-3) = 0.016$, and $r_1(3) = -0.157$. Conceptually, after these transformations, data has the following format for the first study participant.

```
#CD4 data, estimated mean and residuals / first study participant
w           wbar        r            argvals       subj
6.306       6.852       -0.546       -9            1
6.795       6.779       0.016        -3            1
6.488       6.605       -0.157       3             1
```

The next step is to obtain all the cross products of residuals withing the same study participant. Because there are 3 observations, there are $3^2 = 9$ combinations of these observations. The data containing the cross products of residuals has the structure shown below. Consider the first row, which corresponds to the cross product $r_1(-9)r_1(-9) = (-0.546)^2 = 0.298$. The corresponding values for the variable sj1 is -9 and for variable sj2 is -9. This indicates that the cross product corresponds to (s_{1j_1}, s_{1j_2}), where $j_1 = j_2 = 1$ and $s_{1j_1} = s_{1j_2} = -9$. Similarly, the second row corresponds to the cross product $r_1(-9)r_1(-3) = (-0.546) \times (0.016) = -0.009$. The corresponding values for the variable sj1 is -9 and for variable sj2 is -3. This indicates that the cross product corresponds to (s_{1j_1}, s_{1j_2}), where $j_1 = 1$, $j_2 = 2$, $s_{1j_1} = -9$, and $s_{1j_2} = -3$. If a vector ri contains the residuals for study participant i, then the vector of cross products, Kwi, and of time points can be obtained as follows.

```
#Construct the residual cross products
#The i at the end indicates study participant i
Kwi <- as.vector(kronecker(ri, t(ri)))
```

```
#Cross products of residuals / first study participant
Kwi        sj1        sj2
 0.298     -9         -9
-0.009     -9         -3
 0.086     -9          3
-0.009     -3         -9
 0.000     -3         -3
-0.003     -3          3
 0.086      3         -9
-0.003      3         -3
 0.025      3          3
```

Once the data are obtained for each study participant level, the study participant matrices of cross product residuals (Kwi) are bound by rows. The final matrix of cross products is $11{,}463 \times 3$ dimensional and is stored in Kw. This happened because we consider all mutual products (for a total of 11,463) compared to the total number of observed data points (1,877). Figure 2.13 displays the location of each pair of observations (s_{ij_1}, s_{ij_2}) (column sj1 versus column sj2). Points are jittered both on the x and y axis by up to three days. The color of each point depends on the size of the absolute value of the product of residuals, $|r_i(s_{ij_1})r_i(s_{ij_2})|$, (the absolute value of the Kw). Darker colors correspond to smaller cross products. The most surprising part of the plot is that it is quite difficult to see a pattern. One may identify higher residual products in the upper-right corner of the plot (higher density of yellow dots), but the pattern is far from obvious. There are two main reasons for that: (1) the products of residuals are noisy as they are not averaged over all pairs of observations; and (2) there is substantial within-person variability. This is a good example of the limit of what can be done via visual exploration and of the need for smoothing.

Figure 2.14 displays the smooth estimator of the covariance function using the face.sparse function. Both the x- and y-axis represent time from sero-conversion and the matrix is symmetric. The surface has larger values towards the upper-right corner indicating increased variability with the time after sero-conversion. The increase is gradual over time and achieves its maximum 40 months after sero-conversion. A small increase in variability is also apparent in the lower-left corner of the plot, though the magnitude is smaller than in the upper-right corner. Covariances are smaller for pairs of observations taken before and after sero-conversion (note the blue strips in the left and lower parts of the plot). These are effects of variability and not of correlation. Indeed, Figure 2.15 displays the smooth correlation function for the CD4 counts data using the face.sparse. The yellow strip indicates that the estimated correlations are very high for a difference of about

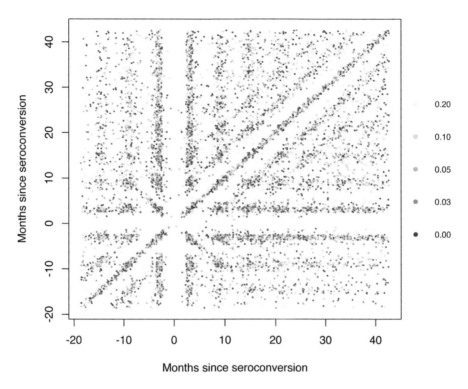

FIGURE 2.13: Product of residuals, $r_i(s_{ij_1})r_i(s_{ij_2})$ after subtracting a smooth mean estimator for every pair of months (j_1, j_2) where data was collected for study participant i. Coloring is done on the absolute value $|r_i(s_{ij_1})r_i(s_{ij_2})|$.

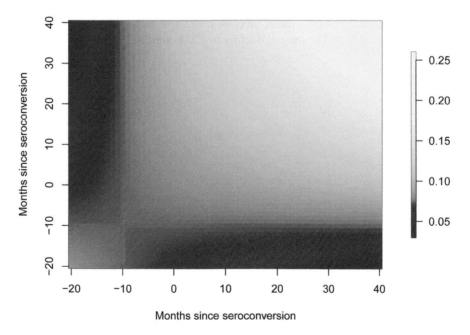

FIGURE 2.14: Estimated smooth covariance function for the CD4 counts data using the `face.sparse` function on data shown in Figure 2.13.

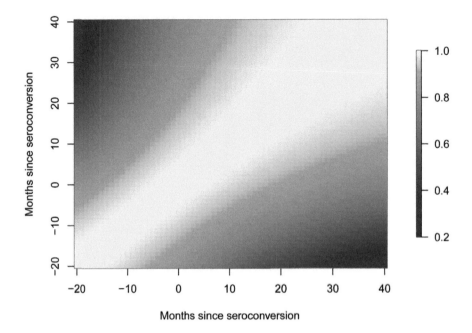

FIGURE 2.15: Estimated smooth correlation function for the CD4 counts data using the `face.sparse` function on data shown in Figure 2.13.

10 months between observations in the first 30 months of the study (days -20 to day 10). For the last 30 months of the study, the high correlations extend to 15 and even 20 months (note the yellow band getting wider in the left-upper corner). So, as time passes, variability of observations around the true mean increases, but the within-person correlations also increase. This is likely due to the overall decline in CD4 counts.

3

Functional Principal Components Analysis

In this chapter we discuss functional principal components analysis (FPCA), an analytic method closely related to PCA for multivariate data that is central to many FDA applications. We will first define FPCA and then highlight connections to classical PCA for multivariate data introduced in Section 2.1.

3.1 Defining FPCA and Connections to PCA

Suppose that we observe the p dimensional functional data $\boldsymbol{W}_i^t = [W_i(s_1), \ldots, W_i(s_p)]$ where $i = 1, \ldots, n$ and $s_j \in S$ for $j = 1, \ldots, p$. Data can be stored in the $n \times p$ dimensional matrix $\mathbf{W}$ with the (i, j) entry equal to $W_i(s_j)$. Each row of $\mathbf{W}$ corresponds to a study participant, i, and each column corresponds to a sampling point in the functional domain, s_j.

One could treat the data matrix $\mathbf{W}$ as multivariate data and use the corresponding analytic tools. However, there are a few important conceptual and practical differences between multivariate and functional data. First, functional data assumes that the grid where data are measured could be much finer or coarser depending on the experimental design and technical limitations. Thus, there is an implicit assumption that one could measure the data "between" s_j and s_{j+1}. This is not the case in multivariate data where there is nothing "between" a person's age and their body mass index. This is the assumption of continuity of the underlying signal. Second, functional data are ordered by the relative geometry of the functional domain, S. For example, if S is time, the observation $W_i(s_{j+1})$ is observed after $W_i(s_j)$ because $s_j < s_{j+1}$. This is the assumption of ordering of the functional domain. Third, functional measurements units $W_i(s_j)$ are on the same scale and have the same interpretation across i and j. For example, the number of deaths due to COVID-19 in the US has the same interpretation on January 1, 2021 and February 19, 2021. This is the assumption of self-consistency. Fourth, observations that are closer in time are likely to be more similar than observations that are farther away (in the distance associated with S). This is the assumption of continuity and smoothness of the underlying signal ("continuous functions that are not too wiggly"). Fifth, the meaning of s_j is the same across all functions. For example, 1 PM means the same thing for each physical activity function measured across individuals. This is the assumption of colocalization.

Given all of these explicit or implicit assumptions about functional data, it is reasonable to wonder whether they make any difference in practice. After all, one could simply decompose the matrix $\mathbf{W}$ using either SVD or PCA, as described in Section 2.1, ignore assumptions, and hope for the best. While this optimistic approach to data analysis sounds a little haphazard, it turns out to be reasonable in many applications, at least as a first step. Moreover, the approach makes sense because the measurement units for $W_i(s_j)$ are the same across i and j and there is no need to standardize the matrix columns or conduct dimension reduction on the correlation operator. We will show that there is a better way of accounting for these assumptions and, especially, for noise.

The conceptual framework for functional PCA starts by assuming that the observed data is

$$W_i(s_j) = X_i(s_j) + \epsilon_{ij} , \tag{3.1}$$

where $\epsilon_{ij} \sim N(0, \sigma_\epsilon^2)$ are independent identically distributed random variables. Here $W_i(s_j)$ are the observed data, $X_i(s_j)$ are the true, unobserved, signals and ϵ_{ij} are the independent noise variables that corrupt the signal to give rise to the observed data. The fundamental idea of functional PCA (FPCA) is to assume that $X_i(\cdot)$ can be decomposed using a low-dimensional orthonormal basis. If such a basis exists and is denoted by $\phi_k(s)$, for $k = 1, \ldots, K$, then model (3.1) can be rewritten as

$$W_i(s_j) = \sum_{k=1}^{K} \xi_{ik} \phi_k(s_j) + \epsilon_{ij} , \tag{3.2}$$

which is a standard regression with orthonormal covariates. Of course, there is the problem of estimating the functions $\phi_k(\cdot)$ and dealing with the potential variability in these estimators [104]. However, the assumptions of smoothness and continuity can be imposed by making assumptions about the underlying signals, $X_i(\cdot)$, or equivalently, about the basis functions $\phi_k(\cdot)$, $k = 1, \ldots, K$. This can be done even though the noise, ϵ_{ij}, may induce substantial variability in the observed data, $W_i(\cdot)$. Ordering, self-consistency, and colocalization are implied characteristics of functional data and will be used to analyze data that can reasonably be fit using model (3.1). The model is especially useful in practice when the number of basis functions in (3.2), K, is relatively small, which can substantially reduce the problem complexity.

Our notation in (3.2) connects FPCA with the discussion of Gaussian Processes in Section 2.2 and to covariance smoothing in Section 2.5. Specifically, the basis $\phi_k(s)$, for $k = 1, \ldots, K$ is obtained through an eigendecomposition of the covariance surface $K_W(s_1, s_2) = \text{Cov}\{W(s_1), W(s_2)\}$ after it has been smoothed to remove the effect of noise and to encourage similarity across adjacent points on the functional domain. Throughout this chapter we will focus on the expression in (3.2) to emphasize the process that generates observed data, but will refer regularly to the underlying covariance surface and its decomposition.

3.1.1 A Simulated Example

To illustrate the use of FPCA, consider the case when data are sampled on a dense and equally spaced grid between 0 and 1. The data generating model is

$$W_i(s) = \sum_{k=1}^{K} \xi_{ik} \phi_k(s) + \epsilon_{is} ,$$

where $K = 4$, $\xi_{ik} \sim N(0, \lambda_K)$, $\epsilon_{is} \sim N(0, \sigma_\epsilon^2)$, ξ_{ik} and ϵ_{is} are mutually independent for all i, k, and s. We set the number of study participants to $n = 50$, the number grid points to $p = 3000$, the variances of the scores to $\lambda_k = 0.5^{k-1}$ for $k = 1, \ldots, 4$, and the standard deviation of the noise to $\sigma_\epsilon = 2$. These parameters can be adjusted to provide different signal to noise ratios. Here we considered a relatively high-dimensional case, $p = 3000$, even though the `fpca.face` [329, 330] function can easily handle much higher dimensional examples. For the orthonormal functions, $\phi_k(\cdot)$, we used the first four Fourier basis functions $\phi_1(s) = \sqrt{2}\sin(2\pi s)$, $\phi_2(s) = \sqrt{2}\cos(2\pi s)$, $\phi_3(s) = \sqrt{2}\sin(4\pi s)$, $\phi_4(s) = \sqrt{2}\cos(4\pi s)$.

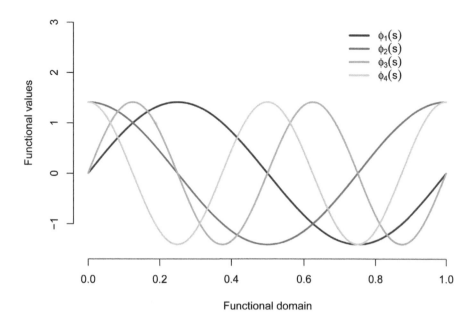

FIGURE 3.1: True eigenfunctions $\phi_k(\cdot)$ with $k = 1, \ldots, 4$ used for simulating the data in model (3.2).

Figure 3.1 displays these four eigenfunctions on $[0, 1]$. The first two eigenfunctions have a period of 1 with slower fluctuations (changes in the first derivative) and the second two eigenfunctions have a period of 2 with faster fluctuations. Note that by construction $\int_0^1 \phi_k^2(s)ds = 1$, but it can be shown that $\int_0^1 \{\phi_3''(s)\}^2 ds = 16 \int_0^1 \{\phi_1''(s)\}^2 ds$, where $f''(\cdot)$ denotes the second derivative of the function $f(\cdot)$. In addition, a straight line has an integral of the square of the second derivative equal to zero. The integral of the square of the second derivative is used extensively in statistics as a measure of functional variability and a penalty on this measure is often referred to as a smoothing penalty. This is different from the mathematical analysis definition of the smoothness of a function which is defined by the number of continuous derivatives. Note that both $\phi_1(\cdot)$ and $\phi_3(\cdot)$ are infinitely continuously differentiable while the integral of the square of the second derivative of $\phi_3(\cdot)$ is 16 times larger compared to the integral of the square of the second derivative of $\phi_1(\cdot)$. This is a quantification of the observed behavior of the two functions, where $\phi_3(\cdot)$ fluctuates more than $\phi_1(\cdot)$. There is substantial confusion between these two concepts and exactly how they are useful in practice. In statistical analysis the number of continuous derivatives is controlled by the choice of the basis function and the addition of a penalty on the square of the second derivative is used to avoid over-fitting. It is this penalty that does the heavy lifting in real-world applications.

3.1.1.1 Code for Generating Data

The R code for simulating the data is shown below. First we specify the number of subjects, dimension of the data, the grid where data are observed, the number of eigenfunctions, the standard deviation of the random noise, the true eigenvalues, and the eigenfunctions. The code here is adapted from the user manual of the function `fpca.face` and all parameters can be modified to produce a variety of data generating mechanisms.

```
#Number of subjects
n <- 50
#Dimension of the data
p <- 3000
#A regular grid on [0,1]
t <- (1:p) / p
#Number of eigenfunctions
K <- 4
#Standard deviation of the random noise
sigma_eps <- 2
#True eigenvalues
lambdaTrue <- c(1, 0.5, 0.25, 0.125)
#Construct eigenfunctions and store them in a p by K matrix
phi <- sqrt(2) * cbind(sin(2 * pi * t), cos(2 * pi * t), sin(4 * pi * t), cos(4 *
pi * t))
```

Functional data with noise is then built by first simulating the scores, $\xi_{ik} \sim N(0, \lambda_K)$. The first line of code below simulates independent identically distributed $N(0,1)$ variables and stores them in an $n \times K$ dimensional matrix. The second line of code re-scales this matrix so that the entries in column k have the true variance equal to λ_k. This could have been done in one line instead of two, but the code would be less pedagogical. The third line of code constructs the signal part of the functional data by multiplying the $n \times K$ dimensional matrix of scores `xi` with the $K \times p$ dimensional matrix `t(phi)` containing the K eigenfunctions. The resulting $n \times p$ dimensional matrix `X` contains the individual functions (without noise) stored by rows. The last step is to construct the observed data matrix `W` by adding the functional signal `X` with an $n \times p$ dimensional matrix with entries simulated independently from a $N(0, \sigma_\epsilon^2)$ distribution.

```
#Simulate an n by K matrix of iid N(0,1)
xi <- matrix(rnorm(n * K), n, K)
#Rescale the matrix by the square root of the eigenvalues
xi <- xi %*% diag(sqrt(lambdaTrue))
#Construct the functional data of size n by p (no noise)
X <- xi %*% t(phi)
#Add the noise scaled by sigma
W <- X + sigma_eps * matrix(rnorm(n * p), n, p)
```

The code shows that model (3.2) is a data generating model, which provides a platform for creating synthetic data sets that could be used for model evaluation and refinement. A first step in this direction is to visualize the type of functions simulated from this model. Recall that functions are stored in the $n \times p$ dimensional matrix `W`.

3.1.1.2   Data Visualization

Figure 3.2 displays the first two rows of the matrix `W`, which correspond to the first two simulated functions from the model (3.2). They both exhibit a high level of noise, though some structure can be observed. Most obvious patterns seem to be associated with the first two true eigenfunctions, which have a frequency equal to 1. Indeed, note how the curve shown in red follows closely a sinusoidal shape with values close to zero at $s = 0$ and $s = 1$ and values close to 3 and -3 at $s = 0.4$ and $s = 0.7$, respectively. The effects of the higher-order eigenfunctions, $\phi_3(\cdot)$ and $\phi_4(\cdot)$, are not as easily recognizable from a direct inspection of the observed data at the individual level.

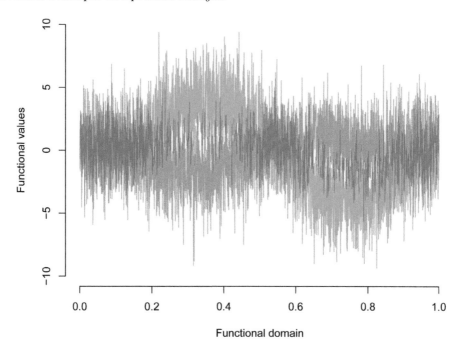

FIGURE 3.2: Two functions simulated from the model (3.2).

It is impractical to plot more than two such functions in the same figure. To partially address this problem, Figure 3.3 displays all 50 functions as a heat map, where each function is displayed on one row and the values of the function are represented as colors (red for low negative values, yellow for values closer to zero, and blue for high positive values). For use of dynamic sorting and visualization of heat maps see [286]. Some structure may become apparent from this heat map, though the noise makes visualization difficult. Indeed, to achieve this level of contrast, colors had to be adjusted to enhance the extreme negative and positive values. Along most rows one can see a wavy pattern with low frequency (functions go from low to high or high to low and back). Some higher frequencies may be observed, but are far less obvious as they are hidden both by the noise and by the higher signal in the lower frequencies.

3.1.1.3   Raw PCA versus FPCA Results

We now compare how multivariate PCA compares to functional PCA, which accounts for noise around the function. Multivariate PCA is obtained directly by using the PCA on the raw data stored in the $n \times p$ dimensional matrix $\mathbf{W}$.

```
#Use PCA on the raw data
results_raw <- prcomp(W)
#Obtain the raw estimated eigenfunctions
raw_eigenvectors <- sqrt(p) * results_raw$rotation
```

Similarly, functional (smooth) PCA is obtained by using the `fpca.face` function.

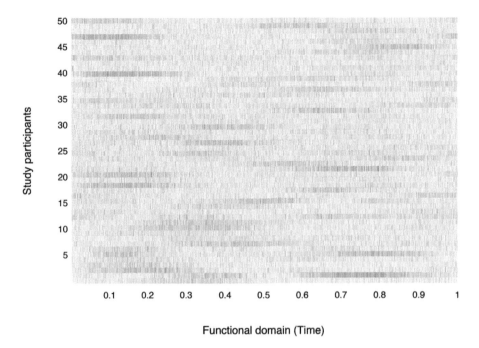

FIGURE 3.3: All 50 functions simulated from the model (3.2). Each function is shown in one row where red corresponds to negative smaller values, yellow corresponds to values closer to zero and blue corresponds to positive larger values. Color has been enhanced for extreme values to improve visualization.

```
#Use FPCA on the raw data
results_smooth <- fpca.face(W, center = TRUE, argvals = t, knots = 100)
#Obtain the smooth estimated eigenfunctions
smooth_eigenvectors <- sqrt(p) * results_smooth$efunctions
```

The calls to the two functions, `prcomp` and `fpca.face`, are similar, which should make the use of `fpca.face` relatively familiar. The difference is mainly internal, where `fpca.face` accounts for the potential noise around the functions and uses the functional structure of the data to smooth the covariance function and its eigenfunctions.

Both the `prcomp` and the `fpca.face` functions produce eigenvectors that need to be re-scaled by multiplication with $\sqrt{p}$ to make them comparable with the true eigenfunctions. Recall that the orthonormality in the functional and vector space are similar up to a normalizing constant (due to the Riemann sum approximation to the integral). Indeed, the functions $\phi_k(\cdot)$ are orthonormal in L_2, but data are actually observed on a grid s_j, $j = 1, \ldots, p$. Thus, we need to work with the vector $\phi_k = \{\phi_k(s_1), \ldots, \phi_k(s_p)\} \in \mathbb{R}^p$ and not with the function $\phi_k : [0,1] \to \mathbb{R}$. Even though the functions $\phi_k(\cdot)$ are orthonormal in L_2, the ϕ_k vectors are not orthonormal in $\mathbb{R}^p$. Indeed, they need to be normalized by $1/\sqrt{p}$ to have norm one. Moreover, the cross products are close to, but not exactly, zero because of numerical approximations. Here L_2 refers to the space of square integrable functions with domain $[0,1]$. It would be more precise to write, $L_2([0,1])$, but we use L_2 for notation simplicity instead.

To better explain this, consider the L_2 norm of any of the functions. It can be shown that $\int_0^1 \phi_k^2(t)dt = 1$ for every k, which indicates that the function has norm 1 in L_2. The integral can be approximated by the Riemann sum

$$1 = \int_0^1 \phi_k^2(t)dt \approx \sum_{j=1}^p (t_j - t_{j-1})\phi_k^2(t_j) = \frac{1}{p}\sum_{j=1}^p \phi_k^2(t_j) =$$

$$= \frac{1}{p}||\phi_k||^2 = \{\phi_k/\sqrt{p}\}^t\{\phi_k/\sqrt{p}\} \, .$$

Here $t_j = j/p$ for $j = 0, \ldots, p$ and $||\phi_k||^2 = \sum_{j=1}^p \phi_k^2(t_j)$ is the Euclidian norm of the vector $\phi_k \in \mathbb{R}^p$. This is different, though closely related to, the L_2 norm in the functional space on $[0,1]$. The norm in the vector space is the Riemann sum approximation to the integral of the square of the function. The two have the same interpretation when the observed vectors are divided by $\sqrt{p}$ (when data are equally spaced and the distance between grid points is $1/p$).

Figure 3.4 displays the first four true eigenfunctions as solid blue lines. For one simulated data set, it also displays the first four estimated eigenfunctions using raw estimators obtained using the function `prcomp` (shown as gray lines) and functional estimators using the function `fpca.face` (shown as dotted red lines). Note that the raw estimators are much noisier than

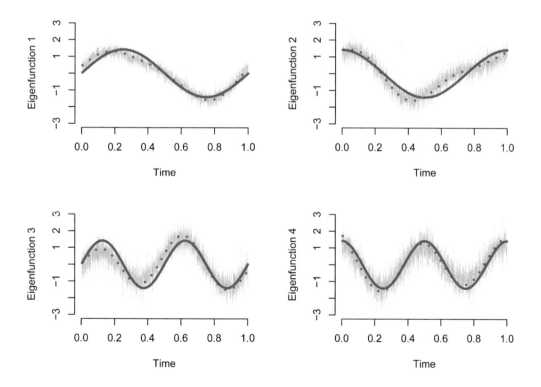

FIGURE 3.4: True (solid blue lines), raw estimated (gray solid lines), and smooth estimated (red dotted lines) eigenfunctions. Raw estimators are obtained by conducting a PCA analysis of the raw data using the `prcomp` function. The smooth estimators are obtained using the `fpca.face` function.

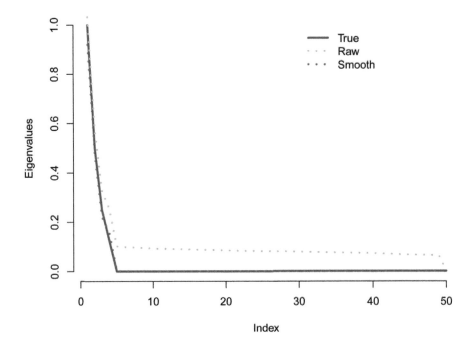

FIGURE 3.5: True (solid blue line), raw estimated (gray solid line), and smooth estimated (red dotted lines) eigenvalues. Raw estimators are obtained by conducting a PCA analysis of the raw data using the `prcomp` function. The smooth estimators are obtained using the `fpca.face` function.

the true functions and would be much harder to interpret in the context of an application. Moreover, it is hard to believe that a true underlying process would change so much for a minuscule change in time. In contrast, the FPCA produces much smoother and more interpretable eigenfunctions that are closer in mean square error to the true eigenfunctions.

Neither the raw nor the smooth estimators of the eigenfunctions are centered on the true eigenfunctions. However, the smooth estimators seem to track pretty closely to an imaginary smooth average through the raw estimators. This happens because we only show results from one simulation. Unbiased estimators would refer to the fact that the average of the red (or gray) curves over multiple simulations is close to the true eigenfunction. This is not required in one simulation, though it is reassuring that the general shape of the function is preserved.

Figure 3.5 displays the true eigenvalues as a solid blue line ($\lambda_k = 0.5^{k-1}$ for $k = 1, \ldots, 4$ and $\lambda_k = 0$ for $k > 5$). The dotted gray line displays the estimated eigenvalues using raw PCA, which ignores the noise in the data. Note that all eigenvalues are slightly overestimated, especially for $k > 5$. Moreover, the decline in the raw estimated eigenvalues is very slow, almost linear after $k > 5$. In applications, such a slow decrease in eigenvalues could be due to the presence of white noise, though, unfortunately, structured noise (with substantial time or space correlations) may also be present. In contrast, the estimated eigenvalues using FPCA track the true eigenvalues much closer. They are not identical due to sampling variability, but indicate much lower bias compared to the raw PCA approach.

3.1.1.4 Functional PCA with Missing Data

So far we have seen the case when all observations are available. However, in many applications, missing data can occur. There are ways of dealing with missing data when conducting PCA on the raw data [13, 68, 289]. A simple solution is to calculate the covariance matrix using the option `use="pairwise.complete.obs"` in the `cov` function. After this step, one can simply use any diagonalization technique. Problems can occur when the dimension of the covariance matrix is too large and cannot be stored in the computer memory. The function `fpca.face` uses the fact that functions are smooth and iterates between smoothing the functions and the covariance operator.

We show how we construct the data set with missing observations and how `fpca.face` is used on the data. We define the missing percentage (in this case 80%), simulate missing data at random, and use the `fpca.face` function. Notice that the call is the same with or without missing data.

```
#Define the percentage of missing data
miss_perc <- 0.8
#Simulate the missing data matrix
NA_mat <- matrix(rbinom(p * n, 1, miss_perc), ncol = p)
#Construct the data with missing observations
W_w_NA <- W
W_w_NA[NA_mat == 1] <- NA
#Apply FPCA to the missing data matrix
results_NA <- fpca.face(W_w_NA, center = TRUE, argvals = t, knots = 100)
```

Figure 3.6 displays the true (solid blue lines), and smooth estimated (red dotted lines) eigenfunctions with 80% missing data. Results illustrate that the FPCA recovers the overall structure of the true eigenfunctions even when 80% of the data are missing. Moreover, the eigenfunction estimators are quite smooth and comparable to the ones obtained from the complete data. Part of the reason is that every function is observed at $p = 3000$ data points and even with 80% missing data there still remain ~ 600 data points per function. This many points are enough to recover the overall shape of each function in spite of the low signal-to-noise ratio. The iteration between functional and covariance smoothing is specific to functional data and relies strongly on the assumptions of smoothness of the underlying signal. Such assumptions would be highly tenuous in a non-functional context, where information is not smooth across covariates. Indeed, consider the case when the data matrix contains age, body mass index (BMI), and systolic blood pressure (SBP) arranged in this order. In this case, age is "not closer" to BMI than to SBP, as ordering of covariates is, in fact, arbitrary.

When missing data are pushed to extremes, say to 99.9% missing, this approach no longer works appropriately and needs to be replaced with methods for sparse functional data. Indeed, with only 2 to 4 observations per function there is not enough information to recover the functional shape by smoothing the individual functional data first. Moreover, the FACE approach to smoothing individual curves relies on the assumption that missing data are not systematically missing, particularly on the boundaries. For example, for a curve that has the first 500 observations missing, FACE uses linear extrapolation to estimate this portion of the curve. If a large number of curves have this type of missingness, other approaches may be more appropriate. We discuss FPCA for sparse and/or irregularly sampled data in Section 3.3.

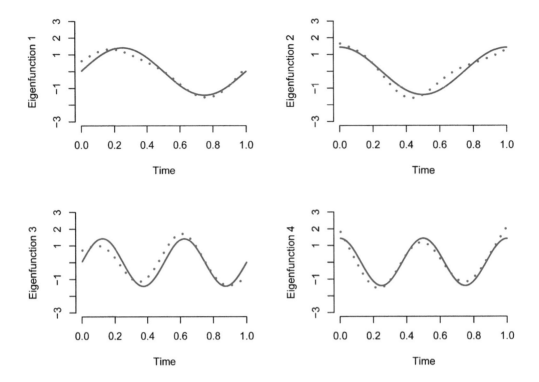

FIGURE 3.6: True (solid blue lines), and smooth estimated (red dotted lines) eigenfunctions with 80% missing data. Smooth estimators are obtained using the `fpca.face` function, but with 80% missing data.

3.1.2   Application to NHANES

So far we have investigated the differences between PCA and FPCA on simulated data. We now investigate these differences in a real-world application by focusing on the minute-level PA data from NHANES.

3.1.2.1   Data Description

The National Health and Nutrition Examination Survey (NHANES) is a large, ongoing study which provides a nationally representative sample of the non-institutionalized US population. NHANES is a study conducted by the Centers for Disease Control (CDC) designed in 2-year waves whose goal is to provide a snapshot of the health and nutrition of the US population. Specific components of the study vary from wave to wave, with wearable accelerometers deployed in the 2003–2004, 2005–2006, 2011–2012, and 2013–2014 waves.

The 2003-2006 accelerometry component of the NHANES study involved participants wearing a waist-worn accelerometer during waking hours. A guide to analyzing these data is provided in [168] and an R data package, `rnhanesdata` [165], is publicly available for download on Github at https://github.com/andrew-leroux/rnhanesdata. The 2011–2014 accelerometer data, released in 2021, are provided in multiple resolutions: subject level, minute level, and raw sub-second level. Here, we use the minute-level MIMS data, which was downloaded directly from the CDC website. All NHANES accelerometry data, including the NHANES 2011-2014 data used in this book, are available through the website `http://www.FunctionalDataAnalysis.org`.

The NHANES data used in this application come from minute-level "activity profiles," representing the average MIMS across days of data with estimated wear time of at least 95% (1,368 minutes out of the 1,440 in a day). Specifically, if we let $W_{im}(s)$ denote the ith individual's recorded MIMS at minute s on day m, and T_{im} be their estimated wear time in minutes for day m, then the average activity profile for participant i can be expressed as $W_i(s) = |\{m : T_{im} \geq 1368\}|^{-1} \sum_{\{m:T_{im} \geq 1368\}} W_{ij}(s)$. For this application we use the NHANES 2011-2014 physical activity data and include all participants aged 5 to 80 with at least one day of accelerometry data with $T_{im} \geq 1368$, resulting in total of 11,820 participants.

Note that although NHANES is a nationally representative sample, obtaining representative estimates for population quantities and/or model parameters requires the use of survey weights and/or survey design [156, 185, 274]. Because the intersection of survey statistics and functional data analysis is a relatively new area of research [32, 33, 34, 226] with few software implementations available, we do not account for survey weights in the data application here or in subsequent chapters.

3.1.2.2 Results

Figure 3.7 displays the first four eigenfunctions estimated using FPCA (left panel) and PCA (right panel). The number of PCs is coded by color: PC1 red, PC2 green, PC3 blue, and PC4 magenta; see the legend in the right-upper corner of the figure. A quick comparison indicates that the two approaches provide almost identical patterns of variability. This is reassuring as both approaches are estimating the same target. Results are also consistent with our simulation results in Figure 3.4 where the FPCA and PCA results showed very similar average patterns, but the PCA results were much noisier.

In NHANES, the PCA results are also noisier, but the noise is not as noticeable, at least at first glance. This is likely due to the fact that the results in Figure 3.4 are based on 50

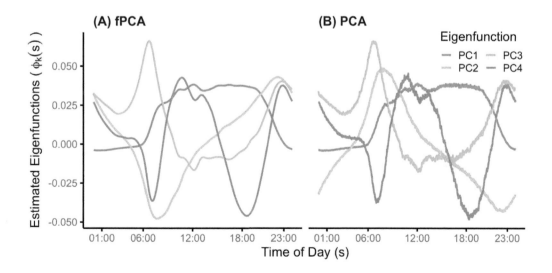

FIGURE 3.7: First four eigenfunctions estimated using FPCA (left panel) versus PCA (right panel) in NHANES. The number of the PC is coded by color: PC1 red, PC2 green, PC3 blue, and PC4 magenta.

functions, whereas results in Figure 3.7 are based on more than 12,000 functions. However, even in this case it would be difficult to use the PCA results because one cannot explain the small, but persistent, variations around the overall trends. Indeed, during a quick inspection of the right panel in Figure 3.7, it seems reasonable to discard the small variations around the pretty clearly identified patterns.

We will have a closer look at the first two eigenfunctions. The first PC (red) is slightly negative during the night (midnight to around 7 AM), strongly positive during the day (from around 9 AM to 9 PM), and decreasing in the evening (10 PM to 12 AM). The second PC (green) is strongly positive between 9 AM and 3 PM and strongly negative during late afternoon and night.

In spite of the wide acceptance of PCA and SVD as methods for data reduction, their widespread use in applications is limited by difficulties related to their interpretation. Indeed, one often needs to conduct PCA and then see if the results suggest simpler, easier to understand summaries based directly on the original data. Figure 3.8 is designed to further explore the interpretation of PCs. The left top panel displays the physical activity trajectories of 10 individuals in the 90th percentile of scores on PC1 (thin red lines). The thick red line is the average trajectory over individuals whose scores are in the 90th percentile. These trajectories are compared to physical activity trajectories of 10 individuals in the 10th percentile of scores on PC1 (thin red lines). The thick blue line is the average trajectory over individuals whose scores are in the 10th percentile. In this example (PC1, shown in left top panel) the difference between the trajectories shown in red and in blue is striking. Indeed, the trajectories shown in red indicate much higher physical activity during the day compared to those shown in blue. Moreover, some of the trajectories shown in blue correspond to higher activity during the night. It may be interesting to follow up and investigate whether the higher physical activity during the night is due to systematic differences or, as suggested by the plot, to a few outlying trajectories.

The other three panels show similar plots for PC2, PC3, and PC4, respectively. Differences between groups continue to make sense, though they are less striking. This happens because the higher numbered PCs explain less variability. However, results for PC2 are still quite obvious: trajectories corresponding to the 90th percentile (red) correspond to much higher physical activity during the night, much lower in early morning to about 2 PM and much higher activity in the evening compared to trajectories corresponding to the 10th percentile (blue). Notice, again, that the difference between the thick red line (average of the trajectories with scores in the upper decile of PC2) and thick blue line (average of the trajectories with scores in the lower decile of PC2) is reflected in the shape of PC2 displayed in Figure 3.7. The difference is that Figure 3.8 provides the contrast between observed groups of trajectories of physical activity.

3.2 Generalized FPCA for Non-Gaussian Functional Data

So far we have discussed FPCA for the case when data are continuous and Gaussian. However, in many applications the functional observations may not be continuous or may exhibit strong departures from Gaussianity. For example, in NHANES, MIMS is a continuous measure of physical activity intensity at every minute of the day. However, for interpretation purposes, MIMS can be transformed via thresholding into (1) binary data corresponding to active/inactive, sleep/wake, wear/non-wear, or walking/non-walking periods; (2) multinomial data corresponding to sedentary, low intensity physical activity (LIPA), and moderate to vigorous physical activity (MVPA); and (3) count data corresponding to the number

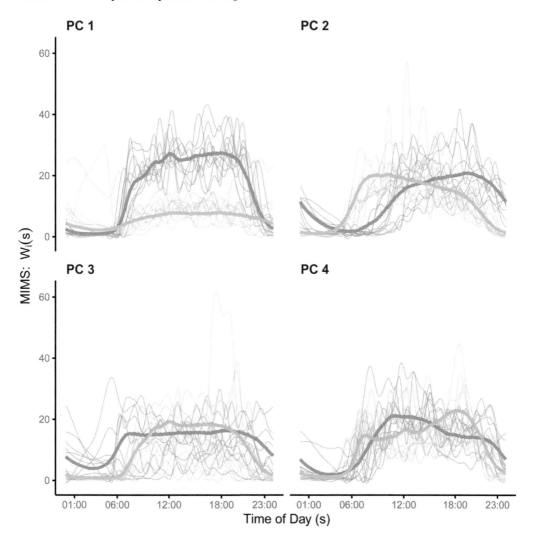

FIGURE 3.8: Average of MIMS profiles with scores in the highest 90th (thick red lines) and lowest 10th (thick blue lines) percentiles for each of the first four principal components in NHANES. For each group, a sample of 10 individual MIMS profiles for those individuals in the 90th (thin red lines) and 10th (thin blue lines) of PC scores are also shown.

of steps or active seconds in every minute. In many other applications, data are collected directly in discrete format (e.g., binary, multinomial, counts) [91, 145, 147, 267, 268, 288] or have strong departures from normality [92, 280].

The question is whether the FPCA ideas can still be used in this context. This is not straightforward, as multivariate PCA and FPCA are developed specifically for Gaussian data and provide a decomposition of the observed variability along orthogonal directions of variation. The corresponding separability of the sum of squares (variance) along directions of variation (eigendirections) does not extend to non-Gaussian outcomes and a different strategy is needed. Here we discuss three approaches for conducting such analyses, highlight their practical implementations in R, and compare computational feasibility and scalability.

3.2.1   Conceptual Framework

When data are continuous but not Gaussian, it may be possible to apply a set of transformations $g_s : \mathbb{R} \to \mathbb{R}$ such that $g_s\{W_i(s)\}$ is reasonably Gaussian for every $s \in S$. In this case PCA and FPCA can be applied to the transformed data. However, such transformations may not be available, interpretable, or widely accepted; for example, there are no generally accepted transformations for the case when some values of $W_i(s)$ are positive and some are negative at the same point s. Moreover, different transformations may be needed for different points in the functional domain; for example, when the empirical distributions of $W_i(s_1)$ and $W_i(s_2)$ for $i = 1, \ldots, n$ are left and right skewed, respectively. When data are not continuous (e.g., binary) transformations do not solve the underlying problem.

To better understand the structure of the problem, consider the case when $W_i(s)$, $s \in S$ are non-Gaussian observed data. We assume that $W_i(s_j)$ have a distribution with mean $E\{W_i(s_j)\} = \mu_i(s_j)$, where

$$\eta_i(s_j) = h\{\mu_i(s_j)\} = X_i(s_j) = \beta_0(s_j) + \sum_{k=1}^{K} \xi_{ik}\phi_k(s_j) \,, \tag{3.3}$$

$\eta_i(\cdot)$ is the linear predictor, $h(\cdot)$ is a link function, $\beta_0(\cdot)$ is a function that varies along the domain S, $\phi_k(\cdot)$ are assumed to be orthonormal functions in the L_2 norm, and ξ_{ik} are mutually uncorrelated random variables. The distribution of $W_i(s_j)$ could belong to the exponential family or any type of non-Gaussian distribution. The expectation is that the number of orthonormal functions, K, is not too large and the functions $\phi_k(\cdot)$, $k = 1, \ldots, K$, are unknown and will be estimated from the data. The advantage of this conceptual model is that, once the functions $\phi_k(\cdot)$ are estimated, the model becomes a generalized linear mixed model (GLMM) with a small number of uncorrelated random effects. This transforms the high-dimensional functional model with non-Gaussian measurements into a low-dimensional GLMM that can be used for estimation and inference. The random effects, ξ_{ik}, have a specific structure, as they are assumed to be independent across i and k. These coefficients can be viewed as independent random slopes on the orthonormal predictors $\phi_k(s_j)$. In practice, often $\phi_1(\cdot) \approx 1$, which is, strictly speaking, a random intercept. Thus, the model can be viewed as a GLMM with one random intercept and $K-1$ random slopes, but this distinction is not necessary in our context. The independence of random effects, orthonormality of $\phi_k(\cdot)$, and the small number of random effects, K, are important characteristics of the model that will be used to ensure that methods are computationally feasible.

A general strategy is to estimate $\eta_i(s_j) = h\{\mu_i(s_j)\}$ and then conduct FPCA on these estimators. We now describe three possible ways of achieving this and illustrate them using the binary active/inactive profiles from the NHANES 2011-2014 data. To obtain binary active/inactive profiles, we first threshold participants' daily MIMS data as $W_{im}^*(s) = 1\{W_{im}(s) \geq 10.558\}$, where $W_{im}(s)$ corresponds to the ith individual's MIMS unit on day $m = 1, \ldots, M_i$ at minute s. We then define their active/inactive profile as $W_i^*(s) = \text{median}\{W_{im}^*(s) : j = 1, \ldots, M_i\}$. For example, if $M_i = 7$, $W_i^*(s)$ is 0 if study participant i was inactive at time s for at least 4 days and 1 otherwise. When M_i is even, say $M_i = 2K_i$, the median is defined as the $(K_i + 1)^{\text{th}}$ largest observation. The threshold for active/inactive on the MIMS unit scale was 10.558, as suggested in [142]. In this section, we use the notation $W_i^*(s)$ to represent the estimated binary active/inactive profiles in NHANES at every s. In general, we have used $W_i(s)$ to denote the continuous MIMS observations at every point or to refer to a general functional measurement.

Figure 3.9 displays the binary active/inactive data for four study participants as a function of time starting from midnight and ending at midnight. Each dot corresponds to one minute of the day with zeros and ones indicating inactive and active minutes, respectively. The red lines show the smoothed estimates of these binary data as a function of the

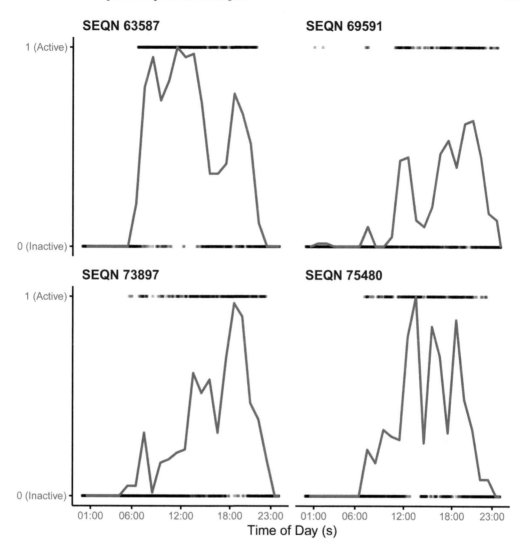

FIGURE 3.9: A sample of 4 individual active/inactive profiles. The red lines represent smoothed estimates of the profiles obtained by binning the data into 60-minute windows and taking the binned mean, corresponding to the estimated probability of being active at a given period of the day.

time of day, with smoothing done by binning the data into 60-minute windows and taking binned averages. These estimates do not borrow strength across study participants, are not specifically designed for binary outcome data, and are highly dependent on the smoothing procedure (i.e., choice of bin width). Moreover, many study participants will have an estimated probability of being active of zero, which is unlikely and cannot be mapped back to the linear scale of the predictor (logit of zero does not exist). For these reasons we are interested in fitting models of the type (3.3) that can extract a small number of directions of variation in the linear predictor space and borrow strength across study participants. In the next sections we review three different approaches to achieve this: fast GFPCA using local mixed effects, binary PCA using the Expectation Maximization (EM) algorithm, and Functional Additive Mixed Models (FAMM).

3.2.2   Fast GFPCA Using Local Mixed Effects

Here we follow the ideas introduced in a recent paper [167], which starts by fixing a value $s \in S$ and a neighborhood N_s around s. Assuming that $W_i^*(s_j)$ has a distribution with mean $\mu_i(s_j)$, one can then fit the following random intercept model

$$\eta_i(s_j) = h\{\mu_i(s_j)\} = \beta_0(s) + X_i(s) \ \text{ for } \ s_j \in N_s, \tag{3.4}$$

where $\beta_0(s)$ is the fixed intercept and $X_i(s) \sim N(0, \sigma_s^2)$ are independent random intercepts for $s \in N_s$ and $i = 1, \ldots, n$, and $\eta_i(s) = \beta_0(s) + X_i(s)$ is the linear predictor. Here we use the $W_i^*(s_j)$ notation because we use the binary example of active/inactive profiles in NHANES, though, in general, we will revert to the $W_i(s_j)$ notation when referring to a functional process, Gaussian or non-Gaussian.

Once this model is fit at one value s, one can move to the next value in the grid and so on using either overlapping or non-overlapping intervals. When observations are close to the boundary asymmetric neighborhoods around the observation can be used. Alternatively, one could fit the models where symmetric neighborhoods are available and extrapolate the mean based on the closest estimable one. In the NHANES example, this is not a problem, as the data are periodic and the beginning of the time interval (say 12 AM) is artificial. Indeed, a neighborhood for midnight could be symmetric around midnight, which allows us to borrow information from data observed before and after midnight.

Applied iteratively, this approach will produce estimators of $X_i(s)$ at every value of $s \in S$. FPCA can then be applied to smooth these estimators across s. An advantage of this approach is that it can be used with any distribution of $W_i^*(s_j)$ as long as the single-level mixed effects model can be fit. Another advantage is that we can add fixed covariate effects to model (3.4), which would allow us to conduct FPCA conditional on these covariates, or, equivalently, FPCA of the residual variability after accounting for fixed effects. Finally, the model could be extended to multilevel or longitudinal functional observations.

The size of the neighborhood, N_s, could influence the quality of estimation; neighborhoods that are too large can lead to over-smoothing while neighborhoods that are too small may lead to under-smoothing and excessive shrinkage toward the population mean (because there are too few observations per bin at the participant level). In specific applications one may want to try several different neighborhood sizes and compare results across the different choices and with the observed data. For example, in the NHANES data we have considered neighborhoods of length 5, 10, and 20 minutes out of a total of 1,440 minutes per day.

The R code for fitting this method is shown below using a neighborhood of 20 minutes. The code assumes that the data are stored in long format. Most steps are relatively straightforward, with the most important step being the fit of the binary GLMMs (option `family=binomial`) using the `glmer` function in the `lme4` package [10]. The structure `Z ~ 1 + (1|SEQN)` indicates that we fit a fixed effects mean with a random intercept model and provides indications of how the model could be expanded to include additional covariates or accommodate multilevel structures. The predicted random effects are stored in the matrix `etamat` where rows correspond to study participants and columns correspond to location in the functional domain. The function `fpca.face` in the `refund` package is used to obtain smooth estimators of the eigenfunctions $\phi_k(s)$, for $k = 1, \ldots, K$. In this case K is chosen so that the proportion of variance explained is `pve=0.95`. We note that choice of K remains an open problem, and in practice, using a proportion of variance explained metric seems to select K that is too large. For that reason we set $K = 4$ in the final estimation procedure.

```
#Fit the local mixed effects models
#Define the window length
w <- 20
#Add column for bin indicator
df_gfpca_local <- df_gfpca %>%
                mutate(sind_w = floor(sind / w) * w + w / 2)
#Fit separate GLMMs for each bin
fit_gfpca_local <-
                df_gfpca_local %>%
                nest_by(sind_w) %>%
                mutate(fit = list(glmer(Z ~ 1 + (1|SEQN),
                data = data, family = binomial))) %>%
                summarize(y_i = coef(fit)$SEQN[[1]]) %>%
                ungroup()
#Re-arrange predicted log odds to wide format
etamat <- matrix(fit_gfpca_local$y_i, nid, 1440 / w + 1)
#Estimate FPCA
fPCA_local <- fpca.face(etamat, pve = 0.95)

#Fit the local mixed effects models
#Fix the number of eigenfunctions used
K <- 4
#get the midpoints of the bins used in the local models
sind_bin <- sort(unique(df_gfpca_local$sind_w))
#Linear interpolation of eigenfunctions on original grid
phi_interp <- matrix(NA, 1440, 4)
colnames(phi_interp) <- paste0("phi_", 1:K)
for(k in 1:K) {
                phi_local <- fPCA_local$eigenfunctions[,k]
                interp_k <- approx(sind_bin, phi_local, sind)
                phi_interp[,k] <- interp_k$y
                }
phi_interp <- data.frame(phi_interp)
phi_interp$sind <- sind
#Add these interpolated eigenfunctions to the long format dataframe
df_gfpca <- left_join(df_gfpca, phi_interp, by = "sind")
#Estimate the GLMM
fit_gfpca <-
                bam(X ~ s(sind, bs = "cc", k = 20) +
                s(SEQN, by = phi_1, bs = "re") +
                s(SEQN, by = phi_2, bs = "re") +
                s(SEQN, by = phi_3, bs = "re") +
                s(SEQN, by = phi_4, bs = "re"),
                method = "fREML", discrete = TRUE, data = df_gfpca,
                family=binomial)
```

The next step of fitting model (3.3) is to interpolate the functions $\phi_k(\cdot)$ on the original grid of the data. This is necessary when the previous steps are applied in non-overlapping intervals because the function is then estimated at a subgrid of points. If overlapping intervals are used, this step is not necessary. For simplicity of presentation, below we use linear interpolation. In practice this can be made more precise using the same B-spline basis used for estimating the covariance operator by `refund::fpca.face`. This is the approach implemented in the `fastGFPCA` package. The estimated eigenfunctions are stored in the $1{,}440 \times K$ dimensional matrix `phi_local`. The final step is to fit model (3.3) conditional on the estimated $\phi_k(s)$ whose random effects are assumed to be uncorrelated. This is implemented in the `bam` function [180, 320, 321] in the `mgcv` package, which allows nonparametric modeling of the fixed intercept $\beta_0(s)$ while accounting for random effects. Note that the function `bam` uses the expression `s(SEQN, by=phi_1, bs="re")` to specify the random effect for the first eigenfunction $\phi_k(\cdot)$. In this implementation the `SEQN` variable (the indicator of study participants) needs to be a factor variable.

One could think of the last step as refitting model (3.3), as the model is initially fit locally to obtain initial estimators of $\beta_0(s)$ and $X_i(s)$. However, this is different because we condition on the eigenfunctions, $\phi_k(s)$, which are unavailable at the beginning of the algorithm. One could stop after applying FPCA to the local estimates and interpolate the predicted log-odds $\widehat{\eta}_i(s_l) = h\{\widehat{\mu}_i(s_j)\}$ obtained from FPCA to obtain subject-specific predictions. While this approach provides good estimates of $\phi_k(\cdot)$, the estimators of the subject-specific coefficients, $\widehat{\xi}_{ik}$, and of the corresponding log-odds tend to be shrunk too much due to the binning of the data during estimation; see [167] for an in-depth description of why this extra step is essential.

The last step of the fitting algorithm can be slower, especially for large data sets. One possibility to speed up the algorithm is to fit the model (3.3) to sub-samples of the data, especially when the number of study participants is very large. For reference we will call the method applied to the entire data `fastGFPCA` and the method applied to four subsets of the data as `modified fastGFPCA`. In Section 3.2.5 we will show that the two methods provide nearly indistinguishable results when applied to the NHANES data. As we have shown, the method can be implemented by any user familiar with mixed model software, though the R package `fastGFPCA` [324] provides a convenient implementation. In a follow-up paper [341] it was shown that using Bayesian approaches may substantially speed up the last step of the algorithm. This is due to the smaller number of random effects, independence of random effects, and orthonormality of the functions $\phi_k(s)$, $k = 1, \ldots, K$.

Figure 3.10 displays the same data as Figure 3.9, where the dots still represent the data and the red lines provide a loess smoothing of the binary data. The blue lines indicate the estimates of the logit$\{\Pr(W_i^*(s) = 1)\}$ based on local GLMMs, while the gold lines represent the same estimators after joint modeling conditioning on the eigenfunction estimators. When applied to these four study participants, the procedure seems to produce sensible results.

Figure 3.11 displays the distribution of estimated probabilities (top panels) and log-odds (bottom panels) of being active as a function of time of day for all study participants in this analysis. Results for the complete fast-GFPCA approach are shown in the left panels, while the binned estimation procedure using local GLMMs are shown in the right panels. For every 30-minute interval a boxplot is shown for the values of all study participants. Note that during the night, when the probability of being active is much lower, the complete approach (right panels) provides more reasonable results as there is better and continuous separation of individuals by their probability of being active during the night. A few study participants are more active during the night (note the dots above 0.5 probability of being active during the night), though fewer individuals are identified following the smoothing procedure. Increased activity probability during the night could correspond to night-shift work, disrupted circadian rhythms, or highly disrupted sleep. During the day, irrespective of method, the median probability of being active culminates around 0.5 somewhere between 11 AM and 12 PM, stays high during the day and decreases in early evening. This agrees with what is known about physical activity, but it is useful to have this quantification and confirmation.

The local estimation method for GFPCA discussed in this section is novel and methods work remains. Nevertheless, the method is (1) grounded in the well-understood GLMM inferential methodology; (2) straightforward to implement for researchers who are familiar with standard mixed model software; and (3) computationally feasible for extremely large

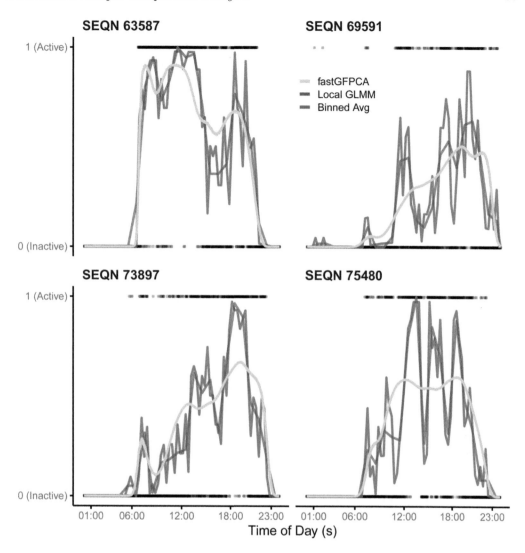

FIGURE 3.10: Results from the local method of estimating GFPCA on the NHANES data using 30-minute bins. Red lines correspond to 60-minute binned averages of the original data, while blue and gold lines present the inverse-logit transformed estimates of logit$\{\Pr(W_i^*(s) = 1)\}$ of the point-wise and smoothed (final fast GFPCA) estimates, respectively.

datasets. As a result, we recommend this as a first-line approach and potential reference for other methods.

3.2.3 Binary PCA Using Exact EM

A fast Expectation Maximization (EM) [61] algorithm for estimating FPCA for functional binary data has been implemented in the `registr` package [323] as part of a registration algorithm [325]. This algorithm requires users to provide the number of eigenfunctions to be estimated and the number of B-splines basis functions for estimating the population

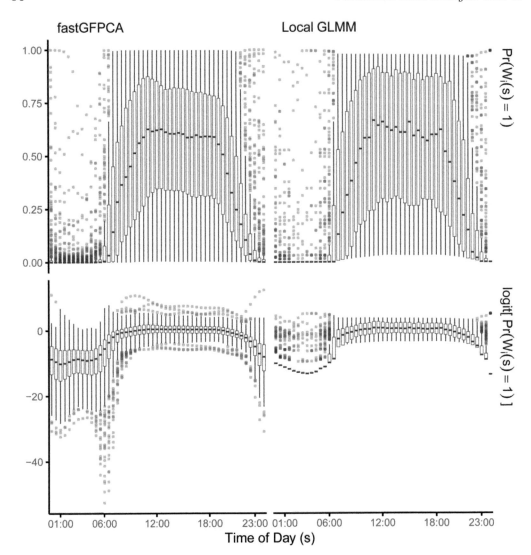

FIGURE 3.11: Population distribution of probability of being active (top panels) and log odds (bottom panels) obtained from fast-GFPCA (left panels) and the binned estimation procedure fit using local GLMMs (right panels).

mean function, $\beta_0(s)$. The results can be sensitive to these choices and options should be explored beyond the current default of one eigenfunction (npc=1) and eight B-spline basis functions for estimating $\beta_0(s)$ (Kt=8). For comparison purposes, in our NHANES application with binary active/inactive profiles we use the registr::bfpca() function with four eigenfunctions and eight basis functions for estimating $\beta_0(s)$. We refer to this method as the variational Bayes FPCA, or vbFPCA.

3.2.4   Functional Additive Mixed Models

A powerful alternative is to use Functional Additive Mixed Models (FAMM) [110, 263, 302] to fit the model

$$\eta_i(s_j) = h\{\mu_i(s)\} = \beta_0(s) + X_i(s) \,, \tag{3.5}$$

where $X_i(\cdot) \sim \mathrm{GP}(0, \mathbf{\Sigma})$ is a zero-mean Gaussian Process with covariance $\mathbf{\Sigma}$. FAMM uses penalized splines to model the study participant functional effects, $X_i(s)$. This approach works with a variety of response families, uses the entire data set to estimate $X_i(s)$, and conducts estimation, smoothing, and inference simultaneously. The approach can also incorporate additional covariates and random effects and was implemented in R using the connection between nonparametric regression models and mixed effects models. An important contribution of the FAMM methodology was to show how nonparametric functional models can be transformed into nonparametric regression models that can then be fit using the `mgcv` function.

To illustrate the procedure, consider a subset of the NHANES data (FAMM cannot be fit on the full data) and fit a Gamma regression for the functional outcomes $W_i(s) \sim \mathrm{Gamma}\{1, \eta_i(s)\}$ and $E\{W_i(s)|X_i(s)\} = \eta_i(s) = \beta_0(s) + X_i(s)$. This is the default R implementation of Gamma regression in `mgcv`, which models the shape parameter using a fixed scale parameter of 1. Under the FAMM approach, this can be fit using the following syntax based on the `bam` function in the `mgcv` package.

```
#Fit the FAMM to NHANES with Gamma outcomes
fit_gfpca_gamma <- bam(W ~ s(index, bs = "cc", k = 10) +
              ti(index, id_fac, bs = c("cc", "re"),
              mc = c(F, T), k = c(10,5)), data = df_gfpca,
              family = Gamma, method = "fREML", discrete = TRUE)
```

Here the functional data are stored in long format. Specifically, the data are contained in the vectors W (functional data), `index` (functional domain), and `id_fac` (factor variable corresponding to subject identifier). The `ti` term syntactically specifies a tensor product smoother of the functional domain and subject identifier, which adds the following subject-specific random effect to the linear predictor

$$X_i(s) = \sum_{k=1}^{K} \xi_{ik} B_k(s) \, .$$

Here $\{B_k(s) : k = 1, \ldots, K\}$ is a spline basis with K basis functions and the coefficients are estimated as $\widehat{\xi}_{ik} = E[\xi_{ik}|\boldsymbol{W}]$, the conditional expectation of ξ_{ik} given the observed data. The functional argument `mc=c(F,T)` in the `ti` function imposes the identifiability constraint $\sum_{i=1}^{n} X_i(s) = 0$ for all $s \in S$ instead of the less stringent default constraint $\sum_{i=1}^{n} \sum_s X_i(s) = 0$. The `mc` argument is shorthand for "marginal constraint" and should be used with caution, while the argument `k=c(10,5)` specifies that $K = 10$ marginal basis functions should be used for estimating $X_i(s)$. The memory requirements and computational time increase substantially with even small increases in K, which could be problematic when the subject-specific random effects have complex shapes.

Once $X_i(s)$ are estimated, one can apply FPCA on these estimators and then proceed as in Section 3.2.2. This was not the intended use of FAMM, but it provides an alternative to the local mixed effects models if we want to conduct dimension reduction on the linear predictor scale for non-Gaussian data. Given the estimates $\widetilde{X}_i(s)$, we may estimate $\mathrm{Cov}\{X_i(s), X_i(u)\}$ using the moment estimator of $\widehat{\mathbf{\Sigma}} = \mathrm{Cov}\{\widetilde{X}_i(s), \widetilde{X}_i(u)\}$. This covariance matrix can be decomposed using either SVD or PCA to obtain the eigenvectors of $\widehat{\mathbf{\Sigma}}$. The model can be re-fit using the first K of these eigenvectors, assuming random effects are independent. One potential limitation of FAMM is that its computational complexity is cubic in the number of subjects. Thus, scaling up the approach could be difficult, though fitting the models on subsets or subsamples of the data may alleviate this problem. This approach can be implemented using standard software for estimating GLMMs as discussed in Section 3.2.2. Here we use the `gamm4` [276] function, which is based on the `gamm` function from the `mgcv` [314, 319] package, but uses `lme4` [10] rather than `nlme` [230] as the fitting method.

```
sm <- smoothCon(s(index, bs = "cc", k = 5), data = df_gfpca_s, absorb.cons =
FALSE)
#Extract the eigenvectors and add them to the data frame
for (k in 1:4){
            df_gfpca_s[[paste0("Phi_s", k)]] <- sm[[1]]$X[, k]
            }
#Fit a nonparametric mean and eigenvectors as random intercepts
fit_glmm <- gamm4(Y_sc ~ s(index, bs = "cc"),
        random = ~ (0 + Phi_s1 + Phi_s2 + Phi_s3 + Phi_s4 | id_fac),
        family = Gamma, data = df_gfpca_s)
```

3.2.5   Comparison of Approaches

Based on the binary active/inactive physical activity data from NHANES 2011-2014 [167] we compare the fast-GFPCA approach based on local mixed effects (`fastGFPCA`) and the binary PCA using the exact EM algorithm `vbFPCA`. Results indicated that as long as the number of spline bases used to model the principal components is sufficiently large for the EM algorithm, the two approaches performed similarly in terms of estimating the linear predictor. However, in our implementation as of the writing of this book, the EM algorithm resulted in biased estimates for both the population mean and eigenfunctions when the population mean function was non-zero. This suggests that if the focus is on dimension reduction and/or interpretation of principal sources of variability in the latent process, the fast-GFPCA approach should be preferred. In contrast, if one is interested only in prediction, either approach may be reasonable. In practice, we recommend estimating GFPCA using multiple approaches and comparing the results, attempting to explain any discrepancies that arise.

Figure 3.12 displays the estimated population means $\widehat{\beta}_0(s)$ (Figure 3.12A), and first four eigenfunctions, $\widehat{\phi}_k(s)$ $k = 1, \ldots, 4$, (Figure 3.12B) of the latent process. Each column of Figure 3.12 corresponds to a different model fit, where color indicates approach (vbFPCA in red, columns 1-2, and fast-GFPCA in blue, columns 3-8). The modified fast-GFPCA estimates a population mean function for each of four randomly selected sub-samples, which are then averaged to produce the overall $\widehat{\beta}_0(s)$. These sub-sample estimates of $\beta_0(s)$ are shown as dashed lines in Figure 3.12A, columns 3-8. As the modified fast-GFPCA uses the population level $\widehat{\phi}_k(s)$, there are no corresponding dashed plots for the eigenfunctions.

The population mean is fairly stable across sub-samples for the modified fast-GFPCA approach (Figure 3.12A, columns 3-8). There was also excellent agreement between the estimated linear predictor of the modified and unmodified fast-GFPCA approaches; results not shown. Moreover, $\widehat{\beta}_0(s)$ and $\widehat{\phi}_k(s)$ for $k = 1, \ldots, 4$ are similar across the fast-GFPCA fits (overlapping versus non-overlapping windows, and bin widths), suggesting that the fast-GFPCA algorithm is fairly robust to the choice of input parameters in the NHANES data.

However, there are substantial differences between vbFPCA and fast-GFPCA both in terms of the estimated population mean function and first four eigenfunctions. The largest differences in $\widehat{\beta}_0(s)$ occur around 6 AM-8 AM and 9 PM-12 AM, where vbFPCA respectively over- and under-estimates $\beta_0(s)$ relative to fast-GFPCA. The vbFPCA estimator for $\widehat{\beta}_0(s)$ suggests that, for participants with $X_i(s) = 0$, the probability of being active between 6 AM and 8 AM is around 80-90% compared to around 50% from fast-GFPCA. This result is unexpected and does not match the results in the data, where we observe approximately 50% probability of being active during this time. One potential explanation could be that

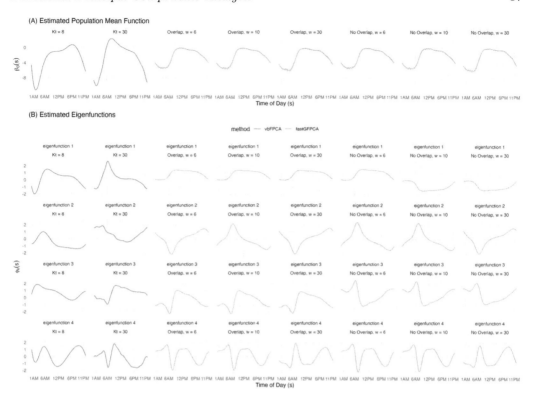

FIGURE 3.12: Estimated population mean function (first row) and the first four estimated eigenfunctions (rows 2–5) in NHANES. Model estimates are presented as red (vbFPCA) or blue lines (fast-GFPCA). The two leftmost columns correspond to vbFPCA (Kt=8 in column 1 and Kt=30 in column 2). The six rightmost columns correspond to fast-GFPCA with different input parameters (overlapping versus non-overlapping windows) and window sizes ($w = 6, 10, 30$). Estimates of the population mean function based on the modified fast-GFPCA are displayed as dashed lines.

vbFPCA provides biased estimators of the mean and that the bias may affect the latent random process variation. Simulation results in [167] further reinforce this hypothesis. Updates to the `registr` package [323] may fix this problem in the future.

3.2.6 Recommendations

We suggest using the fast-GFPCA as the first-line approach to estimating GFPCA, exploring sensitivity to choice of window size, and overlapping versus non-overlapping windows. The `fastGFPCA` function [324] is available, the step-by-step description in this section can be used as a guide, and one can also check the associated website http://www.FunctionalDataAnalysis.com to see specific implementations. A modified version of the approach, where sub-samples of the data are used in the last step of the fitting algorithm can substantially increase computational feasibility. Other approaches, including vbFPCA and FAMM, can also be used and compared to fast-GFPCA to see if they provide similar results. Subsampling of individuals and undersampling or smoothing of observations within individuals may help reduce computational burden and produce results that can be compared with existing solutions.

3.3   Sparse/Irregular FPCA

Thus far we have discussed methods for when data are regularly observed with allowances
for missing data. However, in all applications discussed so far, there are many pairs of
observations for the same distance in time and space. For example, many study participants
have observations at, say, $(0.2, 0.3)$. Sparse or irregularly observed functional data has few
or just one pair of observations at a particular pair of points in the functional domain. Many
applications collect data on participants sparsely (i.e., few observations per curve) and/or
at irregular intervals (i.e., not all participants' data are observed at the same points in the
domain). In this scenario, methods based on smoothing method of moments estimators of
the covariance operator developed for densely observed data do not work.

In Section 2.5.3 we discussed how to recover the information using products of residuals,
as suggested by [283]. The structure of the data in this case is $\{s_{ij}, W_i(s_{ij})\}$, where $i =
1, \ldots, n$ and $j = 1, \ldots, p_i$. The model considered is

$$W_i(s_{ij}) = \beta_0(s_{ij}) + X_i(s_{ij}) + \epsilon_{ij} , \tag{3.6}$$

where $\beta_0(\cdot)$ is a smooth population mean, $X_i(s)$ is a zero mean Gaussian Process with
covariance Σ, and ϵ_{ij} are independent $N(0, \sigma_\epsilon^2)$ random variables. An estimator of the
mean function, $\widehat{\beta}_0(s)$, can be obtained by smoothing the pairs $\{s_{ij}, W_i(s_{ij})\}$ under the
independence assumption using any type of smoother. We prefer a penalized spline smother,
but most other smoothers would work as well. Residuals can then be calculated as $r_i(s_{ij}) =
W_i(s_{ij}) - \widehat{\beta}_0(s_{ij})$.

To estimate and diagonalize Σ, the idea is to consider every pair of observations
(s_{ij_1}, s_{ij_2}) and calculate the products $r_i(s_{ij_1})r_i(s_{ij_2})$ and then smooth these products using
a bivariate smoother. An important methodological contribution was provided by [328],
who developed a fast method for covariance estimation of sparse functional data and implemented it in the `face` package via the `face::face.sparse()` function [329]. We introduce
the motivating dataset, discuss how to apply `face.sparse()` to such data sets, and provide
the context for the methods.

3.3.1   CONTENT Child Growth Data

The CONTENT study [131] was described in Section 1.2.4. Here we provide a summary
of the main features of the study. The CONTENT study was conducted in two peri urban
shanty towns located on the southern edge of Lima city in Peru. Data on weight, length, and
other body measures were collected for the first two years of life. A key goal of the study was
to identify risk factors for growth stunting, a risk factor for poor future health outcomes.
Here we focus on weight and length measurements for 197 children. These measurements
were age- and sex-standardized relative to World Health Organization reference values,
creating Z-scores [20] for length for age (zlen), weight for age (zwei). The study protocol
involved measuring children at semi-regular intervals, with more frequent measurements
closer to birth.

Figure 1.9 in Chapter 1 displays the sampling trajectories for all children in the study.
The figure shows the z-score for length (zlen, first column) and z-score for age (zwei, second
column) for males (first row) and females (second row) as a function of day from birth.
Across the 197 children, there are 4,405 observations for an average of 22 sampling points
per child, with 346 unique sampling points (days from birth). In such cases, when data are
measured at highly irregular intervals, the method of moments (MoM) estimators for the
covariance function do not work. A potential workaround could be to bin the data along the

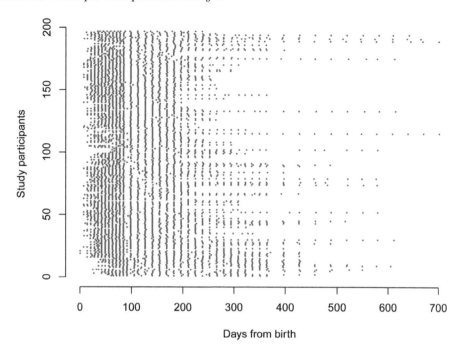

FIGURE 3.13: Distribution of sampling points for each study participant in the CONTENT study. Each study participant is shown on one row and each red dot indicates a time from birth when the child was measured.

functional domain and then use MoM estimators. However, results may depend on the bin size and binning strategy because data are sparse and/or unequally sampled in particular areas of the functional domain.

Figure 3.13 displays the days from birth when each study participant was measured. Each red dot represents a time of a visit from birth and each study participant is shown on a different row. The higher density of points in the first three months is due to the design of the study. The intention was to collect weekly measurements in the first three months of the study. After the first three months observations become sparser and only few children have observations for more than one year. Observations for each child are quite regular, reflecting the design of experiment, though there are missing visits and visits are not synchronized across children.

Figure 3.14 displays the z-scores for length (zlen, shown in blue) and weight (zwei, shown in red) for four study participants. The data for all study participants was displayed before in Figure 1.9, though here we show a different perspective. Indeed, note that for Subject 49 the two time series decrease almost continuously from close to 0 at first measurement to around -2 sometime after day 600. The decrease seems to be synchronized indicating that the baby did not grow in length or weight as fast as other babies, at least relative to the WHO standard. Subject 73 has increasing trajectories both for zlen and zwei, where the baby started around -1.5 for both curves and ended around 0, or close to the WHO mean around day 500 from birth. The increase in zwei (z-score for weight) was faster for the first 200 days, while the increase in zlen (z-score for length) was more gradual. The data for Subject 100 was only collected up to day 250 from birth and indicates that the baby was born longer than the WHO average, grew in length relative to the WHO average in the first 150 days, though the last couple of measurements indicate that the baby's length went back

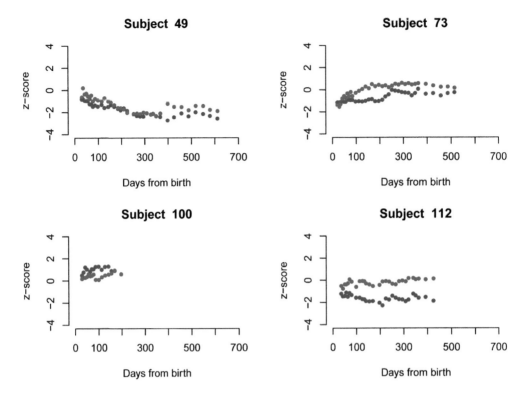

FIGURE 3.14: Example of trajectories of z score for length (zlen, blue points) and weight (zwei, red points) for four children in the CONTENT data.

to the WHO average. The z-score for weight seemed to have an overall increasing pattern, with the exception of the last measurement. Finally, Subject 112 had data collected up to around day 500, had a roughly constant z-score for length (baby was close to the WHO average at every age), but a much lower than average z-score for weight (baby was much below the WHO average for weight at every age). A couple of times, around day 200 and 330, there were increases in z-score for weight, but they were followed by further decreases. While far from complete, the data for these four babies provides an indication of the complexity of the growth trajectories and a glimpse into the close and possibly heterogeneous interactions between the length and weight processes.

3.3.2 Data Structure

Unlike FPCA for regularly measured Gaussian data, software for estimating sparse/irregular FPCA generally require data in long format. This is an efficient way of storing sparse functional data and is similar to how data are stored for fitting traditional linear mixed effects models. The CONTENT data structure is shown below.

```
#CONTENT data structure
id    ma1fe0    weightkg    height    agedays    cbmi     zlen     zwei
1     0         5.62        56.0      61         17.91    -0.53    0.70
1     0         5.99        56.2      65         18.97    -0.62    1.05
1     0         5.97        56.5      71         18.71    -0.75    0.81
...   ...       ...         ...       ...        ...      ...      ...
2     1         5.44        55.5      54         17.66    -1.09    0.13
...   ...       ...         ...       ...        ...      ...      ...
```

The variable `id` is the identifier of individuals in the data and is repeated for every row that corresponds to an individual. The column labeled `malfe0` is the variable that indicates the sex of the child with males being indicated by a 1 and females indicated by a 0. For example, the first study participant `id=1` is a female and the second study participant `id=2` is a male. The sex indicator is repeated for each study participant. The third column `weightkg` is the weight in kilograms and the fourth column `height` is the height (or length) of the baby in centimeters. The fifth column `agedays` is the age in days of the baby. In our models this is t_{ij} and it depends on the subject i (`id=i`) and the observation number j for that baby. For example, $t_{11} = 61$ days indicating that the first measurement for the first baby was taken when she was 61 days old. Similarly, $t_{13} = 71$ days, indicating that the third measurement for the first baby was taken when she was 71 days old. The sixth column `cbmi` is the child's bmi. The seventh column `zlen` is the z-score for length of the child, while the eighth column `zwei` is the z-score for weight. The data set contains two additional variables that were omitted for presentation purposes.

3.3.3 Implementation

There are several implementations of FPCA for sparse/irregular data in R. In particular, the `face` package contains the `face.sparse()` function, which is easy to use, fast, and scalable. It also produces prediction and confidence intervals. Assume that the CONTENT data is contained in the `content_df` variable. The code below shows how to create the data frame and fit `face::face.sparse` for the `zlen` variable.

```
#Prepare the data for sparse PCA
id <- content_df$id
t <- content_df$agedays
w <- content_df$zlen
data <- data.frame(y = w,argvals = t, subj = id)
#Use face to conduct sparse FDA
fit_face <- face.sparse(data, argvals.new = (1:701))
```

This fit takes only seconds on a standard laptop and provides automatic nonparametric smoothing of the covariance operator, which includes optimization over the smoothing parameter. Below we show how to extract the estimated population mean, covariance, variance, and correlation. We also show how to produce pointwise prediction intervals for observations.

```
#Extracting the population mean estimator
m <- fit_face$mu.new
#Extract the covariance estimator
Cov <- fit_face$Chat.new
#Extract the variance estimator
Cov_diag <- diag(Cov)
#Extract the correlation estimator
Cor <- fit_face$Cor.new
#Compute the pointwise prediction interval
m_p_2sd <- m + 2 * sqrt(Cov_diag)
m_m_2sd <- m - 2 * sqrt(Cov_diag)
```

Figure 3.15 displays these estimators. The panel in the first row and column displays the estimated population mean function (m in R code and $\beta_0(s)$ in the statistical model). The trajectory of the mean function increases between day 0 and 200 (from around -0.6 to around -0.3), decreases between day 200 and 450 (from -0.3 to -0.4), and increases again from day 450 to day 700 (from around -0.4 to -0.25). The mean is negative everywhere indicating that the average length of babies was lower than the WHO average. The increasing trend indicates that babies in this study get closer in length to the WHO average as a

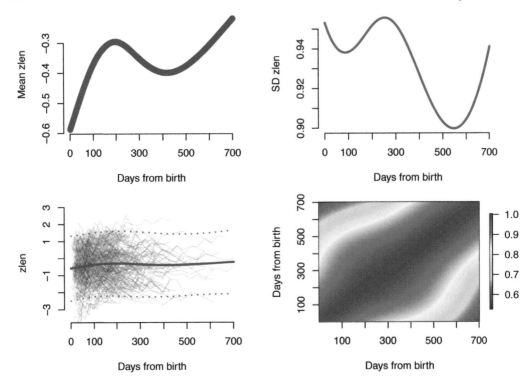

FIGURE 3.15: Some results using `face::face.sparse` function using the CONTENT data. Panel (1,1): Smooth estimator of the population mean as a function of time from birth. Panel (1,2): Smooth estimator of the population standard deviation as a function of time. Panel (2,1): Complete CONTENT data (gray lines), mean and mean ± 2 standard deviations; Panel (2,2): Smooth estimator of the correlation function.

function of time. The apparent decrease in the mean function between day 200 and 450 could be real, but may also be due to the increased sparsity of the data after day 200. Indeed, it would be interesting to study whether babies who were lighter in the first 100 days were more likely to stay in the study longer. The panel in the first row second column displays the estimated pointwise standard deviation function (`sqrt(Cov_diag)`) as a function of time from birth. Note that the function is pretty close to 1 across the functional domain with a slight dip from around 0.94 in the first 200 days to 0.90 around day 500. This is remarkable as the data is normalized with respect to the WHO population, indicating that the within population variability of the length of children is pretty close to the WHO standard. The panel in the second row first column displays all the CONTENT data as gray lines, the mean function in blue, and the mean plus and minus two standard deviations in red. The prediction interval seems to capture roughly 95% of the data, which suggests that the pointwise marginal distributions are not too far from Gaussian distributions. The panel in the second row second column displays the smooth estimate of the correlation function (`Cor`). The plot indicates high positive correlations even 700 days apart, which is consistent with biological processes of child growth (babies who are born longer tend to stay longer). Correlations are in the range of 0.9 and above for time differences of 100 days or less, between 0.8 and 0.9 for time differences between 100 to 400 days, and below 0.8 for time differences longer than 400 days.

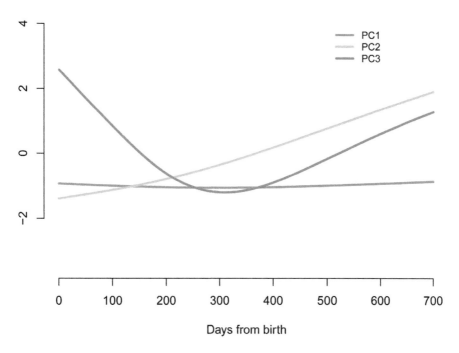

FIGURE 3.16: CONTENT data: first three principal components estimated using sparse FPCA.

The next step is to obtain the eigenfunctions, which in R is simply

```
#Extracting the eigenfunctions
ef <- fit_face$eigenfunctions
```

Figure 3.16 displays the first three principal components (PCs) estimated based on the CONTENT data. These components explain more than 99% of the observed variability after smoothing (90% for PC1, 8.6% for PC2, and 1.4% for PC3). PC4 has an eigenvalue which is an order of magnitude smaller than the eigenvalue for PC3.

The first PC (shown in blue) is very close to a random intercept, with a curvature that is barely visible. The second component (shown in green) is very close to a random slope, though some curvature is still present in this component. It is remarkable that 98.6% of the random effects variability (excluding the error of the random noise) can be captured by a random intercept and a random slope. This is good news for someone who analyzes these data using such a model. However, the sparse FPCA allowed us to quantify how much residual variability is explained (or lost) by using such a model. Moreover, the third component allows for trajectories to bend somewhere between day 200 and 400 from birth. This is an important feature that allows the model to better adapt to the observed data. Indeed, let us have a second look at the plot in Figure 3.14 and inspect the z-score for length (zlen, shown as red dots) for each of the four study participants shown. For Subjects 49, 73, and 112 it is quite apparent that allowing for a change of slope between day 200 and 400 would visually improve the fit. This is not a proof, as we are looking at only 4 study participants, but it provides a practical look at the data guided by sparse FPCA.

Another way to think about the problem is that PC3 could easily be approximated by a linear or quadratic spline with one knot around day 300 from birth. Thus, the sparse FPCA model suggests a linear mixed effects model with a random intercept, one random slope on time, and one random slope on a quadratic spline with one knot at 300. This becomes a standard mixed effects model. Such a direction could be considered for modeling and it is supported by the sparse FPCA analysis. While we do not pursue that analysis here, we note that running first sparse FPCA and then learning the structure of the random effects could be an effective strategy for fitting simpler, traditional models without worrying too much about what could be lost.

We now show how to produce predictions for one study participant at a given set of points using the `face.sparse` function.

```
#The days where we predict
seq <- 1:701
#Length of the predicted sequence
k <- length(seq)
#Indicator of data for subject 1
sel <- which(id == 1)
#Extract data for subject 1
dati <- data[sel,]
```

This code sets up the grid where the function is predicted (`seq`) and selects the data corresponding to a particular study participant. We now set up the data structure required as input for prediction.

```
#Prepare the data frame for prediction
dati_pred <- data.frame(y = rep(NA, nrow(dati) + k),
                        argvals = c(rep(NA, nrow(dati)), seq),
                        subj = rep(dati$subj[1], nrow(dati) + k))
#Fill the first rows of the data frame with the subject data
dati_pred[1:nrow(dati),] <- dati
```

The `dati_pred` variable is a data frame with the same column structure as the original data used for model fitting. The data frame has a number of rows equal to the number of observations for the subject being predicted, `nrow(dati)`, plus the number of grid points used for prediction, `k`. The first `nrow(dati)` rows contain the observed data for the subject. The last `k` rows correspond to the data that will be predicted. For example, the last `k` rows of the `argvals` variable contain the grid of points for prediction `seq`. The last `k` rows of the `w` variable contains `NA` because these values are predicted. The last `k` rows of the variable `subj` are just repeats of the subject's ID. Once this structure is complete, predictions are obtained using the following code.

```
#Obtain the predictions
what2 <- predict(fit_face, dati_pred)
#Get the indicator where predictions are stored
Ord <- nrow(dati) + 1:k
#Get the predictions
mean_pred <- what2$y.pred[Ord]
#Get the standard error of the predictions
se_pred <- what2$se.pred[Ord]
```

The predictions are stored in the variable `mean_pred` while the standard errors are stored in `se_pred`.

Figure 3.17 displays the z-score for length for the same study participants as in Figure 3.14 (data shown as blue dots). However, the plot also displays the prediction of the individual curves (solid red lines) as well as the pointwise 95% confidence intervals for the subject-specific mean function (dashed red lines). This is not the same as the 95% pointwise

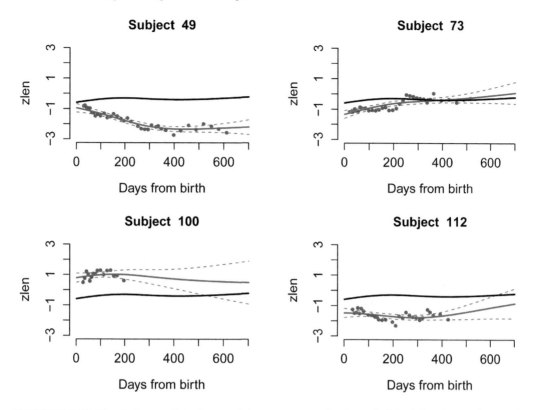

FIGURE 3.17: Predictions of study participants z-score for length (zlen) level together with pointwise 95% confidence intervals for four study participants.

prediction intervals, which would be larger. The population mean function is also displayed as a solid black line. The fits to all four data sets appear reasonable and display some amount of curvature that is quite different from what would be obtained from a random intercept random slope model. Consider, for example, study participant 49, whose z-score for length decreases steadily (linearly) from birth to around day 400 close to a very low −3. However, after that the z-scores plateau around −2.5, the model captures this change relatively well, which is most likely due to the third principal component.

Sparse FPCA produces predictions both outside and inside the range of the observed data for a particular baby. Predictions outside the range are useful to provide future predictions based on the data available for the specific baby while borrowing strength from the data available for the other babies. Moreover, the model can be used to estimate the data at birth or during the period before the first measurement. Consider, for example, study participant 100. The first observation was at 28 days and the last observation was at 196 days from birth. Our model predicts that the z-score for length for this baby was 0.85 on the first day and 0.49 on day 700 from birth.

Prediction within the range of the observations is also very useful. For example, study participant 100 had observations 1.27, 0.99, and 1.25 on days 99, 112, and 126, respectively. The model predicts that the z-scores increased very slowly from day 100 to day 130 from 0.96 to 1.0. In contrast, study participant 112 had observations −1.61, −1.61, −1.74 on days 98, 113, 125, respectively. The model predicts that during this period the scores for this baby decreased slowly from −1.52 to −1.56. Note that data for study participants 100 and 112 are not collected on the same days from birth. However, the model allows us to compare the z-scores on every day from birth. This is also helpful if we are interested in prediction. Suppose that we would like to predict a specific outcome based on the data up to day 150 for all babies. It makes sense to use sparse FPCA to produce predictions of the z-scores for every day between birth and day 150 and then use these inferred trajectories for prediction of outcomes. This becomes a standard scalar on function regression (SoFR) or function-on-function regression (FoFR) depending on whether we predict a scalar or functional outcome. We will discuss these implications in more detail in Chapters 4 and 6, respectively.

3.3.4 About the Methodology for Fast Sparse FPCA

For a full description of this methodology, see [328], and for extensions to multivariate sparse data, see [173]. The idea starts with focusing on the residuals $r_i(s_{ij}) = W_i(s_{ij}) - \widehat{\beta}_0(s_{ij})$, just as described by [283]. However, the approach diverges because it uses a bivariate penalized spline smoother and estimates the covariance operator $\mathbf{\Sigma}$, the noise variance σ_ϵ^2, and the smoothing parameter simultaneously. These choices substantially reduce computation times and use nonparametric smoothing via penalized splines, which avoids manual tuning of the smoothing parameter. Moreover, prediction is conducted very efficiently using the evaluation of the B-spline basis functions at the desired time points, which is computationally simple.

3.4 When PCA Fails

So far we have the discussed how powerful and popular PCA and FPCA are and illustrated them using multiple applications. However, it is important to know what the limits of these approaches are and when one could expect these methods not to perform well. Consider the case when the functions are generated as completely independent white noise $W_i(s) \sim N(0, \sigma^2)$. We simulate data with $i = 1, \ldots, n = 250$ and s equally observed on $p = 3000$ grid points between $[0, 1]$. The simulated data are stored in an $n \times p$ dimensional matrix $\mathbf{W}$ and this matrix is decomposed using PCA.

The four panels in Figure 3.18 correspond to the first four estimated eigenfunctions associated with the first four largest eigenvalues. From the inspection of the graph one cannot see any structure in the eigenfunctions, which substantially reduces the interpretability of the results. We also investigated the magnitude of the corresponding eigenvalues. Figure 3.19 displays the first 50 eigenvalues from the largest to smallest, which indicate an almost linear decrease of the eigenvalues as a function of their index. If such plots are seen in practice, they could indicate that data are highly noisy and functional PCA may need to be used instead of traditional PCA. Note that in Figure 3.5, a similar slow linear decrease in eigenvalues was observed for eigenvalues that correspond to noise; see the dotted gray line in that figure.

Consider now a different model that does not contain noise at all and simulate data from the model

$$W_i(s) = \frac{1}{\sqrt{2\pi}\sigma} \exp \left\{ -\frac{(s - \mu_i)^2}{2\sigma^2} \right\} ,$$

that is, $W_i(s)$ is the probability density function (pdf) of a $N(\mu_i, \sigma)$. The only randomness in the model is induced by sampling the means $\mu_i \sim \text{Uniform}[0, 1]$. For this application we chose $\sigma = 0.01$, though other values could be chosen and/or could be allowed to be subject-specific. Just as in the previous example, we simulated $n = 250$ functions evaluated at an equally spaced grid of $p = 3000$ points between $[0, 1]$.

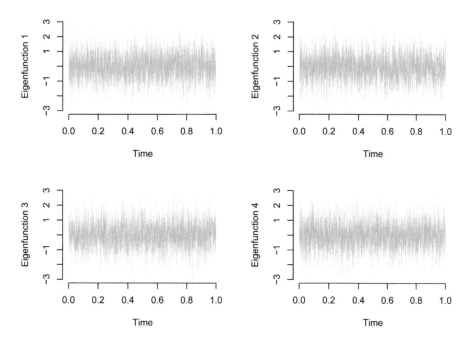

FIGURE 3.18: The first four estimated eigenfunctions when data are simulated as independent normal random variables with mean zero $W_i(s) \sim N(0, \sigma_0^2)$.

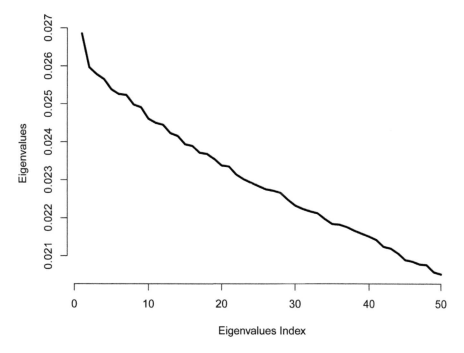

FIGURE 3.19: The first fifty estimated eigenvalues when data are simulated as independent normal random variables with mean zero $W_i(s) \sim N(0, \sigma_0^2)$.

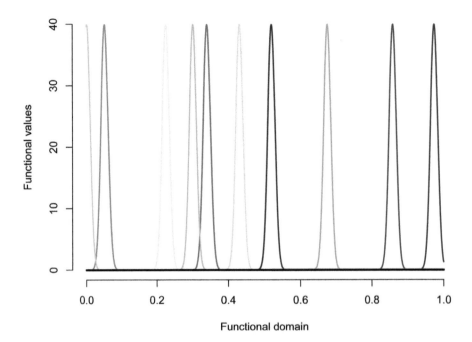

FIGURE 3.20: First ten $N(\mu_i, \sigma^2)$ pdfs, where $\mu_i \sim \mathrm{U}[0,1]$ and $\sigma = 0.01$. Each color corresponds to a different function.

The functions $W_i(s)$ are stored in an $n \times p$ dimensional matrix $\mathbf{W}$, where each row corresponds to one individual and each column corresponds to an observation at that location. PCA was applied to this matrix and the eigenfunctions and eigenvalues were obtained. Figure 3.21 displays the first four eigenfunctions corresponding to the highest four eigenvalues. Knowing the structure of the data used for simulations, the principal components do not seem to capture interesting or important patterns of the data. This is supported by the fact that the first eigenfunction explains 5.7% while the second eigenfunction explains 4.9% of the observed variability. The first eigenvector indicates a peak around 0.4, which is likely due to chance in this particular sample, probably because some bumps may be clustered around this value.

Figure 3.22 displays the first 50 largest eigenvalues of the matrix $\mathbf{W}$ ordered in decreasing order as a function of the eigenvalue index from the largest to the smallest. Just as in the case of random noise data, the decrease in eigenvalues is slow and close to linear as a function of the eigenvalue index. Note that this problem cannot be solved by smoothing the data (FPCA) because data are already perfectly smooth. These examples provide potential explanations for observed behavior of eigenvalues in various practical applications. In general, real data will contain a real signal that can be captured by PCA (typically captured by fast decreasing eigenvalues), noise and de-synchronization of signal, both of which will contribute to slowing down the decrease in eigenvalues.

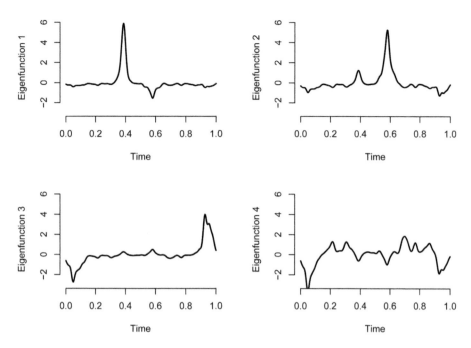

FIGURE 3.21: First four eigenfunctions (corresponding to the highest four eigenvalues) estimated from the data simulated as $N(\mu_i, \sigma^2)$ pdfs, where $\mu_i \sim U[0, 1]$ and $\sigma = 0.01$.

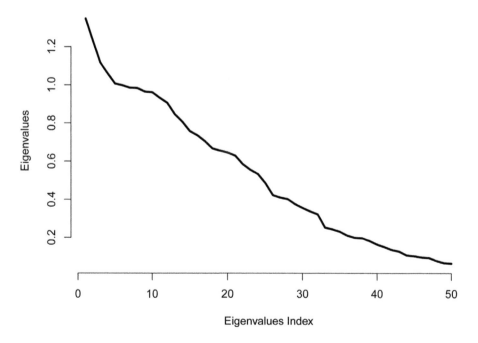

FIGURE 3.22: First 50 eigenvalues in decreasing order (from the highest to the lowest) estimated from the data simulated as $N(\mu_i, \sigma^2)$ pdfs, where $\mu_i \sim U[0, 1]$ and $\sigma = 0.01$. Based on 250 simulated functions, each evaluated at 3000 points.

4

Scalar-on-Function Regression

The basic assumption of FDA is that each observed function is an individual, self-contained and complete "data point." This suggests the need to extend standard statistical methods to include such data; we now consider regression analyses when covariates are functions. Regression models with functional predictors avoid reducing those predictors to single-number summaries or treating observed data within each individual function as a separate, unrelated observation. Instead, this class of regression approaches generally assumes that the association between the predictor and outcome is smooth across the domain of the function to (1) account for the correlation structure within the function; (2) control the change of the association effect for a small change in the functional argument; and (3) allow for enough flexibility to capture the complex association between the scalar outcome and functional predictor. Data analysis where outcomes are scalars and some of the predictors are functions is referred to as scalar-on-function regression (SoFR). This type of models has been introduced by [245, 294, 304], while the first use of the SoFR nomenclature can be traced to [251].

Scalar-on-function regression has been under intense methodological development, which generated many different approaches and publications. Here we will not be able to refer to all these approaches, but we will point out the many types of applications of SoFR: physical activity [56, 75], chemometrics [82, 99, 108, 192, 212, 252, 298], crop yield prediction [224], cardiology [248], intensive care unit (ICU) outcome analysis [93], brain science [102, 124, 182, 188, 253, 339], methylation analysis [174], climate science [12, 83], electroencephalography [24, 186, 222], simulated earthquake data [11], and continuous glucose monitoring (CGM) [194], just to name a few. While these papers are referenced here for their specific area of application, they contain substantial methodological developments that could be explored in detail. Also, this list is neither comprehensive nor does it relate to all aspects of SoFR methodology and applications. The overall goal of this chapter is not to explore the vast array of published methodological tools for SoFR. Instead, we will focus on linear models where the coefficient function is estimated using a basis expansion. We will emphasize the use penalized splines, the equivalence of these models with linear mixed effects models for inference, and the flexibility of well-developed software such as `refund` and `mgcv`. The objective is to ensure that enough readers can get started with fitting SoFR models using stable and reproducible software.

In this chapter we describe the general motivation for SoFR, and build intuition using exploratory analyses and careful interpretation of coefficients. Methods using unpenalized basis expansions are implemented in traditional software, while estimation and inference for SoFR using penalized splines is conducted using the `refund` and `mgcv` packages.

4.1   Motivation and EDA

Much of this chapter will use data from the NHANES to illustrate techniques for modeling scalar outcomes using functional predictors. In particular, we will use MIMS profiles as functional predictors of body mass index (BMI) as a continuous outcome. BMI is defined as the body mass expressed in kilograms divided by the square of the body height expressed in meters. We recognize that BMI is an imperfect measure of adiposity, and that the relationship between BMI and activity is more complex than is captured by the models we show. That said, BMI is often useful in practice and is related to physical activity; this motivates the exploration in this chapter.

We will begin with a simple approach and show how this is related to more general linear scalar-on-function regression models. For each participant, we will obtain the average MIMS value in each of 12 two-hour bins and then use these bin averages as predictors in a standard linear regression model. This has the advantage of allowing some degree of flexibility in the association between MIMS values and BMI over the course of the day, while not requiring non-standard modeling techniques or interpretations. Indeed, in our experience, this kind of approach can be a good starting point in collaborative settings or serve as a useful check on results from more complex approaches.

The code chunk below imports and organizes the processed NHANES data that will be used in this chapter. We read in the data using `readRDS()` from the `base` R package. Using the `mutate()` function, we create a new variable, `death_2year`, which indicates whether the participant died within two years of their inclusion in the study. In the next step, we use `select()` to retain only those variables in the processed NHANES data that will be relevant for this chapter; in addition to BMI and MIMS, we keep the subject ID, age, gender, and our two-year mortality indicator. The variable name `gender` used in this chapter is drawn directly from NHANES to be consistent with the framing of questions in that survey, and is not intended to conflate sex and gender. We rename `MIMS` and `MIMS_sd` to include the suffix `_mat` to indicate these variables are stored as matrices. Finally, we use `filter()` to restrict the dataset to participants 25 years of age or older and ensure that the resulting dataframe has the class `tibble` for better default printing options and other benefits. This dataframe is stored as `nhanes_df` and will be used throughout the chapter.

```
nhanes_df =
  readRDS(
    here::here("data", "nhanes_fda_with_r.rds")) %>%
  mutate(
    death_2yr = ifelse(event == 1 & time <= 24, 1, 0)) %>%
  select(
    SEQN, BMI, age, gender, death_2yr,
    MIMS_mat = MIMS, MIMS_sd_mat = MIMS_sd) %>%
  filter(age >= 25) %>%
  drop_na(BMI) %>%
  tibble()
```

In the next code chunk, we convert MIMS_mat and MIMS_sd_mat to tidyfun objects using tfd(). Note that we use the arg argument in tfd() to be explicit about the grid over which functions are observed: $\left[\frac{1}{60}, \frac{2}{60}, \dots, \frac{1440}{60}\right]$, so that minutes are in $\frac{1}{60}$ increments and hours of the day fall on integers from 1 to 24. Here and elsewhere, the use of the tidyfun package is not required, but takes advantage of a collection of tools designed to facilitate data manipulation and exploratory analysis when one or more variables is functional in nature. Indeed, although we will make extensive use of this framework in this chapter, the use of matrices or other data formats is possible.

```
nhanes_df =
  nhanes_df %>%
  mutate(
    MIMS_tf = matrix(MIMS_mat, ncol = 1440),
    MIMS_tf = tfd(MIMS_tf, arg = seq(1/60, 24, length = 1440)),
    MIMS_sd_tf = matrix(MIMS_sd_mat, ncol = 1440),
    MIMS_sd_tf = tfd(MIMS_sd_tf, arg = seq(1/60, 24, length = 1440)))
```

The next code chunk contains two components. The first component creates a new data frame, nhanes_bin_df, containing average MIMS values in two-hour bins by computing the rolling mean of each MIMS_tf observation with a bin width of 120 minutes, and then evaluating that rolling mean at hours $1, 3, \dots, 23$. The result is saved as MIMS_binned, and for the next step only BMI and MIMS_binned are retained using select(). The second component of this code chunk fits the regression of BMI on these bin averages. The tf_spread() function produces a wide-format dataframe with columns corresponding to each bin average in the MIMS_binned variable, and the call to lm() regresses BMI on all of these averages. We save the result of the regression in an object called fit_binned.

```
nhanes_bin_df =
  nhanes_df %>%
  mutate(
    MIMS_binned =
      tf_smooth(MIMS_tf, method = "rollmean", k = 120, align = "center"),
    MIMS_binned = tfd(MIMS_binned, arg = seq(1, 23, by = 2))) %>%
  select(BMI, MIMS_binned)

fit_binned =
  lm(BMI ~ .,
    data = nhanes_bin_df %>% tf_spread(MIMS_binned))
```

We now show the binned predictors and the resulting coefficients. The first plot generated in the code chunk below shows the MIMS_binned variable for the first 500 rows (other data points are omitted to prevent overplotting). Because MIMS_binned is a tidyfun object, we use related tools for plotting with ggplot. Specifically, we set the aesthetic mapping y = MIMS_binned and use geometries geom_spaghetti() and geom_meatballs() to show lines and points, respectively. We also set the aesthetic color = BMI to color the resulting plot by the observed model outcome. The second plot shows the coefficients for each bin averaged MIMS values. We create this plot by tidying the model fit stored in fit_binned and omitting the intercept term. An hour variable is then created by manipulating the coefficient names, and upper and lower 95% confidence bounds for each hour are obtained by adding and subtracting 1.96 times the standard error from the estimates. We plot the estimates as lines and points, and add error bars for the confidence intervals.

```
nhanes_bin_df %>%
  slice(1:500) %>%
  ggplot(aes(y = MIMS_binned, color = BMI)) +
  geom_spaghetti() +
  geom_meatballs()

fit_binned %>%
  broom::tidy() %>%
  filter(term != "(Intercept)") %>%
  mutate(
    hour = str_replace(term, "MIMS_binned_", ""),
    hour = as.numeric(hour),
    ub = estimate + 1.96 * std.error,
    lb = estimate - 1.96 * std.error) %>%
  ggplot(aes(x = hour, y = estimate)) +
  geom_point() +
  geom_path() +
  geom_errorbar(aes(ymax = ub, ymin = lb), width = .5)
```

The resulting panels are shown in Figure 4.3. The binned MIMS profiles (left panel) show expected diurnal patterns of activity, where there is generally lower activity at night and higher activity during the day. Binning has collapsed some moment-to-moment variability observed in minute-level MIMS profiles, making some patterns easier to observe; here, we see a general trend that participants with higher BMI values had lower levels of observed physical activity during the daytime hours; trends in the morning, evening, and nighttime are less obvious based on this plot. The results of the regression using bin-averaged MIMS values as predictors for BMI (right panel) are consistent with these observed trends. Coefficients for bin averages during the day are generally below zero and some (2-hour intervals between 8-10 AM and 6-8 PM) are statistically significant. Interestingly, coefficients in the morning and parts of the night are positive, suggesting that higher activity in these times is associated with higher BMI values. All bin averages are used jointly in this model, so coefficients can be interpreted as adjusting for the activity at other times of the day; standard errors and confidence intervals also reflect the correlation in activity over the course of the day.

Our choice of 2-hour bins was arbitrary; this may create enough bins to capture changes in the association between MIMS and BMI over the course of the day without becoming overwhelming, but other choices are obviously possible. Indeed, a common choice (in physical activity and other settings) is to average over the complete functional domain, effectively using one bin. Doing so in this setting produces a single measure of total activity which, arguably, could suffice to understand associations between activity and BMI. However, we found that a single average over the day performed notably worse: twelve 2-hour bins produced a fit with an adjusted R^2 of 0.0267, while the single average yielded an adjusted R^2 of 0.0167. Alternatively, one could use more bins rather than fewer. These models are relatively easy to implement through adjustments to the preceding code, and the results are not surprising: as bins become smaller, the coefficients become less similar over time and the trends are harder to interpret. Even in Figure 4.1, it is not clear if the changes in coefficient values in the first three bins reflect a true signal or noise. Conceptually, scalar-on-function regression models are intended to avoid issues like bin size by considering predictors and coefficients as functions and ensuring smoothness.

We will motivate a shift to scalar-on-function regression by recasting the preceding model based on bin averages. Notationally, let $s_j = \frac{j}{60}$ for $1 \leq j \leq 1440$ be the grid over

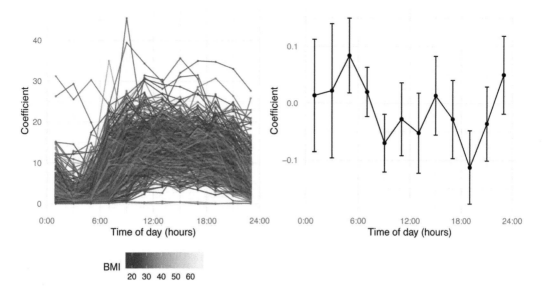

FIGURE 4.1: Left panel: NHANES physical activity profiles averaged in 2-hour intervals as a function of time from midnight (labeled 0) to midnight (labeled 24). Individual trajectories are colored from low BMI (dark blue) to high BMI (yellow). Right panel: Pointwise and 95% confidence intervals of the association between average physical activity measured in every 2-hour interval and BMI. All 2-hour intervals are used as regressors in the same model.

which functions are observed and $X_i(s_j)$ be the observed MIMS value for subject i at time s_j. Additionally, let $\overline{X}_{ib}$ be the average MIMS value for subject i in bin $1 \leq b \leq 12$. For example,

$$\overline{X}_{i1} = \frac{1}{120} \sum_{j=1}^{120} X_i(s_j)$$

is the average MIMS in the first two hours of the day. Additionally, let β_b be the regression coefficient corresponding to bin average b in the model regressing BMI on bin average MIMS values. A key insight is that

$$
\begin{aligned}
\beta_1 \overline{X}_{i1} &= \beta_1 \frac{1}{120} \sum_{j=1}^{120} X_i(s_j) \\
&= \frac{\beta_1}{2} \left[\frac{1}{60} \sum_{j=1}^{120} X_i(s_j) \right] \\
&\approx \int_0^2 \frac{\beta_1}{2} X_i(s)\, ds\ ,
\end{aligned}
\tag{4.1}
$$

where the approximation in the last line results from the numeric approximation to the true integral between hours 0 and 2 of the day. This notation also emphasizes that $X_i(s)$ is conceptually a continuous function, although it is observed over a discrete grid.

Taking this a step further, define a coefficient function $\beta^{\text{step}}(s)$ over the same domain as MIMS values through

$$
\beta^{\text{step}}(s) = \begin{cases}
\frac{\beta_1}{2}, & 0 < s \leq 2\,, \\
\frac{\beta_2}{2}, & 2 < s \leq 4\,, \\
\cdots \\
\frac{\beta_{12}}{2}, & 22 < s \leq 24\,.
\end{cases}
\tag{4.2}
$$

Building on the intuition from (4.1), we have that

$$\sum_{b=1}^{12} \beta_b \overline{X}_{ib} \approx \int_0^{24} \beta^{step}(s) X_i(s) \, ds. \tag{4.3}$$

That is, the model using bin averages can be expressed in terms of a specific functional coefficient and functional predictors by integrating over their product. In this case, the specific functional coefficient is a step function with step heights equal to half of the coefficient in the bin average model. The rescaling that converts bin coefficients to the step function depends on the bin width and on the domain for s (e.g., a two-hour bin on $[0, 24]$ requires halving the coefficient). Assuming $s \in [0, 24]$ is observed on an equally spaced grid with length 1,440 yields time increments of $\frac{1}{60}$, which are used in the numeric approximation to the integral term. Alternatively assuming $s \in [0, 1]$ or $s \in [0, 1440]$, both of which could be reasonable in this case, would require some slight modifications to (4.1) and (4.2) that would affect the scale of $\beta(s)$, but not the value of the approximate integrals.

The code chunk below creates a dataframe that contains the coefficient function defined in (4.2). We use the tidied output of `fit_binned`, again omitting the intercept term and now focusing only on the estimated coefficients in the `estimate` variable. Using `slice()`, we repeat each of these values 120 times. The next steps use `mutate()` to define the method name (a variable that will be used in later sections of this chapter), divide the estimate by 2, and define the `arg` variable in a way that is consistent with the specification of `MIMS_tf` in `nhanes_df`. Finally, we use `tf_nest()` to collapse the long-form data frame into a dataset containing the coefficient function as a `tidyfun` object.

```
stepfun_coef_df =
  fit_binned %>%
  broom::tidy() %>%
  filter(term != "(Intercept)") %>%
  select(estimate, std.error) %>%
  slice(rep(1:12, each = 120)) %>%
  mutate(
    method = "Step",
    estimate = .5 * estimate,
    arg = seq(1/60, 24, length = 1440)) %>%
  tf_nest(.id = method)
```

A plot showing the complete (not binned) `MIMS_tf` trajectories alongside the step coefficient function is shown in Figure 4.2. The difference between this plot and the one shown in Figure 4.1 is that in Figure 4.1 we took the average of the functions in a bin and then regressed the outcome on the collection of bin averages. In Figure 4.2 we obtained the fitted values by integrating the product of predictor functions (left) and the coefficient function (right); these two approaches are identical up to a re-scaling of the regression parameters. Connecting the bin average approach to a truly functional coefficient is an intuitive starting point for the more flexible linear SoFR models considered in the next section.

4.2 "Simple" Linear Scalar-on-Function Regression

We now introduce the linear scalar-on-function regression model in which there is only one functional predictor and no scalar covariates. This is analogous to "simple" linear

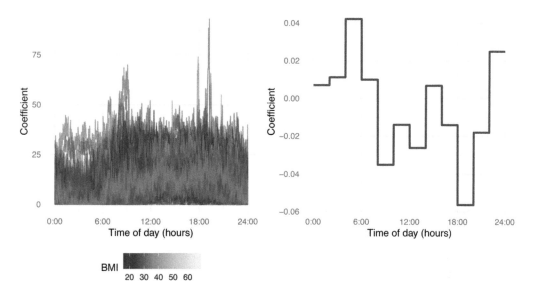

FIGURE 4.2: Left panel: NHANES physical activity profiles averaged in two-hour intervals as a function of time from midnight (labeled 0) to midnight (labeled 24). Individual trajectories are colored from low BMI (dark blue) to high BMI (yellow). Right panel: pointwise estimators of the step-wise association between physical activity and BMI.

regression, and will be useful for introducing key concepts in interpretation, estimation, and inference for models with functional predictors. Later sections will consider extensions of this approach.

4.2.1 Model Specification and Interpretation

For participants $i = 1, \ldots, n$, let Y_i be a scalar response of interest and $X_i : S \to \mathbb{R}$ be a functional predictor observed over the domain S. The simple linear scalar-on-function regression model is

$$Y_i = \beta_0 + \int_S \beta_1(s) X_i(s)\, ds + \epsilon_i \tag{4.4}$$

where β_0 is a population-level scalar intercept, $\beta_1(s) : S \to \mathbb{R}$ is the functional coefficient of interest, and ϵ_i is a residual with mean zero and variance σ_ϵ^2. This model specification generalizes the specific case in (4.3), where a coefficient function constrained to be a step function approximated a regression model based on bin averages. The coefficient function in (4.4) can be more flexible, but the interpretation is analogous: $\beta_1(s)$ defines the weight given to predictor functions at each $s \in S$, and the product of $X_i(s)$ and $\beta_1(s)$ is integrated to obtain the overall contribution of the functional term in the model. At each time point, $\beta(s)$ is the effect on the outcome for a one-unit increase in $X_i(s)$ when all other values $X_i(s')$ $s' \in S, s' \neq s$ are kept unchanged. This interpretation is admittedly somewhat awkward, but unavoidable when regressing a scalar outcome on a functional predictor. Note that the coefficients adjust for effects of all other $s' \in S$.

The interpretation of the coefficient function is illustrated in Figure 4.3. The first column of panels displays the observed profiles for three individuals, $X_i(s)$. The first study participant starts their activity around 6 AM and relatively high activity is maintained throughout the day. The second study participant is less active than the first participant during the day, but has higher levels of activity at night. Finally, the third participant exhibits only

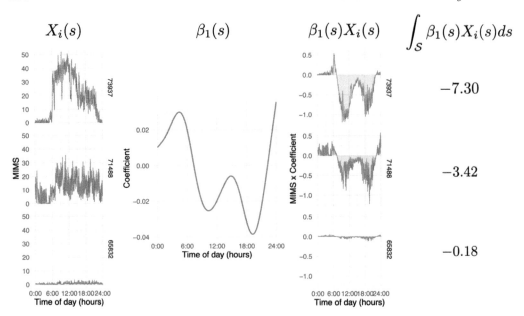

FIGURE 4.3: Interpretation of functional predictors in SoFR. First column of panels indicate observed profiles for three individuals, $X_i(s)$. The middle panel displays $\beta(s)$ along its domain. Final column of panels indicate the pointwise product between $\beta_1(s)X_i(s)$. The shaded area shows to $\int_S \beta_1(s)X_i(s)ds$. For each shaded area, the number to the right indicates the value of the corresponding integral.

very low levels of activity over the 24 hours of observation. These participants were selected because they have different activity trajectories. The middle panel in Figure 4.3 displays the coefficient $\beta(\cdot)$ over its domain; how this estimate was obtained will be discussed in later sections of this chapter. The next column of panels contains the pointwise products $\beta_1(s)X_i(s)$. The shaded areas highlight the integrals $\int_S \beta_1(s)X_i(s)ds$, and the numbers to the right are the values of this integral.

The innovation in scalar-on-function regression, compared to nonfunctional models, is a coefficient function that integrates with covariate functions to produce scalar terms in the linear predictor. The corresponding challenge is developing an estimation strategy that minimizes

$$\min_{\beta_0, \beta_1(s)} \sum_{i=1}^n \left\{ Y_i - \beta_0 - \int_S \beta_1(s)X_i(s)\,ds \right\}^2 \tag{4.5}$$

in a way that is flexible and computationally feasible.

4.2.2   Parametric Estimation of the Coefficient Function

Our first approach expands the coefficient function $\beta_1(s)$ using a relatively low-dimensional basis expansion; doing so leads to a more familiar setting in which scalar basis coefficients are the target of estimation and inference. Specifically, let $\beta_1(s) = \sum_{k=1}^K \beta_{1k}B_k(s)$ where $B_1(s), \ldots, B_K(s)$ is a collection of basis functions. Substituting this expansion into the integral term in (4.4) gives

$$E[Y_i] = \beta_0 + \int_S \beta_1(s)X_i(s)\,ds$$

$$= \beta_0 + \sum_{k=1}^{K}\left[\int_S B_k(s)X_i(s)\,ds\right]\beta_{1k} \tag{4.6}$$

$$= \beta_0 + \mathbf{C}_i^t\boldsymbol{\beta}_1$$

where $C_{ik} = \int_S B_k(s)X_i(s)\,ds$, $\mathbf{C}_i = [C_{i1},\ldots,C_{iK}]^t$, and $\boldsymbol{\beta}_1 = (\beta_{11},\ldots,\beta_{1K})^t$ is the vector of basis coefficients. The result of the basis expansion for the coefficient function, therefore, is a recognizable multiple linear regression with carefully defined scalar covariates and corresponding coefficients. Specifically, let $\mathbf{y} = (y_1,\ldots,y_n)^t$, the matrix $\mathbf{X}$ be constructed by row-stacking vectors $[1, C_{i1},\ldots,C_{iK}]$, and $\boldsymbol{\beta} = [\beta_0, \beta_{11},\ldots,\beta_{1K}]^t$ be the vector of regression coefficients including the population intercept and spline coefficients. The ordinary least squares solution for $\boldsymbol{\beta}$ is given by minimizing the sum of squares criterion

$$\min_{\boldsymbol{\beta}} ||\mathbf{y} - \mathbf{X}\boldsymbol{\beta}||^2 . \tag{4.7}$$

This very familiar expression, which is a practical reframing of (4.5), is possible due to the construction of a design matrix $\mathbf{X}$ that is suitable for scalar-on-function regression.

Functional predictors are actually observed over a finite grid, and the definite integrals that define the C_{ik} are in practice estimated using numeric quadrature. In the illustration constraining $\beta_1(s)$ to be a step function, we approximated this integral using a Riemann sum with bin widths (or quadrature weights) equal to $\frac{1}{60}$; in general, we will use

$$C_{ik} \approx \frac{1}{60}\sum_{j=1}^{1440} X_i(s_j)B_k(s_j) \tag{4.8}$$

throughout this chapter to approximate the integrals needed by our spline expansion. We reiterate here that the choice of S and the implied quadrature weighting can affect the scale of the resulting basis coefficient estimates; being consistent in the approach to numeric integration in model fitting and in constructing subsequent fitted values or predictions for new observations is critical, and failing to do so is often a source of confusion. We also note that, except in specific cases like the step function approach, the basis coefficients β_{1k} $k = 1,\ldots,K$ are not individually interpretable and examining the coefficient function $\beta_1(s)$ provides more insights.

Many options for the basis have been considered in the expansive literature for SoFR. To illustrate the ideas in this section, we start with a quadratic basis and obtain the corresponding estimate of the coefficient function $\beta_1(s)$. We define the basis

$$\begin{cases} B_1(s) = 1\,, \\ B_2(s) = s\,, \\ B_3(s) = s^2\,, \end{cases} \tag{4.9}$$

and, given this, obtain scalar predictors C_{ik} that can be used in standard linear model software. The basis expansion includes an intercept term, which should not be confused with the model's intercept, β_0. The intercept in the basis expansion allows the coefficient function $\beta_1(s)$ to shift as needed, while the population intercept is the expected value of the response when the predictor function is zero over the entire domain. Because the intercept in the basis expansion is integrated with the predictor functions to produce C_{i1}, the basis expansion intercept $B_1(s)$ does not introduce identifiability concerns with respect to the model's intercept β_0.

Continuing to focus on BMI as an outcome and MIMS as a functional predictor, the code chunk below defines the quadratic basis and obtains the numeric integrals in the C_{ik}. For consistency with other code in this chapter, the grid over which functions are observed is called `arg` and is set to $\left[\frac{1}{60}, \frac{2}{60}, \dots, \frac{1440}{60}\right]$. The basis matrix B is defined in terms of `arg` and given column names `int`, `lin`, `quad` for the intercept, linear, and quadratic terms. Next, we construct the data frame `num_int_df`, which contains the numeric integrals. The first step is to multiply the matrix of functional predictors, stored as `MIMS_mat` in the data frame `nhanes_df`, by the basis B. Doing so gives $\sum_{j=1}^{1440} X_i(s_j)B_k(s_j)$ for each i and k, and multiplying by the quadrature weight $\frac{1}{60}$ produces the numeric integrals defining the C_{ik} using (4.8). We retain the row names of the matrix product (inherited from the `MIMS_mat` matrix) in the resulting dataframe, and convert this to a numeric variable for consistency with `nhanes_df`.

```
epoch_arg = seq(1/60, 24, length = 1440)


B = cbind(1, epoch_arg, epoch_arg^2)
colnames(B) = c("int", "lin", "quad")


num_int_df =
  as_tibble(
    (nhanes_df$MIMS_mat %*% B) * (1/60),
    rownames = "SEQN") %>%
  mutate(SEQN = as.numeric(SEQN))
```

The next code chunk implements the regression and processes the results. We first define a new data frame, `nhanes_quad_df`, that contains variables relevant for the scalar-on-function regression of BMI on MIMS trajectories using (4.4) and expand the coefficient function $\beta_1(s)$ in terms of the quadratic basis defined in the previous code chunk. This is created by joining `nhanes_df` and `num_int_df` using the variable `SEQN` as the key to define matching rows. We keep only BMI and the columns corresponding to the numeric integrals C_{ik}. Using `nhanes_quad_df`, we fit a linear regression of BMI on the C_{ik}; the formula specification includes a population intercept `1` to reiterate that the model's intercept β_0 is distinct from the basis expansion's intercept, which appears in C_{i1}. Finally, we combine the coefficient estimates in `fit_quad` with the basis matrix B to obtain the estimate of the coefficient function. For any $s_j \in S$, $\widehat{\beta}_1(s_j) = \sum_{k=1}^{3} \beta_{1k}B_k(s_j) = \mathbf{B}(s_j)\boldsymbol{\beta}_1$, where $\mathbf{B}(s_j) = [B_1(s_j), B_2(s_j), B_3(s_j)]$ is the row vector of basis functions evaluated at s_j. Let $\mathbf{s} = \{s_j\}_{j=1}^{1440}$ be the grid over which functions are observed and $\mathbf{B}(\mathbf{s})$ be the 1440×3 matrix of basis functions evaluated over all entries $s_j \in \mathbf{s}$. This is obtained by stacking the 1×3 dimensional row vectors $\mathbf{B}(s_j)$ over s_j, $1 = 1, \dots, p = 1440$. If $\beta_1(\mathbf{s}) = \{\beta_1(s_1), \dots, \beta_1(s_p)\}^t$ is the $p \times 1$ dimensional vector where the function is evaluated, then $\beta_1(\mathbf{s}) = \mathbf{B}(\mathbf{s})\boldsymbol{\beta}_1$. In the code below, we therefore compute the matrix product of B and the coefficients of `fit_quad` (omitting the population intercept), and convert the result to a `tidyfun` object using the `tfd()` function with the `arg` parameter defined consistently with other `tidyfun` objects in this chapter. The coefficient function is stored in a data frame called `quad_coef_df`, along with a variable `method` with the value `quad`.

```
nhanes_quad_df =
  left_join(nhanes_df, num_int_df, by = "SEQN") %>%
  select(BMI, int, lin, quad)

fit_quad =
  nhanes_quad_df %>%
  lm(BMI ~ 1 + int + lin + quad, data = .)

quad_coef_df =
  tibble(
    method = "Quadratic",
    estimate = tfd(t(B %*% coef(fit_quad)[-1]), arg = epoch_arg))
```

This general strategy for estimating coefficient functions can be readily adapted to other basis choices. The next code defines a cubic B-spline basis with eight degrees of freedom; this is more flexible than the quadratic basis, while also ensuring a degree of smoothness that is absent from the stepwise estimate of the coefficient function. Cubic B-splines are an appealing general-purpose basis expansion and we use them throughout the book, but other bases can be useful, depending on the context. For instance, in this application a periodic basis (e.g., a Fourier or periodic B-spline basis) could be a good choice, since it would ensure that the coefficient function began and ended at the same value.

```
B_bspline = splines::bs(epoch_arg, df = 8, intercept = TRUE)
colnames(B_bspline) = str_c("BS_", 1:8)

num_int_df =
  as_tibble(
    (nhanes_df$MIMS_mat %*% B_bspline) * (1/60),
    rownames = "SEQN") %>%
  mutate(SEQN = as.numeric(SEQN))

nhanes_bspline_df =
  left_join(nhanes_df, num_int_df, by = "SEQN") %>%
  select(BMI, BS_1:BS_8)

fit_bspline =
  lm(BMI ~ 1 + ., data = nhanes_bspline_df)

bspline_coef_df =
  tibble(
    method = "B-Spline",
    estimate =
      tfd(t(B_bspline %*% coef(fit_bspline)[-1]), arg = epoch_arg)
```

Once the basis is defined, the remaining steps in the code chunk mirror those used to estimate the coefficient function using a quadratic basis, with a small number of minor changes. The basis is generated using the `bs()` function in the `splines` package, and there are now eight basis functions instead of three. There is a corresponding increase in the number of columns in `num_int_df`, and for convenience we write the formula in the `lm()` call as BMI ~ 1 + . instead of listing columns individually. The final step in this code chunk constructs the estimated coefficient function by multiplying the matrix of basis functions evaluated over **s** by the vector of B-spline coefficients; the result is stored in a data frame called `bspline_coef_df`, now with a variable `method` taking the value `B-Spline`. The similarity between this model fitting and the one using a quadratic basis is intended to emphasize that the basis expansion approach to fitting (4.4) can be easy to implement for a broad range of basis choices.

We show how to display all coefficient function estimates in the next code chunk. The first step uses `bind_rows()` to combine data frames containing the stepwise, quadratic, and B-spline estimated coefficient functions. The result is data with three rows, one for each estimate, and two columns containing the `method` and `estimate` variables. Because `estimate` is a `tidyfun` object, we plot the estimates using `ggplot()` and `geom_spaghetti()` by setting the aesthetics for y and `color` to `estimate` and `method`, respectively.

```
bind_rows(stepfun_coef_df, quad_coef_df, bspline_coef_df) %>%
  ggplot(aes(y = estimate, color = method)) +
  geom_spaghetti(alpha = 1, linewidth = 1.2)
```

In the resulting plot, shown in Figure 4.4, the coefficient functions have some broad similarities across basis specifications. That said, the quadratic basis has a much higher estimate in the nighttime than other methods because of the constraints on the shape of the coefficient function. The stepwise coefficient has the bin average interpretation developed in Section 4.1, but the lack of smoothness across bins is scientifically implausible. Of the coefficients presented so far, then, the B-spline basis with eight degrees of freedom is our preference as a way to include both flexibility and smoothness in the estimate of $\beta_1(\cdot)$. In this case, diagnostic metrics also slightly favor the B-spline approach: the adjusted R^2 for the B-spline model is 0.0273 compared to 0.0267 for the stepwise coefficient and 0.0252 for the quadratic basis fit.

The degree of smoothness is closely connected to the choice of degrees of freedom in this approach. While it is possible (and in some cases useful) to explore this choice using

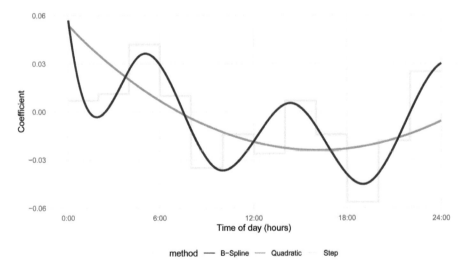

FIGURE 4.4: Estimates of the coefficient function $\beta_1(s)$ in equation (4.4) obtained using B-splines of degree 8 (purple), quadratic (green), and step function with bin sizes of length two hours (yellow). The outcome is the BMI and the predictors are the MIMS profiles during the day.

traditional techniques for model selection in linear models, we will next incorporate explicit smoothness constraints in a penalized likelihood framework for estimation.

4.2.3 Penalized Spline Estimation

Our approach to scalar-on-function regression using smoothness penalties relies on key insights introduced by Goldsmith et al., 2011 [102], that connect functional regression to scatterplot smoothing and mixed models. As we have already seen, by expressing the coefficient function in (4.4) using a spline expansion, it is possible to cast scalar-on-function regression in terms of a linear regression model. While the design matrix for that model has a specific construction, once it is available, usual model fitting approaches can be used directly. Analogously, we will see that the use of penalized splines requires the careful construction of design and penalty matrices, but once these are in place, the techniques for scatterplot smoothing described in Section 2.3.2 can be applied. Critically, this includes casting scalar-on-function regression as a mixed model, which we alluded to in Section 2.3.3. This connection makes it possible to use the rich collection of techniques for mixed model estimation and inference for functional regression. Indeed, while we now favor `mgcv` for estimation, the scalar-on-function regression in [102] was implemented using `nlme`. Leveraging the connection between functional regression models and mixed models has also facilitated Bayesian approaches, allowed models with non-Gaussian outcomes, included random effects in longitudinal models, and extended the framework to many other settings. The widespread use of mixed models in functional regression makes it clear that recognizing this connection is critical in functional data analysis.

Broadly speaking, in functional regression models we prefer to use rich spline basis expansions because they are flexible and numerically stable. We include penalties that encourage smoothness in the target of estimation, which generally take the form of penalties on the overall magnitude of the squared second derivative of the estimand. This combination results in a quadratic penalty with a tuning parameter that controls the degree of smoothness; higher penalization leads to "flatter" coefficient functions, while lower penalization allows more flexible but "wigglier" coefficients. We will now discuss how to construct the appropriate design and penalty matrices, fit the resulting models for fixed values of the tuning parameter, recast FoSR as a mixed model, and estimate the tuning parameter as a ratio of variance components.

As before, let $\beta_1(s) = \sum_{k=1}^{K} \beta_{1k} B_k(s)$ where $B_1(s), \ldots, B_K(s)$ is a collection of basis functions. In contrast to previous examples, though, we now let K be "large" so that estimation of spline coefficients using (4.7) will produce a non-smooth estimated coefficient function. We add a smoothness-inducing penalty of the form $\int_S \{\beta_1''(s)\}^2 ds$ to the minimization criterion. Intuitively, estimates of $\beta_1(s)$ that include many sharp turns will have squared second derivatives $\{\beta_1''(s)\}^2$ with many large values, while smooth estimates of $\beta_1(s)$ will have second derivatives that are close to 0 over the domain S. To implement the squared second derivative penalty, note that

$$
\begin{aligned}
\int_S \{\beta_1''(s)\}^2 ds &= \int_S \beta_1''(s)\beta_1''(s) ds \\
&= \int_S \boldsymbol{\beta}_1^t \mathbf{B}''(s)\mathbf{B}''(s)^t \boldsymbol{\beta}_1 ds \\
&= \boldsymbol{\beta}_1^t \left[\int_S \mathbf{B}''(s)\mathbf{B}''(s)^t ds \right] \boldsymbol{\beta}_1 \\
&= \boldsymbol{\beta}_1^t \mathbf{P} \boldsymbol{\beta}_1
\end{aligned}
\tag{4.10}
$$

where $\mathbf{P}$ is the $K \times K$ matrix with the $(i,j)^{\text{th}}$ entry equal to $\int_S B_i''(s)B_j''(s) ds$.

Let $\mathbf{X}$ be an $n \times (K+1)$ matrix in which the ith row is $[1, C_{i1}, \ldots, C_{iK}]$, and let $\boldsymbol{\beta}$ be the $(K+1)$ dimensional column vector that concatenates the population intercept β_0 and the spline coefficients $\boldsymbol{\beta}_1$. Adding the second derivative penalty to the minimization criterion in (4.7) yields a new penalized sum of squares

$$\min_{\boldsymbol{\beta}} \|\mathbf{y} - \mathbf{X}\boldsymbol{\beta}\|^2 + \lambda \boldsymbol{\beta}^t \mathbf{D} \boldsymbol{\beta} . \tag{4.11}$$

Here $\lambda \geq 0$ is a scalar tuning parameter and $\mathbf{D}$ is the matrix given by

$$\mathbf{D} = \begin{bmatrix} \mathbf{0}_{1 \times 1} & \mathbf{0}_{K \times 1} \\ \mathbf{0}_{1 \times K} & \mathbf{P} \end{bmatrix} ,$$

where $\mathbf{0}_{a \times b}$ is a matrix of zero entries with a rows and b columns. This notation and specification intentionally mimics what was used for penalized scatterplot smoothing in Section 2.3.2, and many of the same insights can be drawn. For fixed values of λ, a closed form solution exists for $\widehat{\boldsymbol{\beta}}$. Varying λ from 0 to ∞ will induce no penalization and full penalization, respectively, and choosing an appropriate tuning parameter is an important practical challenge. As elsewhere in this chapter, though, we emphasize that the familiar form in (4.11) should not mask the novelty and innovation of this model, which implements penalized spline smoothing to estimate the coefficient function in a scalar-on-function regression.

We illustrate these ideas in the next code chunk, which continues to use BMI as an outcome and MIMS as a functional predictor. The code draws on elements that have been seen previously. We first define a B-spline basis with 30 degrees of freedom evaluated over the finite grid `arg`, previously defined to take values $\left[\frac{1}{60}, \frac{2}{60}, \ldots, \frac{1440}{60}\right]$. Using functionality in the `splines2` package, we obtain the second derivative of each spline basis function evaluated over the same finite grid. The elements of the penalty matrix $\mathbf{P}$ are obtained through numeric approximations to the integrals through $\frac{1}{60} \sum_{j=1}^{1440} B_i''(s_j) B_l''(s_j)$. The design matrix $\mathbf{X}$ is obtained by adding a column taking the value 1 everywhere to the terms C_{ik} given by the numeric integration of the predictor functions and the spline basis using (4.8). Next, we construct the matrix $\mathbf{D}$ by adding a row and column taking the value 0 everywhere to the penalty matrix $\mathbf{P}$. The response vector $\mathbf{y}$ is extracted from `nhanes_df` and we choose high and low values for the tuning parameter λ. Given all of these elements, we estimate the coefficient vector $\boldsymbol{\beta}$ through $\widehat{\boldsymbol{\beta}} = (\mathbf{X}^t \mathbf{X} + \lambda \mathbf{D})^{-1} \mathbf{X}^t \mathbf{y}$ using the pre-selected values of λ to obtain `coef_high` and `coef_low`. We note that these include estimates of the population intercept as well as the spline coefficients.

```
B_bspline = splines::bs(epoch_arg, df = 30, intercept = TRUE)
sec_deriv = splines2::bSpline(epoch_arg, df = 30, intercept = TRUE, derivs = 2)
P = t(sec_deriv) %*% sec_deriv * (1/60)

X = cbind(1, (nhanes_df$MIMS_mat %*% B_bspline) * (1/60))
D = rbind(0, cbind(0, P))


y = nhanes_df$BMI


lambda_high = 10e6
lambda_low = 100


coef_high = solve(t(X) %*% X + lambda_high * D) %*% t(X) %*% y
coef_low  = solve(t(X) %*% X + lambda_low  * D) %*% t(X) %*% y
```

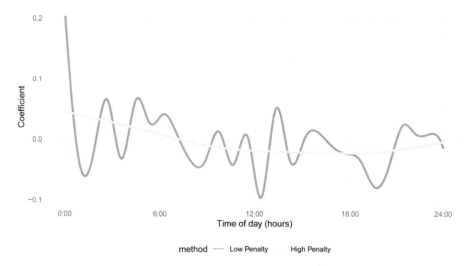

FIGURE 4.5: Estimates of the coefficient function $\beta_1(s)$ in equation (4.4) using penalized B-splines of maximum degree 30 using low penalization (green) and high penalization (yellow). The outcome is BMI and the predictors are the MIMS profiles during the day.

The estimated coefficient functions that correspond to the estimates in `coef_high` and `coef_low` can be produced through simple modifications to the previous code, so we omit this step. Figure 4.5 shows the resulting coefficient functions. Most strikingly, the choice of λ has a substantial impact on the estimated coefficient function. With 30 basis functions and low penalization, the estimated coefficient function (shown in green) is indeed very wiggly – there is a high spike at the beginning of the domain and rapid oscillations over the day. The more highly penalized estimate (shown in yellow), meanwhile, varies smoothly from values above zero in the evening hours and below zero during the day. These can be compared to the coefficient functions seen in Figure 4.4; in general, all coefficient functions suggest temporal variation in the association between BMI and MIMS values, but the model specification and choice of tuning parameter significantly impacts the resulting estimates.

Recasting the penalized sum of squares in (4.11) as a mixed model allows the data-driven estimation of tuning parameters; more broadly, this opens the door to using a wide range of methods for mixed model estimation and inference in functional regression settings. Using an approach similar to that in Section 2.3.3, we note that (4.11) can be re-written as a maximization problem. First, we will separate the population intercept from spline coefficients; let $\mathbf{1}_n$ be a column of length n containing the value 1 everywhere and $\mathbf{C}$ be the $n \times K$ dimensional matrix constructed by row-stacking the vectors $[C_{i1}, \ldots, C_{iK}]$. Letting $\lambda = \frac{\sigma_\epsilon^2}{\sigma_{\beta_1}^2}$ and multiplying by $\frac{-1}{2\sigma_\epsilon^2}$, we now have

$$\frac{-||\mathbf{y} - \mathbf{1}_n\beta_0 - \mathbf{C}\boldsymbol{\beta}_1||^2}{2\sigma_\epsilon^2} - \frac{\boldsymbol{\beta}_1^t \mathbf{P} \boldsymbol{\beta}_1}{2\sigma_{\beta_1}^2} . \tag{4.12}$$

Although we initially developed our objective function as a penalized sum of squares, the same objective arises through the use of maximum likelihood estimation for the model

$$\begin{cases} [\mathbf{y}|\beta_0, \boldsymbol{\beta}_1, \sigma_\epsilon^2] = N(\mathbf{1}_n\beta_0 + \mathbf{C}\boldsymbol{\beta}_1, \sigma_\epsilon^2 \mathbf{I}_p) ; \\ [\boldsymbol{\beta}_1|\sigma_{\beta_1}^2] \quad = \frac{\det(\mathbf{P})^{1/2}}{(2\pi)^{K/2}\sigma_{\beta_1}} \exp\left(-\frac{\boldsymbol{\beta}_1^t \mathbf{P} \boldsymbol{\beta}_1}{2\sigma_{\beta_1}^2}\right) . \end{cases} \tag{4.13}$$

As elsewhere, we are using the notation $[y|x]$ to indicate the conditional probability density function y given x. Using restricted maximum likelihood estimation (REML) for (4.13), we estimate spline coefficients as best linear unbiased predictors; estimate the tuning parameter λ by estimating the variance components σ_ϵ^2 and $\sigma_{\beta_1}^2$; and conduct inference using the mixed effects model framework.

We make a few brief comments on the degeneracy of the conditional distribution $[\beta_1|\sigma_{\beta_1}^2]$. In the model construction we have described, the penalty matrix $\mathbf{P}$ may not be full rank. Intuitively, the issue is that we only penalize departures from a straight line (because all straight lines have a second derivative that is zero everywhere), but our basis spans a space that includes straight lines. There are some solutions to this problem. First, one can mumble vaguely about "uninformative priors" and otherwise ignore the issue; this works quite well in practice but can raise eyebrows among more detail-oriented peers and book coauthors. A second ad hoc solution is to construct the penalty matrix $(\alpha\mathbf{I}_{K\times K}) + \mathbf{P}$, where α is a small value. The resulting penalty is no longer exactly a second derivative penalty because it includes some overall shrinkage of spline coefficients, but it is full rank. The third solution is to extract the intercept and linear terms from the basis and the penalty matrix to create unpenalized and penalized terms that are equivalent to the original basis and penalty. The technical details for this solution are beyond the scope of this book, but can be found in [258, 319]. We first address this issue in Section 2.3.2, and refer to it in multiple chapters throughout the book.

Recognizing that scalar-on-function regression with a penalty on the second derivative of the coefficient function can be expressed as a standard mixed model allows generic tools for estimation and inference to be applied in this setting. Below, we construct the design matrix $\mathbf{C}$ using the numeric integration in (4.8) as elsewhere in this chapter. The penalty matrix $\mathbf{P}$, which contains the numeric integral of the squared second derivative of the basis functions, is reused from a prior code chunk. We pass these as arguments into `mgcv::gam()` by specifying C in the formula that defines the regression structure, and then use the `paraPen` argument to supply our penalty matrix P for the design matrix C. Lastly, we specify the estimation method to be REML. We omit code that multiplies the basis functions by the resulting spline coefficients to obtain the estimated coefficient function, but plot the result below. Previous penalized estimates, based on high and low values of the tuning parameter λ, over- and under-smoothed the coefficient function; the data-driven approach to tuning parameter selection yields an estimate that is smooth but time-varying.

```
C = (nhanes_df$MIMS_mat %*% B_bspline) * (1/60)


fit_REML_penalty =
  gam(y ~ 1 + C, paraPen = list(C = list(P)),
    method = "REML")
```

Figure 4.6 displays the resulting penalized spline fit using a penalized B-spline with a maximum degree of 30 based on REML estimation of the smoothing parameter. This fit is shown as a solid purple line and it is superimposed over the two fits shown in Figure 4.5 with a fixed low penalty (green) and a fixed high penalty (yellow). The figure indicates that the REML estimation is somewhere between the low and high penalization estimates. It also provides a pleasing compromise that is obtained automatically via penalized splines.

The preceding text shows that it is certainly possible to set up the basis matrices defining $\beta_1(s)$ and corresponding penalty matrices, supply those quantities to `mgcv::gam()`, and let the software do the smoothing parameter selection. However, the linear algebra and "time cost" for learning how to set up the model may deter potential users of the method. Luckily, many of the strengths of the `mgcv` package can be leveraged through the `refund`

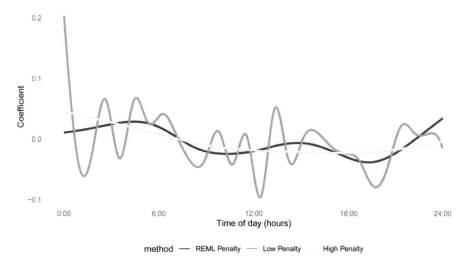

FIGURE 4.6: Estimates of the coefficient function $\beta_1(s)$ in equation (4.4) using penalized B-splines of maximum degree 30 using low penalization (green), high penalization (yellow), and REML penalization (purple). The outcome is BMI and the predictors are the MIMS profiles during the day.

package, which adds functionality, quality-of-life features, and user interfaces relevant to FDA. For SoFR, this means that instead of building models using knowledge of the linear algebra underlying penalized spline estimation, we instead only require correctly specified data structures and syntax for `refund::pfr()`. In the code chunk below, we regress BMI on MIMS using the matrix variable `MIMS_mat` using the linear specification in `lf()`. We additionally indicate the grid over which predictors are observed, and specify the use of REML to choose the tuning parameter. Among other things, the `pfr()` function organizes data so that model fitting can be performed by calls to `gam()`; more details are provided in Section 4.5. The next component of this code chunk extracts the resulting coefficient and structures it for plotting.

```
pfr_fit =
  pfr(
    BMI ~ lf(MIMS_mat, argvals = seq(1/60, 24, length = 1440)),
    method = "REML", data = nhanes_df)

pfr_coef_df =
  coef(pfr_fit) %>%
  mutate(method = "refund::pfr()") %>%
  tf_nest(.id = method, .arg = MIMS_mat.argvals) %>%
  rename(estimate = value)
```

Figure 4.7 displays results obtained using `mgcv::gam()` and `refund::pfr()`; there are some minor differences in the default model implementations and these results do not align perfectly, although they are very similar and can be made exactly the same. For reference, we also show the coefficient function based on an unpenalized B-spline basis with eight degrees of freedom.

However, it is worth pausing and reflecting on the simplicity of the implementation of the function on scalar implementation in `refund::pfr`. Indeed, all one needs to do is to

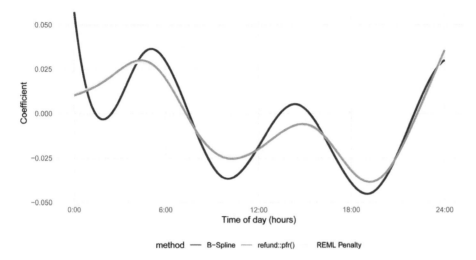

FIGURE 4.7: Estimates of the coefficient function $\beta_1(s)$ in equation (4.4) using penalized B-splines implemented in `mgcv::gam()` and `refund::pfr()`, as well as an unpenalized B-spline with eight degrees of freedom. The outcome is BMI and the predictors are the MIMS profiles during the day.

input the outcome vector, `BMI` in our case, and the predictor matrix, `MIMS_mat` in our case, which contains the observed functions for each study participant per row. As we will show, this implementation easily expands to other types of outcomes, additional scalar covariates, and additional functional predictors. Moreover, the outcome of this fit, `pfr_fit`, contains all the ingredients to extract model estimates and conduct inference, as we will show in this chapter. In some sense, this is a culmination of research conducted by thousands of researchers. Our major accomplishment here was to transform the SoFR into a penalized spline regression, make the connection between FoSR and semiparametric regression, embed this into a mixed effects model framework, and identify the software elements that work well together and correspond to the correct inferential framework. This is not the only way to conduct FoSR, not by a long shot. But it is one of the best ways we know to do it. We hope that this will be helpful to many others.

4.2.4 Data-Driven Basis Expansion

To this point, we have developed unpenalized and penalized estimation strategies using basis expansions constructed independently of observed data – step functions, polynomial bases, B-splines. The use of a data-driven basis obtained through FPCA is a popular alternative, and indeed was among the first approaches to SoFR. This approach, sometimes referred to as "Functional Principal Components Regression" (FPCR), has useful numerical features that can make it an appealing option in some practical settings. For more details see, for example, [253] and references therein.

Borrowing some notation from Chapter 3, suppose the functional predictors can be expanded using

$$X_i(s) = \beta_0(s) + \sum_{k=1}^{K} \xi_{ik}\phi_k(s) , \tag{4.14}$$

where $\beta_0(s)$ is a population mean function, the $\phi_k(s)$ are orthonormal eigenfunctions, and the scores $\xi_{ik} = \int_S \{X_i(s) - \beta_0(s)\} \phi_k(s)ds$ are mean-zero random variables. We will use the same basis to express the coefficient function, so that $\beta_1(s) = \sum_{k=1}^{K} \beta_{1k}\phi_k(s)$. One could, at this point, use numeric approximations to the second derivatives of the $\phi_k(s)$ to pursue the penalized estimation techniques developed in Section 4.2.3. Instead, it is common to assume that the largest directions of variation in the $X_i(s)$ are also the most relevant for the outcome Y_i. This suggests selecting a truncation level K and proceed as in Section 4.2.2.

More specifically, we will use existing approaches for FPCA to define the basis $\phi_1(s), \ldots, \phi_K(s)$. Next, we will let $C_{ik} = \int_S X_i(s)\phi_k(s)ds$ and $\mathbf{C}_i = [C_{i1}, \ldots, C_{iK}]^t$, so that estimating basis coefficients $\boldsymbol{\beta}_1 = [\beta_{11}, \ldots, \beta_{1K}]^t$ relies on OLS estimation using the $\mathbf{C}_i$ as covariates. We again use numeric integration to obtain C_{ik}, but first recall an issue raised in Chapter 3: many implementations of FPCA implicitly use a quadrature weight of 1 and return a basis that needs to be rescaled to have the correct properties under numeric integration. In our case, let $\mathbf{s} = \{s_j\}_{j=1}^{1440} = \left[\frac{1}{60}, \frac{2}{60}, \ldots, \frac{1440}{60}\right]$ be the finite observation grid. For an FPCA implementation that returns the basis $\phi_1^*(s), \ldots, \phi_K^*(s)$ such that $\sum_{j=1}^{1440} \phi_k^*(s_j)\phi_k^*(s_j) = 1$ for all k, we define $\phi_k(s) = \phi_k^*(s)\sqrt{60}$ so that $\frac{1}{60}\sum_{j=1}^{1440} \phi_k(s_j)\phi_k(s_j) = 1$. This basis can be used analogously to the quadratic and B-spline bases seen in Section 4.2.2.

The code chunk below implements the scalar-on-function regression of BMI on MIMS using a data-driven basis. In the first lines of code, we use the function `refunder::rfr_fpca()` to conduct FPCA. This function is under active development, but takes a `tf` vector as an input; for data observed over a regular grid, this serves as a wrapper for `fpca.face()`. We specify `npc = 4` to return $K = 4$ principal components. The remainder of this code chunk is essentially copied from Section 4.2.2. Using naming conventions similar to previous code, we define B_fpca by extracting `efunctions` from nhanes_fpca, and rescaling them to have the desired numerical properties. Given this basis, we compute numerical integrals to define the C_{ik}; merge the resulting dataframe with nhanes_df; retain only the outcome BMI and predictors `efunc_1:efunc_4`; and fit the regression. We save this as `fit_fpcr_int` to indicate that this conducts FPCR using numeric integration to obtain the covariates C_{ik}. We omit code that multiplies the basis coefficients and the basis expansion to produce the estimated coefficient function, since this is identical to code seen elsewhere.

```
nhanes_fpca =
  rfr_fpca("MIMS_tf", data = nhanes_df, npc = 4)


B_fpca = nhanes_fpca$efunctions * sqrt(60)
colnames(B_fpca) = str_c("efunc_", 1:4)


num_int_df =
  as_tibble(
    (nhanes_df$MIMS_mat %>% B_fpca) * (1/60),
    rownames = "SEQN") %>%
  mutate(SEQN = as.numeric(SEQN))


nhanes_fpcr_df =
  left_join(nhanes_df, num_int_df, by = "SEQN") %>%
  select(BMI, efunc_1:efunc_4)


fit_fpcr_int =
  lm(BMI ~ 1 + ., data = nhanes_fpcr_df)
```

We used numerical integration to obtain the C_{ik} in the model exposition and example code, but an important advantage of FPCR is that principal component scores are the projection of centered functional observations onto eigenfunctions. That is, the ξ_{ik} in (4.14) are given by $\xi_{ik} = \int_S \{X_i(s) - \beta_0(s)\} \phi_k(s) ds$. Using this in the context of scalar-on-function regression while expanding the coefficient function using the $\phi_1(s), \ldots, \phi_K(s)$ functions yields the following:

$$
\begin{aligned}
E[Y_i] &= \beta_0 + \int_S \beta_1(s) X_i(s) \, ds \\
&= \beta_0 + \int_S \beta_1(s)\beta_0(s) \, ds + \int_S \beta_1(s) \{X_i(s) - \beta_0(s)\} \, ds \\
&= \beta_0^* + \sum_{k=1}^K \left[\int_S \phi_k(s) \{X_i(s) - \beta_0(s)\} \, ds \right] \beta_{1k} \\
&= \beta_0^* + \boldsymbol{\xi}_i^t \boldsymbol{\beta}_1 \,,
\end{aligned}
\tag{4.15}
$$

where $\boldsymbol{\xi}_i = (\xi_{i1}, \ldots, \xi_{iK})^t$, $\boldsymbol{\beta}_1 = (\beta_{11}, \ldots, \beta_{1K})^t$, and $\beta_0^* = \beta_0 + \int_S \beta_1(s)\beta_0(s) \, ds$ is the population intercept when the covariate functions $X_i(s)$ are centered.

The fact that FPCR can be carried out using a regression on FPC scores directly is a key strength. There are many practical settings where the numeric integration used to construct the design matrices throughout this chapter – for pre-specified and data-driven basis expansions – is not possible. For functional data that are sparsely observed or that are measured with substantial noise, numeric integration can be difficult or impossible. In both those settings, FPCA methods can produce estimates of eigenfunctions and the associated scores and thereby enable scalar-on-function regression in a wide range of real-world settings. At the other extreme, for very high-dimensional functional observations, it may be necessary to conduct dimension reduction as a pre-processing step to reduce memory and computational burdens. The FPCR gives an interpretable scalar-on-function regression in this setting as well. That said, because FPCR is a regression on FPC scores, only the effects of $X_i(s)$ on Y_i that are captured by the directions of variation contained in the $\phi_1(s), \ldots, \phi_K(s)$ functions can be accounted for using this approach. Moreover, the smoothing of the estimated coefficient function depends on the intrinsic choice of the number of eigenfunctions, K. This tends to be less problematic when one is interested in prediction performance, but may have large effects on the estimation of the $\beta_1(s)$ coefficient.

The code chunk below re-implements the previous FPCR specification. Because the underlying FPCA implementation scaled the eigenfunctions $\phi_k^*(s)$ using quadrature weights of 1, we first need to appropriately rescale the principal component scores. Let $\xi_{ik}^* = \int_S \{X_i(s) - \beta_0(s)\} \phi_k^*(s) ds$ be the scores based on the incorrectly scaled eigenfunctions. Multiplying both ξ_{ik}^* and $\phi_k^*(s)$ by $\sqrt{60}$ addresses the scaling of the eigenfunctions; additionally, multiplying by the correct quadrature weight $\frac{1}{60}$ produces scores ξ_{ik} on the right scale. This scaling process is necessary whether the FPCA method uses numeric integration for the ξ_{ik}^* or estimates them using BLUPs in a mixed model. Indeed, both approaches are built around incorrectly scaled eigenfunctions, $\phi_k^*(s)$, and need to account for the quadrature weight used for numeric integration to obtain fitted values from predictor functions. As noted above, being careful about weights for numeric integration can be a point of confusion in SoFR; small inconsistencies can produce coefficient functions with similar shapes but very different scales, with corresponding differences in the fitted values. We have made these mistakes too many times, and we hope that others will benefit from our experience and avoid them.

In the code below, we extract **scores** from the FPCA object **nhanes_fpca** obtained in a previous code chunk. Mirroring code elsewhere, we create a dataframe containing the scores; merge this with **nhanes_df** and retain BMI and the predictors of interest; and fit a linear model, storing the results as **fit_fpcr_score** to reflect that we have performed

FPCR using score estimates. A table showing coefficient estimates from `fit_fpcr_int` and `fit_fpcr_score` is shown after this code chunk. As expected, the intercepts from the two models differ – one is based on centered $X_i(s)$ covariate functions and the other is not – but basis coefficients are nearly identical.

```
C = nhanes_fpca$scores * (sqrt(60) / 60)

colnames(C) = str_c("score_", 1:4)
rownames(C) = nhanes_df$SEQN

nhanes_score_df =
  as_tibble(
    C, rownames = "SEQN") %>%
  mutate(SEQN = as.numeric(SEQN))

nhanes_fpcr_df =
  left_join(nhanes_df, nhanes_score_df, by = "SEQN") %>%
  select(BMI, score_1:score_4)

fit_fpcr_score =
  lm(BMI ~ 1 + ., data = nhanes_fpcr_df)
```

method	(Intercept)	Coef. 1	Coef. 2	Coef. 3	Coef. 4
Numeric Integration	32.402	0.067	-0.008	0.052	0.068
FPCA Score	29.312	0.067	-0.008	0.053	0.068

Throughout this section, we have deferred the important methodological consideration of how to choose the truncation level K. Because our estimation approach does not include any explicit smoothness constraints, the flexibility of the underlying basis controls the smoothness of the resulting coefficient function. For that reason, K is effectively a tuning parameter. Figure 4.8 illustrates this point by showing the coefficient function with $K = 4$ obtained using the code above, as well as the coefficient function with $K = 12$ obtained through straightforward modifications to that code. We also show the coefficient function expressed as a step function for reference. The coefficient function with $K = 4$ is, unsurprisingly, the smoothest, while the coefficient function with $K = 12$ more closely tracks the step coefficient function.

FPCR is a common approach; a thorough review and comparison for selecting K is outside the scope of this book, although we will briefly note some options. One can select K by a percent variance explained threshold in FPCA, which uses information only about the functional space, not the outcome. The hope is that the components necessary to explain, for example, 95% of the variation in the $X_i(s)$ provide a sufficient but not too rich basis. A serious drawback of this approach is that the estimated coefficient depends on the proportion of variance explained and, implicitly, on the shape of all principal components. As the number of PCs increases, the shape of the functional coefficient can change substantially, even for small increases in the number of PCs. For example, in NHANES $K = 30$ components are needed to explain 95% of the variation in $X_i(s)$, which leads to obvious under-smoothing of the functional coefficient.

An alternative would be to choose K based on a combination of information about the outcome and predictor space. Such criteria for estimating K include AIC, BIC and cross validation [205, 213, 250, 252], but rarely account for the variability associated with the choice of K, which can be substantial in the decision area. The effect on the parameter

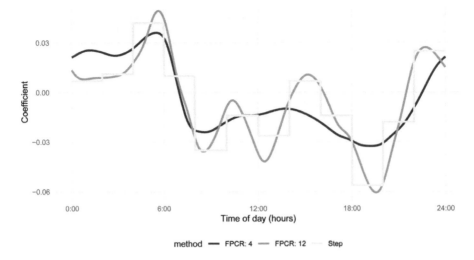

FIGURE 4.8: Estimates of the coefficient function $\beta_1(s)$ in equation (4.4) using FPCR with the first four (green) and twelve (purple) smooth eigenfunctions of the predictor space. Also shown is the estimated coefficient using a piece-wise constant spline every two hours (yellow). The outcome is BMI and the predictors are the MIMS profiles during the day.

estimate can be very large, as increasing or decreasing the number of components by one can significantly impact the shape of the estimated functional parameter.

Yet another approach could be to start with a large value for K and use a variable selection criterion that retains only the PCs that are predictive of the response. Such criteria could be p-values, cross-validation, or adjusted R-square. An important drawback of these approaches is that they become much more complicated when one incorporates additional functional or scalar covariates, random coefficients, non-parametric components, and non-Gaussian outcomes. It is not the complexity of an individual component, but the joint complexity of all components of the method that reduces the overall appeal of these methods.

The expression in (4.14) requires some additional discussion. Throughout this chapter we have worked with the notation $X_i(s)$, which typically refers to the underlying true functional observation. Functional data are often measured with noise and the observed functional process is denoted by $W_i(s)$. The standard functional model connecting the observed and true underlying processes is

$$W_i(s) = X_i(s) + \epsilon_i(s)$$

$$= \beta_0(s) + \sum_{k=1}^{K} \xi_{ik} \phi_k(s) + \epsilon_i(s) \, ,$$

(4.16)

which raises methodological questions about how and whether to account for measurement error in functional predictors. This issue is not limited to data-driven basis expansions, although it arises naturally in this setting. There can also be debate about whether a smooth, unobserved $X_i(s)$ should be considered the "true" predictor instead of the observed data in $W_i(s)$. In practice, the most popular strategies have been to (1) ignore the measurement error and induce smoothness in the functional coefficient; or (2) to pre-smooth functional $W_i(s)$ using FPCA or another smoothing method applied to each predictor. Throughout much of this chapter, we have taken the first approach by using observed MIMS trajectories to construct necessary model terms. Effectively, this strategy assumes that the measurement error does not substantially impact the numeric integration and trusts that smoothness in the basis expansion for the coefficient function is sufficient. The use of a data-driven basis arguably pre-smooths functional predictors: although observed data are used to estimate FPC scores, only those scores are included as predictors and any variation in the functional predictor not accounted for in the first K FPCs is omitted. For more formal treatments of addressing measurement error in scalar-on-function regression see, for example, [290, 291, 338].

4.3 Inference in "Simple" Linear Scalar-on-Function Regression

Section 4.2 focused on estimation of the coefficient function $\beta_1(s)$ in model (4.4); we now turn our attention to inference. First, we will develop pointwise confidence intervals that are not adjusted for correlation and multiplicity (not CMA), as defined in Section 2.4. We refer to these as "unadjusted" confidence intervals, even though they are adjusted for other scalar and/or functional covariates. In Section 4.5.2 we will discuss correlation (to account for the correlation of $\widehat{\beta}_1(s_j)$) and multiplicity (to account for the multiple locations s_j, $j = 1, \ldots, p$) adjusted confidence intervals. We refer to this double adjustment as correlation and multiplicity adjusted (CMA) inference.

4.3.1 Unadjusted Inference for Functional Predictors

Our approach to unadjusted inference assumes Normality and constructs $(1 - \alpha)$ confidence intervals of the form
$$\widehat{\beta}_1(s) \pm z_{1-\alpha/2}\sqrt{\mathrm{Var}\{\widehat{\beta}_1(s)\}}$$
where z_α is the α quantile of a standard Normal distribution. Recall that the coefficient $\widehat{\beta}_1(s)$ was expanded in a basis $\widehat{\beta}_1(s) = \mathbf{B}(s)\boldsymbol{\beta}_1$. Therefore, the variance $\mathrm{Var}\{\widehat{\beta}_1(s)\}$ at any point $s \in S$ can be obtained as

$$\mathrm{Var}\{\widehat{\beta}_1(s)\} = \mathbf{B}(s)\mathrm{Var}(\widehat{\boldsymbol{\beta}}_1)\mathbf{B}^t(s) , \qquad (4.17)$$

which casts inference for $\widehat{\beta}_1(s)$ in terms of inference for the estimated basis coefficients, $\widehat{\boldsymbol{\beta}}_1$. This is very useful because $\beta_1(s)$ is infinite dimensional, whereas the dimension of the basis coefficient vector $\widehat{\boldsymbol{\beta}}_1$ is finite and quite low dimensional.

When using a fixed basis and no penalization, the resulting inference is familiar from usual linear models. After constructing the design matrix $\mathbf{X}$ and estimating all model coefficients, $\boldsymbol{\beta}_1$, using ordinary least squares and the error variance, σ_ϵ^2, based on model residuals, $\mathrm{Var}(\widehat{\boldsymbol{\beta}}_1)$ can be extracted from $\mathrm{Var}(\widehat{\boldsymbol{\beta}}_1) = \widehat{\sigma}_\epsilon^2 (\mathbf{X}^t\mathbf{X})$. Indeed, this can be quickly illustrated using previously fit models; in the code chunk below, we use the `vcov()` function to obtain $\mathrm{Var}(\widehat{\boldsymbol{\beta}})$ from `fit_fpca_int`, the linear model object for FPCR with $K = 4$. We remove the row and column corresponding to the population intercept, and then pre- and post-multiply by the FPCA basis matrix $\mathbf{B}(\mathbf{s})$ stored in `B_fpca`. The resulting covariance matrix is 1440×1440, and has the variances $\mathrm{Var}\{\widehat{\beta}_1(s)\}$ for each value in the observation grid $s \in \mathbf{s}$ on the main diagonal. The final part of the code chunk creates the estimate $\widehat{\beta}_1(s)$ as a `tf` object, uses similar code to obtain the pointwise standard error (as the square root of

entries on the diagonal of the covariance matrix), and constructs upper and lower bounds
for the 95% confidence interval.

```
var_basis_coef = vcov(fit_fpcr_int)[-1,-1]
var_coef_func = B_fpca %*% var_basis_coef %*% t(B_fpca)


fpcr_inf_df =
  tibble(
    method = c("FPCR: 4"),
    estimate = tfd(t(B_fpca %*% coef(fit_fpcr_int)[-1]), arg = epoch_arg),
    se = tfd(sqrt(diag(var_coef_func)), arg = epoch_arg)
  ) %>%
  mutate(
    ub = estimate + 1.96 * se,
    lb = estimate - 1.96 * se)
```

The process for obtaining confidence intervals for penalized spline estimation is conceptually similar. Again, inference is built on (4.17), and the primary difference is in the
calculation of $\text{Var}(\widehat{\beta}_1)$. The necessary step is to perform inference on fixed and random
effects in a mixed model. While the details are somewhat beyond the scope of this text, for
Gaussian outcomes and fixed variance components (or, equivalently, fixed tuning parameters), a closed form expression for the covariance matrix is available. Earlier in this chapter,
we made use of the connection between penalized spline estimation for scalar-on-function
regression and mixed models. Inference can similarly be conducted directly via the connection to mixed effects models using appropriate software, and we pause to appreciate the
impact of coupling approaches to estimation and inference in SoFR with high-performance
implementations for mixed effects models. In later sections and chapters, we will explore
this connection in detail; for now, we will use the helpful wrapper `pfr()` for inference.
Indeed, the object `pfr_coef_df`, obtained in a previous code chunk using `coef(pfr_fit)`,
already includes a column `se` containing the pointwise standard error. In the code chunk
below, we use this column to construct upper and lower bounds of a 95% confidence interval.

```
pfr_inf_df =
  pfr_coef_df %>%
  mutate(
    ub = estimate + 1.96 * se,
    lb = estimate - 1.96 * se)
```

Our next code chunk will plot the estimates and confidence intervals created in the
previous code. We combine the dataframes containing estimates and confidence bounds
for the penalized spline and FPCR methods contained in `pfr_inf_df` and `fpcr_inf_df`,
respectively, which corresponds to using $K = 4$ PCs. In code not shown, the dataframe
`fpcr_inf_df` used to construct this plot was updated to also include estimates and confidence bounds for FPCR using $K = 8$ and $K = 12$, respectively. The result is passed into
`ggplot()`, where we set the aesthetic `y = estimate` to plot the estimated coefficient functions using `geom_spaghetti()`. The next line uses `geom_errorband()` to plot the confidence
band; this requires the aesthetics `ymax` and `ymin`, which map to columns `ub` and `lb` in our
dataframe. Finally, we use `facet_grid()` to create separate panels based on the estimation
approach.

```
bind_rows(pfr_inf_df, fpcr_inf_df) %>%
  ggplot(aes(y = estimate)) +
  geom_spaghetti() +
  geom_errorband(aes(ymax = ub, ymin = lb)) +
  facet_grid(.~method)
```

Figure 4.9 shows the estimates and confidence intervals for the penalized spline and FPCR methods using $K = 4$, 8, and 12 FPCs in separate panels. We saw in Figure 4.8 that the shape of coefficient functions estimated using FPCR depending on the choice of K, which serves as a tuning parameter; we now see that the corresponding confidence intervals are sensitive to this choice as well. When $K = 4$, the 95% confidence bands are narrow and exclude zero almost everywhere, suggesting strong associations between the predictor and response over the functional domain. Meanwhile, when $K = 12$ the confidence bands are wide and include zero except in a few regions. The results obtained through penalized splines with data-driven tuning parameter selection avoids this sensitivity, and leads to results that are stable and reproducible. The estimate and confidence band from `refund::pfr()` indicates significant negative associations between MIMS and BMI in the mid-morning and evening, as well as significant positive associations in early morning. These findings are qualitatively consistent with results obtained using other models, including the binned regression approach and several choices of K in FPCR implementations. Our experience is that the penalized spline model has good inferential properties without the need to select important tuning parameters by hand, and we generally recommend this approach.

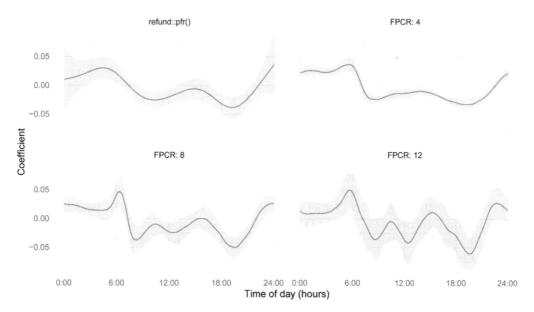

FIGURE 4.9: Estimated coefficient function $\beta_1(s)$ in equation (4.4) and confidence intervals obtained using `refund::pfr()` (top right panel) and FPCR based on the first four, eight, and twelve eigenfunctions (remaining panels). The outcome is BMI and the predictors are the MIMS profiles during the day.

4.4   Extensions of Scalar-on-Function Regression

Our discussion of the "simple" linear scalar-on-function regression in Section 4.2 was deliberate. Beginning from a binning-based approach to build intuition, we introduced parametric expansions for the coefficient function and demonstrated that this reduced the model to a familiar linear regression. From there, we introduced penalized spline methods; although these can be fit "by hand," we took advantage of the key insight that penalized approaches to scalar-on-function regression are connected to mixed models, which allows the use of standard software for tuning parameter selection, estimation, and inference. In general, we favor the use of the `pfr()` function in the `refund` package to conduct this model fitting.

In this section, we will consider several obvious and necessary extensions of the "simple" linear model. We will build on the careful exposition in the previous section to move more quickly through technical details, and focus on implementation of models using `pfr()`. Note that this section only deals with cross-sectional settings; multilevel and longitudinal functional data are the subject of Chapter 8. The emphasis here is on the fact that these extensions can be implemented using small changes to the familiar code of functional regression and attempts to make SoFR models as easy to fit as linear and semiparametric models.

4.4.1   Adding Scalar Covariates

Most real-data analyses that involve a functional predictor will also include one or more scalar covariates. All methods in the previous section can easily be adapted to this setting. To be explicit, we are now interested in fitting models of the form

$$Y_i = \beta_0 + \int_S \beta_1(s) X_i(s)\, ds + \boldsymbol{Z}_i^t \boldsymbol{\gamma} + \epsilon_i \,, \tag{4.18}$$

where $\boldsymbol{Z}_i$ is a $Q \times 1$ dimensional vector of scalar covariates for subject i and $\boldsymbol{\gamma}$ is the vector of associated coefficients. For non-penalized approaches, the design matrix $\mathbf{X}$ appearing in (4.7) is augmented so that it contains an intercept, columns for the scalar predictors, and numeric integrals stemming from the basis expansion. This can be fit using ordinary least squares, and is easy to implement using `lm()`. For penalized approaches, we similarly add columns containing scalar predictors to the design matrix and avoid penalization by expanding the 0 entries of the matrix $\mathbf{D}$ in (4.11). Inference for scalar covariates is analogous to that in non-functional settings, and relies on entries on the diagonal of the covariance matrix of all coefficients.

The code chunk below regresses `BMI` on `MIMS_mat` as a functional predictor and `age` and `gender` as scalar covariates using `pfr()`; recall that `pfr()` expects functional predictors to be structured as matrices. Except for the addition of `age` and `gender` in the formula, all other elements of the model fitting code have been seen before. Moreover, all subsequent steps build directly on previous code chunks. Functional coefficients and standard errors can be extracted using `coef` and combined to obtain confidence intervals, and then plotted using tools from `tidyfun` or other graphics packages. Estimates and inference for non-functional coefficients can be obtained using `summary()` on the fitted model object. Each of these steps is straightforward enough that they are omitted here.

```
pfr_adj_fit =
  pfr(
    BMI ~ age + gender + lf(MIMS_mat, argvals = seq(1/60, 24, length = 1440)),
    data = nhanes_df)
```

The results of this model indicate age is a significant predictor of BMI after adjusting for gender and MIMS as a functional predictor; the coefficient estimate is -0.039 with p-value < 0.001. Gender is also a significant predictor, indicating that female participants have, on average, BMI values 1.68 units larger than males ($p < 0.001$), keeping age and MIMS fixed. While these are highly significant, the effect sizes are small. As a result, the effect of MIMS on BMI is not notably changed after adjustment for age and gender. The coefficient function and 95% confidence bands are broadly similar to those shown in Figure 4.9. In Chapter 5, we will use age and gender as predictors of MIMS in a function-on-scalar regression (FoSR); combining that analysis with the scalar-on-function regression (SoFR) presented here can provide a detailed and complementary view of the mutual relationships among these variables.

Because models can potentially include a large number of scalar (and, we will see in Section 4.4.2, functional) predictors, model building strategies are a necessary component of scalar-on-function regression. That said, the strategies that work in non-functional models also work in those with functional predictors. Our preference is to determine *a priori* a collection of confounding variables based on subject-matter expertise, and focus inference on variables where examining statistical significance is most relevant.

4.4.2 Multiple Functional Coefficients

It is frequently the case that more than one functional predictor is available; often the functions are observed over the same domain, but this is not necessary. The goal is to fit a model of the form

$$Y_i = \beta_0 + \sum_{r=1}^{R} \int_{S_r} \beta_r(s) X_{ir}(s) \, ds + \boldsymbol{Z}_i^t \boldsymbol{\gamma} + \epsilon_i \tag{4.19}$$

where $X_{ir}(s)$, $1 \leq r \leq R$ are functional predictors observed over domains S_r with associated coefficient functions $\beta_r(s)$. The interpretation of the coefficients here is similar to that elsewhere; it is possible to interpret the effect of $X_{ir}(s)$ on the expected value of Y_i through the coefficient $\beta_r(s)$, keeping all other functional and non-functional predictors fixed. As when adding scalar predictors to the "simple" scalar-on-function regression model in Section 4.4.1, the techniques described previously can be readily adapted to this setting. Each coefficient function $\beta_r(s)$ can be estimated using any of the above approaches by creating an appropriate design matrix $\mathbf{X}_r$. If more than one coefficient function is penalized, it is necessary to construct a block diagonal penalty matrix $\mathbf{D}$ with diagonal elements that implement that penalty and estimate the tuning parameter for each coefficient function. This structure is analogous to the penalized approach to multivariate regression splines that appears in equation (2.14) described in Section 2.3.1.3.

We will again use `pfr()` to implement an example of model (4.19). The code chunk below regresses BMI on scalar covariates `age` and `gender`. The functional predictor `MIMS_mat` is familiar from many analyses in this section and is used here. We add the functional predictor `MIMS_sd_mat`, which is the standard deviation of MIMS values taken across several days of observation for each participant, also stored as a matrix. `MIMS_sd_mat` is included in the formula specification exactly as `MIMS_mat` has been in previous code chunks, and the subsequent extraction of estimates and inference is almost identical – with the key addition of specifying `select = 1` or `select = 2` in the call to `coef()` to extract values for `MIMS_mat` and `MIMS_sd_mat`, respectively.

```
pfr_mult_fit =
  pfr(
    BMI ~ age + gender +
      lf(MIMS_mat, argvals = seq(1/60, 24, length = 1440)) +
      lf(MIMS_sd_mat, argvals = seq(1/60, 24, length = 1440)),
    data = nhanes_df)
```

`MIMS_mat` and `MIMS_sd_mat` are intuitively related, in that participants with higher physical activity also tend to have higher variability in their physical activity. In univariate analyses, the shape of coefficient functions for `MIMS_mat` and `MIMS_sd_mat` are similar as well. The results of the multivariate analysis, shown in Figure 4.10, suggest that the effect of `MIMS_mat` is largely attenuated by adjusting for `MIMS_sd_mat`, while the coefficient function for `MIMS_sd_mat` is similar to coefficients seen throughout this chapter. That is, participants with higher variability in physical activity during the nighttime hours tended to have higher BMI values, possibly suggesting that less consistent sleep is associated with higher BMI. On the other hand, higher daytime variability was associated with lower BMI. The effect of `MIMS_mat` after adjusting for `MIMS_sd_mat` is somewhat less intuitive, indicating that higher activity values in the nighttime hours are associated with lower BMI, while keeping other variables fixed.

Again, our primary purpose in this book is not to provide a detailed approach to model building. When selecting scalar and functional variables, we suggest to pre-specification of a confounding structure and identification of primary hypotheses of interest.

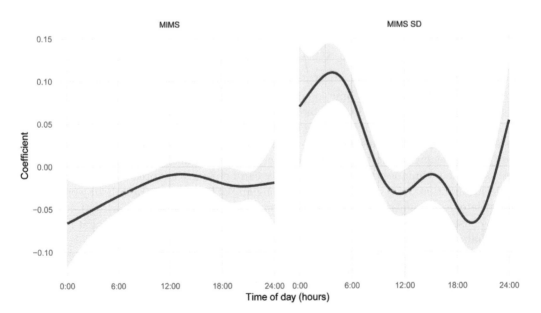

FIGURE 4.10: The left and right panels show coefficient functions for `MIMS_mat` and `MIMS_sd_mat`, respectively, in a model with both functional predictors, BMI as an outcome, and adjusting for age and gender. Coefficients are estimated using penalized splines through `pfr()`.

4.4.3 Exponential Family Outcomes

Outcomes that follow binomial, Poisson, and other exponential family distributions are common in practice, and require the use of a generalized linear model framework for estimation and inference. This is, of course, a natural extension to the estimation methods we favor. For example, given a binary outcome Y_i, scalar predictors $\mathbf{Z}_i$, and a functional predictor $X_i(s)$, one can pose the logistic scalar-on-function regression $E[Y_i|\{X_i(s) : s \in S\}, \mathbf{Z}_i] = \mu_i$, where

$$g(\mu_i) = \beta_0 + \int_S \beta_1(s)X_i(s)ds + \mathbf{Z}_i^t \boldsymbol{\gamma} \,, \tag{4.20}$$

where $g(\cdot)$ is the logit link function. Model parameters can be interpreted as log odd ratios or exponentiated to obtain odds ratios. Expanding the coefficient function in terms of a basis expansion again provides a mechanism for estimation and inference. Rather than minimizing a sum of squares as in (4.7) or the penalized equivalent in (4.11), one now minimizes a (penalized) log likelihood. Tuning parameters for penalized approaches can be selected in a variety of ways, but we continue to take advantage of the connection between roughness penalties and mixed model representations. That is, by viewing spline coefficients for $\beta_1(s)$ as random effects, we can estimate the degree of smoothness as a variance component and estimate it using the associated mixed effects framework. We therefore can take advantage of standard mixed model implementations to fit a broad range of scalar-on-function regression models. Non-penalized approaches can be fit directly using `glm()`, while penalized models are available through `mgcv::gam()` or `pfr()`.

In the code chunk below, we fit a logistic scalar-on-function regression in the NHANES dataset. Our binary outcome is two-year mortality, with the value 0 indicating that the participant survived two years after enrollment. We adjust for age, gender, and BMI, and focus on MIMS as a functional predictor. The model specification using `pfr()` sets the argument `family` to `binomial()`, but other aspects of this code are drawn directly from prior examples. Extracting estimated coefficient functions and conducting inference can be accomplished using the `coef()` function to obtain estimates and pointwise standard errors. Post-processing to construct confidence intervals is direct, and these can be inverse-logit transformed to obtain estimates and inference as odds ratios. Because this code is unchanged from earlier examples, it is omitted here.

```
pfr_mort_fit =
  pfr(
    death_2yr ~ age + gender + BMI +
      lf(MIMS_mat, argvals = seq(1/60, 24, length = 1440)),
    family = binomial(), method = "REML", data = nhanes_df)
```

The results of this analysis suggest that higher MIMS values in the daytime hours are statistically significantly associated with a reduction in the risk of death within two years, keeping other variables fixed. A more detailed analysis of this dataset using survival analysis is presented in Chapter 7.

4.4.4 Other Scalar-on-Function Regression Models

Scalar-on-function regression is the subject of a literature too extensive to survey here, and we recommend reviews of functional regression broadly [205] and scalar-on-function regression specifically [250] to interested readers. Instead, we will simply note several extensions of and alternatives to the linear model with coefficient function estimated through a basis expansion that was the emphasis of this chapter.

An important class of models allows for non-linearity. Single index models extend the model (4.4) by including a smooth function around the integral term [2, 70, 85, 179], while additive models estimate a bivariate coefficient surface in place of the univariate coefficient function [198, 210]. Many other basis expansions are possible; using a wavelet basis (often with variable selection methods for estimation) can be suitable when the coefficient function is not expected to be smooth [191, 309, 340]. Variable selection methods have also been used with spline bases to encourage sparsity and build interpretability in the estimated coefficient [135]. Scalar-on-function regression has been extended to quantile regression, which is necessary to study elements of the response distribution other than the expected value [31, 40]. Quantile regression can be helpful when data are skewed, a setting also considered in [176]. The association between scalar outcomes and functional predictors can be assessed through non-parametric estimation techniques, for example through a functional Nadaraya-Watson estimator or a reproducing kernel Hilbert space [81, 84, 233].

4.5   Estimation and Inference Using mgcv

Estimation

Although `refund::pfr()` is simple, easy to use, and efficient for estimating SoFR models, advanced readers may want to build on the `refund` framework to extend the basic SoFR model to more complex variants. Or, curious readers may want to know exactly what `refund::pfr()` is doing "under the hood" when it calls functions from the `mgcv` family. To that end, here we describe in precise detail how to estimate SoFR using `mgcv` and extract quantities of interest (e.g., point estimates and uncertainty estimates). The basic procedure to estimation using `mgcv` mirrors the procedure used for `refund::pfr()`. Specifically, the first step in this process is setting up the data inputs, which we discuss in detail below. The statistical details are followed immediately with corresponding R code.

Estimation of the linear SoFR model in `refund::pfr()` (and thus `mgcv::gam()`) does not assume that the functional linear term has a convenient closed form solution. Instead, the term $\int_S \beta_1(s)X_i(s)ds$ is approximated numerically. The key insight for estimation is that, when no closed form solution exists for $\int_S \beta_1(s)X_i(s)ds$, we approximate the functional term using numeric integration, resulting in the approximation

$$E[Y_i|\boldsymbol{X_i}] = \beta_0 + \int_{\mathcal{S}} \beta_1(s)X_i(s)ds \tag{4.21}$$

$$\approx \beta_0 + \sum_{j=1}^{p} q_j X_i(s_j)\beta_1(s_j) , \tag{4.22}$$

where q_j are the quadrature weights associated with a particular numeric approximation of the integral. For example, if s_j, $j = 1, \ldots, p$ are equally spaced between 0 and 1, $q_j = 1/p$. Note that the sum $\sum_{j=1}^{p}\{q_j X_i(s_j)\}\beta_1(s_j)$ can be viewed as a sum over s_j of $\beta_1(s_j)$ weighted by $q_j X_i(s_j)$. It turns out that `mgcv` can fit such structures using the "linear functional terms" (see `?linear.functional.terms`). To estimate this model, a user must specify that the outcome, Y_i, is a smooth function of the functional domain, s_j, multiplied by the product of the quadrature weight used for numeric integration, q_j, and the functional predictor at the corresponding point on the domain, $X_i(s_j)$. This quantity is summed up over the observed domain and then added to the linear predictor. A smoothness penalty is then automatically applied, much in the same way as was shown in Section 4.2.3 using the `paraPen` argument, but without the need for the user to manually derive a penalty matrix and basis expansion for $\beta_1(s)$.

Estimating this model then boils down to the construction of the appropriate data inputs to supply to `mgcv::gam()` and identifying the syntax associated with the model we would like to fit or penalized log likelihood we would like to optimize over. First consider construction of the data inputs to `mgcv::gam()`. We require: (1) the vector of responses, $\mathbf{y} = [y_1, \dots, y_n]^t$; (2) the matrix associated with the functional domain $\mathbf{s} = \mathbf{1}_n \otimes \mathbf{s}^t$ where $\mathbf{s}^t = [s_1, \dots, s_p]$ is the row vector containing the domain of the observed functions (in the case of the hourly NHANES data, $p = 1440$ and $\mathbf{s}^t = [1, \dots, 1440]/60$) and $\otimes$ denotes the Kronecker product; and (3) a matrix containing the quadrature weights associated with each functional predictor $\mathbf{P} = \mathbf{1}_n \otimes \mathbf{Q}^t$, where $\mathbf{Q}^t = [q_1, \dots, q_p]$ is the row vector containing the quadrature weights; and lastly (4) a matrix containing the element-wise product of the functional predictor and the quadrature weights $\mathbf{X}_L = \mathbf{X} \odot \mathbf{P}$, where $\odot$ denotes the element-wise product (Hadamard product) of two matrices. The code below constructs these matrices and puts them in a data frame that we will pass to `mgcv`.

```
#Functional predictor
X <- nhanes_df$MIMS_mat
#Number of participants
N <- nrow(nhanes_df)
#Vector containing functional domain of observed data
s_vec <- seq(1/60, 24, length = 1440)
#Matrix containing domain for each person (row)
S <- kronecker(matrix(1, N, 1), t(s_vec))
#Vector quadrature weights (Simpson's rule)
q <- matrix((s_vec[length(s_vec)] - s_vec[1]) / length(s_vec) / 3 *
              c(1, rep(c(4, 2), len = 1440 - 2), 1),
            1440, 1)
#Matrix containing quadrature weights for each person (row)
L <- kronecker(matrix(1, N, 1), t(q))
#Functional predictor multiplied by quadrature weights, elementwise
X_L <- X * L
df_mgcv <-
  data.frame(
    X_L = I(X_L),
    S = I(S),
    y = nhanes_df$BMI
  )
```

The data frame for fitting the model directly using `mgcv` then contains

- y: an $n \times 1$ dimensional vector containing the response, denoted as y;

- $\mathbf{X}_L$: an $n \times p$ matrix corresponding to the functional predictor multiplied by the quadrature weights for numeric integration, denoted as X_L;

- s: an $n \times p$ dimensional matrix, where each row consists of $\mathbf{s}^t$, the $1 \times p$ dimensional row vector containing the domain of the functional predictor, denoted as S.

The SoFR model can then be fit via the function call below.

```
gam_fit = gam(y ~ s(S, by = X_L, bs = "tp", k = 10), method="REML", data =
df_mgcv)
```

We now connect the syntax in the one line of code above, to the SoFR model formulation. Again, this connection is the key insight which allowed for the development of functional regression methodology based on the highly flexible penalized spline framework implemented in the `mgcv` package. As with the `refund::pfr()` function, the first quantity specified is the response variable in vector format, `y`, followed by the syntax which specifies the linear predictor, separated by a tilde ($\sim$). The `s()` function specifies that the response is a smooth function of the variable(s) supplied as unnamed arguments (in the SoFR model, this is `s`). When the variable supplied is a matrix, `mgcv` adds to the linear predictor $\sum_{j=1}^{p} \beta_1(s_{ij})$ for each row (response unit) $i = 1, \ldots, N$ (in the SoFR model with regular data, $s_{ij} = s_{i'j} = s_j$ for all $i, i' \in 1, \ldots, N$). The variable supplied to the `by` argument, $\boldsymbol{X_L}$, which must have the same dimension as the variables, indicates that the smooth terms should be multiplied elementwise by the items in this matrix, $q_j X_i(s_j)$. Combining these two facts leads to a linear predictor of the form shown in equation (4.22). The remaining arguments relate to specification of the basis expansion for $\beta_1(\cdot)$. The argument `bs="tp"` specifies the use of thin plate regression splines [313], while `k=10` sets the dimension of the basis to be 10.

We emphasize that the default behavior of the `refund::lf()` function used by `refund::pfr()` to fit SoFR in Section 4.2.3 is to use the default arguments of `mgcv::s()`, which is to use a thin plate regression spline basis (`bs="tp"`) of dimension `k=10`. In practice, users should explore sensitivity to results to increasing `k`, until the estimated function is well approximated and remains stable when further increasing `k`. For most applications a value of `k` of 10-30 is enough, with some applications requiring larger values. This can be readily checked by examining the `edf` column associated with $\widehat{\beta}(s)$ in the `summary.gam` output. In the case of the NHANES data, the function is well approximated by a smooth of degree 6.9, indicating a degree 10 penalized smooth is likely sufficient. Indeed, increasing the number of basis functions in this example does not appreciably change the shape of the estimated coefficient.

Quantities of interest (e.g., point estimates and standard errors) can be obtained from this fitted object. In addition, calling the `summary.gam()` method on the fitted object provides a host of useful information related to model fit. R users will find the structure of the summary output very similar to what is obtained from (generalized) linear regression fits using the R functions `lm()` and `glm()`. Moreover, this software structure should immediately indicate the extraordinary flexibility and wider implications of this approach. Specifying the type of spline, number of knots, estimation method for smoothing parameters, adding scalar and functional covariates, and changing the outcome distribution, follow immediately from this `mgcv` implementation.

4.5.1   Unadjusted Pointwise Inference for SoFR Using `mgcv`

As we described in Section 4.3.1, confidence intervals can easily be obtained using `refund::pfr()`. We now show how to do the same thing using directly `mgcv::gam()`. Recall that unadjusted confidence intervals refers to the fact that the confidence intervals do not account for correlation and multiplicity of tests. More precisely, we focus on constructing confidence intervals for $\beta_1(s)$ using $\widehat{\beta}_1(s)$ at some fixed $s \in S$. One straightforward approach is to construct Wald type confidence intervals of the form $\widehat{\beta}_1(s) \pm 1.96 \times \text{SE}\{\widehat{\beta}_1(s)\}$. Pointwise p-values corresponding to the null hypothesis $H_0 : \beta_1(s) = 0$ for $s \in S$ can be obtained using the large sample result that, for large N, $\widehat{\beta}_1(s)/\text{SE}\{\widehat{\beta}_1(s)\}$ has an approximately standard normal distribution under H_0. This approach implicitly conditions on the smoothing parameter selected by the smoothing parameter selection criteria (e.g., REML, GCV). Further, note that technically the standard errors produced by `mgcv::gam()` are based on the Bayesian posterior distribution of the spline coefficients. See [316] for a discussion of the implementation of standard error calculation in `mgcv`.

Suppose that we wish to construct pointwise 95% confidence intervals on a set of points over the functional domain, concatenated in a column vector denoted as $\mathbf{s}_{\text{pred}}$. The number of points we want to obtain predictions on is then $|\mathbf{s}_{\text{pred}}|$, the length of the vector $\mathbf{s}_{\text{pred}}$. Continuing the notation for basis matrices used previously in this chapter, let $\mathbf{B}(\mathbf{s}_{\text{pred}})$ be the $|\mathbf{s}_{\text{pred}}| \times K$ matrix containing the basis used for estimating $\beta_1(s)$ evaluated at $\mathbf{s}_{\text{pred}}$. Then $\widehat{\beta}_1(\mathbf{s}_{\text{pred}}) = \mathbf{B}(\mathbf{s}_{\text{pred}})\widehat{\boldsymbol{\beta}}_1$ is the estimated coefficient function evaluated at $\mathbf{s}_{\text{pred}}$. It follows that $\text{SE}\{\widehat{\beta}_1(\mathbf{s}_{\text{pred}})\} = \sqrt{\text{diag}\{\mathbf{B}(\mathbf{s}_{\text{pred}})\text{Var}(\widehat{\boldsymbol{\beta}}_1)\mathbf{B}^t(\mathbf{s}_{\text{pred}})\}}$. This quantity can be obtained directly from `mgcv::predict.gam` using the appropriate data inputs. There are multiple ways to do this, but the most straightforward one is to supply `predict.mgcv` with a data frame that contains $q_j X_i(s_j) = 1$ and $\mathbf{s} = \mathbf{s}_{\text{pred}}$. The code below shows how to do this for the case where $\mathbf{s}_{\text{pred}}$ is a regular grid of length 100 on $[0, 1]$.

```
#Set up grid for prediction
s_pred <- seq(0,24,len=100)
#Put required data inputs in a data frame
df_pred <- data.frame(S = s_pred, X_L = 1)
#Call predict.gam
coef_est <- predict(gam_fit, newdata = df_pred, type = "terms", se.fit = TRUE)
#Extract point estimates and pointwise standard errors
beta_hat <- coef_est$fit[,1]
se_beta_hat <- coef_est$se.fit[,1]
#Construct lower and upper bounds for pointwise 95% confidence intervals
beta_hat_LB <- beta_hat - qnorm(0.975) * se_beta_hat
beta_hat_UB <- beta_hat + qnorm(0.975) * se_beta_hat
```

The `mgcv` package contains functionality which allows users to obtain point estimates and standard errors for linear functions of each term included in the model, separately. This applies to non-linear functions estimated using penalized splines, as is the case with $\beta_1(s)$. To obtain the point estimates $\widehat{\beta}_1(s)$ and their corresponding standard errors using `mgcv::gam()`, the correct data inputs must be supplied to the `predict.gam()` function. Specifically, levels for all predictors used in model fitting must be supplied. In the SoFR model we fit, the only predictor in the model was the linear functional term $\int_S \beta_1(s)X_i(s)ds$, so only predictors associated with this term need be included. Recall that the syntax used to specify the linear function term was `s(S, by = X_L, bs = "tp", k = 10)`. The data inputs for this term are the objects `S` and `X_L`, the matrices containing the functional domain ($\mathbf{s}$) and elementwise product of the quadrature weights and the functional predictor ($\mathbf{X}_L$), respectively. The `predict.gam` function will evaluate each term of the model at the values supplied to the `newdata` argument. In the example above, we specify that we wish to evaluate $\widehat{\beta}_1(s)$ at $\mathbf{s}_{\text{pred}}$ with $\mathbf{X}_L = 1$. Returning to the numeric approximation approach to estimating the linear functional term, we have

$$E[Y_i|\{X_i : s \in S\}] \approx \beta_0 + \sum_{j=1}^{p} q_j X_i(s_j)\beta_1(s_j) \, .$$

So if for a fixed s_j, we obtain predictions with $X_i(s_j)q_j = 1$, we obtain $\widehat{\beta}_1(s_j)$. By setting $\mathbf{X}_L = 1$ in the data frame `df_pred`, we do exactly this. Thus, `mgcv` will provide point estimators and their standard errors for $\beta_1(s)$ for fixed points on the domain (hence pointwise confidence intervals). The lower and upper bounds for the 95% pointwise confidence interval constructed using the code above can be used to exactly re-create the upper left panel of Figure 4.9, the point estimate and 95% confidence intervals obtained from `refund::pfr()`.

Having supplied the correct data inputs to `predict.gam()` with the argument `se.fit=TRUE`, the object returned is a list with two elements. The first element contains a matrix with the point estimates for each term in the model, the second contains a matrix of the same dimension with the corresponding standard errors for the point estimates contained in the first returned element. In our example we only had one term in the model, so we extracted the first column. In a model with more terms, one would need to either manually extract the correct column based on the order in which the linear functional term was specified, or by using regular expressions.

4.5.2   Correlation and Multiplicity Adjusted (CMA) Inference for SoFR

The unadjusted pointwise confidence intervals presented Section 4.5.1 are useful as a fast, initial approach to interpreting the direction and strength of the association at various points on the domain S. It is tempting to interpret the coefficient as "statistically significant" where the confidence intervals do not cross 0. Indeed, this is commonly done in practice. However, this approach ignores that these confidence intervals and corresponding statistics are correlated and they are obtained at many different points along the functional domain.

To address these problems we focus on the correlation and multiplicity adjusted (CMA) inference described in Section 2.4. We consider testing the null hypothesis

$$H_0 : \beta_1(s) = 0, \quad s \in \mathbf{s}_{\text{pred}} \ ,$$

which is a null hypothesis that requires $|\mathbf{s}_{\text{pred}}|$ tests, one for every $s \in \mathbf{s}_{\text{pred}}$. This number of tests depends on the grid of points we obtain predictions for and can be increased arbitrarily. In the current example, there are 100 tests conducted because we have chosen to evaluate $\widehat{\beta}_1(s)$ on an evenly spaced grid of length 100 along S. However, this number could, in principle, have been chosen to be 10 or 10 billion due to the continuity assumption on the functional coefficient. When the number of tests is very large, approaches for controlling the family-wise error rate (FWER), such as the Bonferonni correction, or the Benjamini-Hochberg false discovery rate (FDR) [15] are inadequate. Indeed, ignoring the correlation among tests and applying such corrections can lead to arbitrarily long confidence intervals and tests that are too conservative and never reject the null hypothesis irrespective of the strength of the signal. This is a "continuity paradox," because the number of tests could be anything between 1 and infinity, depending on our definition of $\mathbf{s}_{\text{pred}}$. Thus, if we want to address the problem of multiplicity, there is no way around the problem of correlation, at least in the context of functional data. The question is then twofold: how dense should $\mathbf{s}_{\text{pred}} \subset S$ be to sufficiently capture the entire domain, and how to address the fact that tests are correlated?

Some methods exist for addressing this problem. In particular, the `summary` method for `mgcv::gam()` fitted objects contains a p-value associated with this null hypothesis [318], though we urge caution in using these p-values in the context of functional regression, as they do not always seem to be consistent with p-values obtained from alternative approaches and with the visual inspection of results. We will discuss this in more detail, but, as far as we are concerned, this area of research is definitely still wide open.

To address the problem of correlation and multiplicity adjusted (CMA) confidence intervals, three methods were introduced in Section 2.4. The first method discussed in Section 2.4.1 introduced the joint confidence intervals based on multivariate normality. This method requires the calculation of the exact quantiles of a multivariate normal distribution. This is reasonable using existing software such as `mvtnorm::qmvnorm()` for a random vector of length 100 as is used in our example. This approach involves a slight jittering of the covariance operator which seems to work well in practice, but does not as of writing have a theoretical justification. The other two approaches are simulation based, joint confidence intervals based on parameter simulations introduced in Section 2.4.2 and joint confidence intervals based on the max absolute statistics introduced in Section 2.4.3. We also discuss some potential pitfalls associated with using the PCA or SVD decomposition of the covariance estimator of $\widehat{\beta}_1(\mathbf{s}_{\mathrm{pred}})$.

Simulations from the Spline Parameter Distribution

For sufficiently large sample sizes, conditional on the estimated smoothing parameter, the joint distribution of the spline coefficients is approximately normal with mean $\widehat{\boldsymbol{\beta}}_1$ and variance $\mathrm{Var}(\widehat{\boldsymbol{\beta}}_1)$. Noting that $\widehat{\beta}_1(\mathbf{s}_{\mathrm{pred}}) = \mathbf{B}(\mathbf{s}_{\mathrm{pred}})\widehat{\boldsymbol{\beta}}_1$, where $\mathbf{B}(\mathbf{s}_{\mathrm{pred}})$ is the collection of basis functions used to estimate $\beta_1(\cdot)$ evaluated at $\mathbf{s}_{\mathrm{pred}}$. Therefore, simulating $\widehat{\beta}_1(\mathbf{s}_{\mathrm{pred}})$ is as simple as simulating $\widehat{\boldsymbol{\beta}}_1$, which is low dimensional, and pre-multiplying it with the $\mathbf{B}(\mathbf{s}_{\mathrm{pred}})$ matrix. The simulations from the spline parameter distribution are described in the Algorithm below, and follow directly from the approach presented in [57].

Algorithm 1 Algorithm for Simulations from the Spline Parameter Distribution: SoFR

Input
$B, \mathbf{B}(\mathbf{s}_{\mathrm{pred}}), \widehat{\boldsymbol{\beta}}_1, \mathrm{Var}(\widehat{\boldsymbol{\beta}}_1), Z = \max\{|\widehat{\beta}_1(s)|/\mathrm{SE}\{\widehat{\beta}_1(s)\} : s \in \mathbf{s}_{\mathrm{pred}}\}$

Output
$p_{\mathrm{gCMA}} = p(\max_{s \in \mathbf{s}_{\mathrm{pred}}}\{|\widehat{\beta}_1(s)|/\mathrm{SE}\{\widehat{\beta}_1(s)\}\} \ge |Z||H_0 : \beta_1(s) = 0, \forall s \in \mathbf{s}_{\mathrm{pred}})$
$d^b, b = 1, \ldots, B$ simulations from the distribution of $\max_{s \in \mathbf{s}_{\mathrm{pred}}}\{|\widehat{\beta}_1(s)|/\mathrm{SE}\{\widehat{\beta}_1(s)\}|H_0\}$

for $b = 1, \ldots, B$ **do**
$\quad \boldsymbol{\beta}_1^b \sim N\{\widehat{\boldsymbol{\beta}}_1, \mathrm{Var}(\widehat{\boldsymbol{\beta}}_1)\}$
$\quad \beta_1^b(\mathbf{s}_{\mathrm{pred}}) = \mathbf{B}(\mathbf{s}_{\mathrm{pred}})\boldsymbol{\beta}_1^b$
$\quad d^b = \max_{s \in \mathbf{s}_{\mathrm{pred}}}|\beta_1^b(s) - \widehat{\beta}_1(s)|/\mathrm{SE}\{\widehat{\beta}_1(s)\}$
end for
$p_{\mathrm{CMA}} = \max\{B^{-1}, B^{-1}\sum_{b=1}^{B} d^b > \max\{|\widehat{\beta}_1(s)|/\mathrm{SE}\{\widehat{\beta}_1(s)\}| : s \in \mathbf{s}_{\mathrm{pred}}\}\}$

The simulated test statistic d^b, $b = 1, \ldots, B$ is a sample from the distribution of the maximum of the point-wise Wald statistic over $\mathbf{s}_{\mathrm{pred}}$. The $1 - \alpha$ quantile of $\{d^b : 1 \le b \le B\}$, denoted as $q(\mathbf{C}_\beta, 1 - \alpha)$, then represents the value such that

$$\Pr_{H_0:\beta_1(s)=0}[\{|\widehat{\beta}_1(s)|/\mathrm{SE}\{\widehat{\beta}_1(s)\} \ge q(\mathbf{C}_\beta, 1 - \alpha) : s \in \mathbf{s}_{\mathrm{pred}}\}] \approx \alpha,$$

with the approximation due to the variability associated with the simulation procedure and the normal approximation of the distribution of $\widehat{\boldsymbol{\beta}}$. Here we have used the notation $q(\mathbf{C}_\beta, 1 - \alpha)$ to indicate the dependence of the quantile on the correlation matrix $\mathbf{C}_\beta$ of $\widehat{\beta}_1(\mathbf{s}_{\mathrm{pred}})$ and to keep the notation consistent with the one introduced in Section 2.4.

It follows that the $1 - \alpha$ level correlation and multiplicity adjusted (CMA) confidence interval is

$$\widehat{\beta}_1(s) \pm q(\mathbf{C}_\beta, 1 - \alpha)\mathrm{SE}\{\widehat{\beta}_1(s)\}$$

for all values $s \in \mathbf{s}_{\mathrm{pred}}$. Conveniently, CMA confidence intervals can be inverted to form both pointwise and global CMA p-values, allowing for straightforward evaluation of inference accounting for the correlated nature of tests along the domain. First, consider the procedure for constructing pointwise CMA p-values. For a fixed s, we can simply find the smallest value of α for which the above interval does not contain zero (the null hypothesis is rejected). We denote this probability by $p_{\mathrm{pCMA}}(s)$ and refer to it as the pointwise correlation and multiplicity adjusted (pointwise CMA) p-value. To calculate the global pointwise correlation and multiplicity adjusted (global CMA) p-value we define the minimum α level at which at least one confidence interval $\widehat{\beta}_1(s) \pm q(\mathbf{C}_\beta, 1 - \alpha)\mathrm{SE}\{\widehat{\beta}_1(s)\}$ for $s \in \mathbf{s}_{\mathrm{pred}}$ does not contain zero. We denote this p-value by $p_{\mathrm{gCMA}}(\mathbf{s}_{\mathrm{pred}})$. As discussed in Section 2.4.4, it can be shown that

$$p_{\mathrm{gCMA}}(\mathbf{s}_{\mathrm{pred}}) = \min\{p_{\mathrm{pCMA}}(s) : s \in \mathbf{s}_{\mathrm{pred}}\} \ .$$

Calculating both the pointwise and global CMA adjusted p-values requires only one simulation to obtain the distribution of d^b, though a large number of bootstrap samples, B, may be required to estimate extreme p-values. Here we set $B = 10^7$ to illustrate a point comparing p-values obtained using different methods when p-values are very small. In general, if we are only interested in cases when the p-value is < 0.001, $B = 10^4$ simulations should be enough.

Even with the relatively large number of simulations, the entire procedure is fairly fast as it does not involve model refitting, simply simulating from a multivariate normal of reasonable dimension, in our case K dimensional. Below we show how to conduct these simulations and calculate the CMA adjusted confidence intervals and global p-values. An essential step is to extract the covariance matrix of $\widehat{\beta}_1$ from the `mgcv` fit. This is accomplished in the expression Vbheta <- vcov(gam_fit)[inx_beta,inx_beta], where inx_beta are the indices associated with the spline coefficients used to estimate $\beta_1(\cdot)$. The code chunk below focuses on obtaining the CMA confidence intervals at a particular confidence interval, in this case $\alpha = 0.05$.

```
#Get the design matrix associated with s_pred
lpmat <- predict(gam_fit, newdata = df_pred, type = "lpmatrix")
#Get the column indices associated with the functional term
inx_beta <- which(grepl("s\\(S\\):X_L\\.[0-9]+", dimnames(lpmat)[[2]]))
#Get the design matrix A associated with beta(s)
Bmat <- lpmat[,inx_beta]
#Get Var spline coefficients (beta_sp) associated with beta(u,s)
beta_sp <- coef(gam_fit)[inx_beta]
Vbeta_sp <- vcov(gam_fit)[inx_beta,inx_beta]
#Number of bootstrap samples (B)
nboot <- 1e7
#Set up container for bootstrap
beta_mat_boot <- matrix(NA, nboot, length(s_pred))
#Do the bootstrap
for(i in 1:nboot){
  beta_sp_i <- MASS::mvrnorm(n = 1, mu = beta_sp, Sigma = Vbeta_sp)
  beta_mat_boot[i,] <- Bmat %*% beta_sp_i
}

#Find the max statistic
dvec <- apply(beta_mat_boot, 1, function(x) max(abs(x - beta_hat) / se_beta_hat))
#Get 95% global confidence band
Z_global <- quantile(dvec, 0.95)
beta_hat_LB_global <- beta_hat - Z_global * se_beta_hat
beta_hat_UB_global <- beta_hat + Z_global * se_beta_hat
```

We now show how to invert these confidence intervals and obtain the pointwise adjusted CMA p-values, $\{p_{\mathrm{pCMA}}(s) : s \in \mathbf{s}_{\mathrm{pred}}\}$, as well as the global CMA adjusted p-value $p_{\mathrm{gCMA}}(\mathbf{s}_{\mathrm{pred}})$. In practice all p-values we estimate are limited by the number of simulations we use.

```
#Set up a vector of quantiles to search over
qs_vec <- seq(0.90, 1, len = 10000)
#Get the corresponding quantiles of the max statistic
qs_vec_d <- quantile(dvec, qs_vec)
#Loop over quantiles of max statistic, evaluate which values result
# in global confidence bands covering 0 over all regions of interest
cover_0 <-
        vapply(qs_vec_d, function(x){
        all((beta_hat - x * se_beta_hat) < 0 & (beta_hat + x * se_beta_hat > 0))
        }, numeric(1))

#Get the p-value
inx_sup_p <- which(cover_0 == 1)
p_val_g <- ifelse(length(inx_sup_p) == 0, 1 / nboot, 1 - min(qs_vec[inx_sup_p]) )
#Find the max statistic
dvec <- apply(beta_mat_boot, 1, function(x) max(abs(x - beta_hat) / se_beta_hat))
#Get 95% global confidence band
Z_global <- quantile(dvec, 0.95)
beta_hat_LB_global <- beta_hat - Z_global_np * se_beta_hat
beta_hat_UB_global <- beta_hat + Z_global_np * se_beta_hat
```

The pointwise adjusted p-values are stored in the vector `p_val_lg` while $p_{\mathrm{pCMA}}(\mathbf{s}_{\mathrm{pred}})$ is stored in the scalar (`p_val_g`). Based on $B = 10^7$ simulations we obtain a global CMA p-value of 1×10^{-5} for testing the null hypothesis of 0 effect. Figure 4.11 displays the results of the CMA inference. The left panel of Figure 4.11 presents both the unadjusted and pointwise CMA adjusted p-values $\{p_{\mathrm{pCMA}}(s) : s \in \mathbf{s}_{\mathrm{pred}}\}$ in dark and light gray, respectively. The dashed gray line corresponds to probability 0.05. The right panel of Figure 4.11 presents the unadjusted and pointwise CMA adjusted 95% confidence intervals, again in dark and light gray, respectively. The dashed gray line in this panel corresponds to zero effect. We find that after adjusting for correlations and multiple comparisons, the three broad periods of time identified as significant by the unadjusted pointwise inference persist (early morning, late morning, and early evening), however the exact periods of time in which activity is significantly associated with BMI is reduced due to the widening of confidence intervals.

Non-parametric Bootstrap of the Max Absolute Statistic

The procedure for performing the non-parametric bootstrap approach to inference does not condition on estimated model parameters. Instead, study participants are sampled with replacement and the joint variability is assessed using these resamples. Specifically, the non-parametric bootstrap (1) re-samples units (in this case individuals) with replacement; (2) re-estimates the model within each re-sampled data set; (3) extracts the estimated coefficient on the grid of interest; and (4) calculates the quantiles of the distribution of the maximum statistic to obtain global inference. In contrast to sampling from the distribution of $\widehat{\beta}_1$, this procedure does not require the Normality of $\widehat{\beta}_1$, nor does it condition on the estimated smoothing parameter obtained from the initial model fit. However, the non-parametric bootstrap implicitly requires an assumption of spherical symmetry of the distribution of $\widehat{\beta}_1(\mathbf{s}_{\mathrm{pred}})$. As of this writing, it remains an open problem how to perform inference when this implicit assumption is violated. The approach was proposed by [258] in the context of nonparametric smoothing and was adapted for functional inference by [53].

To be precise, the non-parametric bootstrap procedure is as follows. Fix the grid on which we conduct global inference on, and denoted it, once again, as $\mathbf{s}_{\mathrm{pred}}$. First, we obtain $b = 1, \ldots, B$ non-parametric bootstrap estimates of $\widehat{\beta}_1(\mathbf{s}_{\mathrm{pred}})$, denoted as $\widehat{\beta}_1^b(\mathbf{s}_{\mathrm{pred}})$. Then

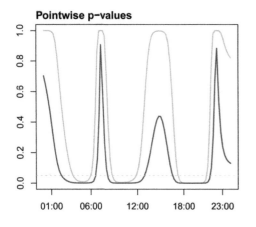
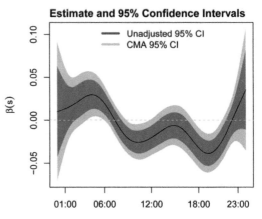

FIGURE 4.11: Pointwise CMA inference for SoFR based on simulations from the distribution of spline coefficients. BMI is the outcome and PA functions are the predictors. Left panel: estimated pointwise unadjusted (dark gray) and CMA (light gray) p-values denoted by $p_{\mathrm{pCMA}}(s)$. Right panel: the 95% pointwise unadjusted (dark gray) and CMA confidence intervals for $\beta(s)$.

$\{\widehat{\beta}_1^b(s) : 1 \leq b \leq B, s \in \mathbf{s}_{\mathrm{pred}}\}$ is the collection of bootstrap estimates over the grid $\mathbf{s}_{\mathrm{pred}}$. As with the simulations from the spline parameter distribution, we then obtain

$$d^b = \max_{s \in \mathbf{s}_{\mathrm{pred}}} |\widehat{\beta}_1^b(s) - \widehat{\beta}_1(s)| / \mathrm{SE}\{\widehat{\beta}_1(s)\} \ .$$

This procedure requires extracting the variances of $\widehat{\beta}_1(\mathbf{s}_{\mathrm{pred}})$, but not the entire covariance matrix. However, it requires refitting the model B times, which increases the computational complexity, increasing the time to obtain extreme p-values.

The code below illustrates how to perform the non-parametric bootstrap with repeated model estimation done using `mgcv::gam()` instead of `refund::pfr()`. For SoFR, either function is equally easy to use for estimation given the wide format storage of the data used as input for both functions (as compared to function-on-function regression; see Chapter 6). However, as of this writing, `pfr` is less straightforward to use, as it requires a function call to `predict.gam`. This is not an inherent limitation of `pfr`, but rather a design decision made when designing the `predict.pfr` method.

Given the matrix of non-parametric resampled estimated $\widehat{\beta}_1^b(\mathbf{s}_{\mathrm{pred}})$ contained in the matrix `beta_mat_boot_np` above, the calculation of pointwise CMA adjusted and global p-values (along with their corresponding confidence intervals) proceeds exactly the same as with the non-parametric bootstrap and thus the code to do so is omitted here. In this particular example, the non-parametric bootstrap yields qualitatively similar results when compared to the simulations from the spline parameter distribution. Figure 4.12 presents the results in the same format as Figure 4.11. We note that the confidence intervals are slightly wider, resulting in the early morning associations not being statistically significant at the 0.05 level when conducting CMA inference. In addition, the late morning period loses much of its statistical significance.

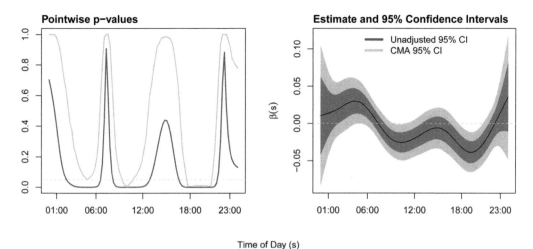

FIGURE 4.12: Pointwise CMA inference for SoFR based on the nonparametric bootstrap of the max statistic. BMI is the outcome and PA functions are the predictors. Left panel: estimated pointwise unadjusted (dark gray) and CMA (light gray) p-values denoted by $p_{\mathrm{pCMA}}(s)$. Right panel: the 95% pointwise unadjusted (dark gray) and CMA confidence intervals for $\beta(s)$.

```
#Set up container for bootstrap
beta_mat_boot_np <- matrix(NA, nboot, length(s_pred))
#Number of participants to resample
N <- nrow(df_mgcv)
#Do the bootstrap
for(i in 1:nboot){
   inx_i <- sample(1:N, size = N, replace = TRUE)
   df_i <- df_mgcv[inx_i,]
   gam_i <- gam(y ~ s(S, by = X_L, bs = "tp", k = 10), method = "REML", data = df_i)
   beta_mat_boot_np[i,] <- predict(gam_i, newdata = df_pred, type = "terms")
}

#Find the max statistic
dvec_np <- apply(beta_mat_boot_np, 1, function(x) max(abs(x - beta_hat) /
se_beta_hat))
#Get 95% global confidence band
Z_global <- quantile(dvec, 0.95)
beta_hat_LB_global_np <- beta_hat - Z_global_np * se_beta_hat
beta_hat_UB_global_np <- beta_hat + Z_global_np * se_beta_hat
```

Analytic Solution to the Parametric Bootstrap Approximation

In Section 2.4.4 we described an approach based on the PCA decomposition of the covariance operator, which may seem like a good practice. However, we will point out problems that can occur in this context and we do not recommend this approach at this time. The lesson here is that not all methods that seem similar are similar, and in practice, one should compare approaches and identify why methods may provide discordant results.

Consider the case when the distribution of $\widehat{\beta}_1$ is well approximated by a multivariate normal distribution; this can be justified either based on the distribution of the data or by asymptotic considerations. We have already seen that in this application, this assumption is probably incorrect. Regardless, we would like to find an analytic solution to this problem and compare the results with those obtained via simulations from the spline coefficients distribution. Under the assumption of normality of $\widehat{\beta}_1$ it follows that $\widehat{\beta}_1(\mathbf{s}_{\mathrm{pred}}) = \mathbf{B}(\mathbf{s}_{\mathrm{pred}})\widehat{\beta}_1$ has a multivariate normal distribution with covariance matrix $\mathrm{Var}\{\widehat{\beta}_1(\mathbf{s}_{\mathrm{pred}})\} = \mathbf{B}(\mathbf{s}_{\mathrm{pred}})\mathrm{Var}(\widehat{\beta}_1)\mathbf{B}^t(\mathbf{s}_{\mathrm{pred}})$. Using this fact, one could proceed with a decomposition approach to obtain CMA confidence intervals as well as pointwise and global p-values for testing the null hypothesis: $H_0 : \beta_1(s) = 0, \forall s \in \mathbf{s}_{\mathrm{pred}}$.

Using the same argument discussed in Section 2.4.1

$$[\{\widehat{\beta}_1(s) - \beta(s)\}/\mathrm{SE}\{\widehat{\beta}_1(s)\} : s \in \mathbf{s}_{\mathrm{pred}}] \sim N(0, \mathbf{C}_\beta) \, ,$$

where $\mathbf{C}_\beta$ is the correlation matrix corresponding to the covariance matrix $\mathrm{Var}\{\widehat{\beta}_1(\mathbf{s}_{\mathrm{pred}})\} = \mathbf{B}(\mathbf{s}_{\mathrm{pred}})\mathrm{Var}(\widehat{\beta}_1)\mathbf{B}^t(\mathbf{s}_{\mathrm{pred}})$. As we discussed in Section 2.4.1, we need to find a value $q(\mathbf{C}_\beta, 1-\alpha)$ such that

$$P\{q(\mathbf{C}_\beta, 1 - \alpha) \times \mathbf{e} \leq \mathbf{X} \leq q(\mathbf{C}_\beta, 1 - \alpha) \times \mathbf{e}\} = 1 - \alpha \, ,$$

where $\mathbf{X} \sim N(0, \mathbf{C}_\beta)$ and $e = (1, \ldots, 1)^t$ is the $|\mathbf{s}_{\mathrm{pred}}| \times 1$ dimensional vector of ones. Once $q(\mathbf{C}_\beta, 1 - \alpha)$ is available, we can obtain a CMA $1 - \alpha$ level confidence interval for $\beta_1(s)$ as

$$\widehat{\beta}_1(s) \pm q(\mathbf{C}_\beta, 1 - \alpha)\mathrm{SE}\{\widehat{\beta}_1(s)\} : \forall \, s \in s \in \mathbf{s}_{\mathrm{pred}} \, .$$

Luckily, the function `qmvnorm` in the R package `mvtnorm` [96, 97] is designed to extract such quantiles. Unluckily, the function does not work for matrices $\mathbf{C}_\beta$ that are singular and very high dimensional. Therefore we need to find a theoretical way around the problem.

Indeed, $\widehat{\beta}_1(\mathbf{s}_{\mathrm{pred}}) = \mathbf{B}(\mathbf{s}_{\mathrm{pred}})\widehat{\beta}_1$ has a degenerate normal distribution, because its rank is at most K, the number of basis functions used to estimate $\beta_1(s)$. Since we have evaluated $\widehat{\beta}_1(s)$ on a grid of $|\mathbf{s}_{\mathrm{pred}}| = 100$, the covariance and correlation matrices of $\widehat{\beta}_1(\mathbf{s}_{\mathrm{pred}})$ are 100 dimensional. To better understand this, consider the case where $\widehat{\beta}_1(s) = c + s$ for $c \in \mathbb{R}$. Then $\mathrm{rank}(\mathbf{C}_\beta) \leq 2$ for any choice of $\mathbf{s}_{\mathrm{pred}}$.

We will use a statistical trick based on the eigendecomposition of the covariance matrix. Recall that if $\widehat{\beta}_1(\mathbf{s}_{\mathrm{pred}})$ has a degenerate multivariate normal distribution of rank $m \leq K$, then there exists some random vector of independent standard normal random variables $\mathbf{Q} \in \mathbb{R}^m$ such that

$$\widehat{\beta}_1^t(\mathbf{s}_{\mathrm{pred}}) = \mathbf{Q}^t\mathbf{D} \, ,$$

with $\mathbf{D}^t\mathbf{D} = \mathbf{B}(\mathbf{s}_{\mathrm{pred}})\mathrm{Var}(\widehat{\beta}_1)\mathbf{B}^t(\mathbf{s}_{\mathrm{pred}})$. If we find m and a matrix $\mathbf{D}$ with these properties, the problem is solved, at least theoretically. Consider the eigendecomposition $\mathbf{B}(\mathbf{s}_{\mathrm{pred}})\mathrm{Var}(\widehat{\beta}_1)\mathbf{B}^t(\mathbf{s}_{\mathrm{pred}}) = \mathbf{U}\mathbf{\Lambda}\mathbf{U}^t$, where $\mathbf{\Lambda}$ is a diagonal matrix of eigenvalues and the matrix $\mathbf{U}\mathbf{U}^t = \mathbf{I}_{|\mathbf{U}_{\mathrm{pred}}| \times |\mathbf{s}_{\mathrm{pred}}|}$ is an orthonormal matrix with the kth column being the eigenvector corresponding to the kth eigenvalue. Note that all eigenvalues $\lambda_k = 0$ for $k > m$ and

$$\mathbf{B}(\mathbf{s}_{\mathrm{pred}})\mathrm{Var}(\widehat{\beta}_1)\mathbf{B}^t(\mathbf{s}_{\mathrm{pred}}) = \mathbf{U}_m\mathbf{\Lambda}_m\mathbf{U}_m^t \, ,$$

where $\mathbf{U}_m$ is the $|\mathbf{s}_{\mathrm{pred}}| \times m$ dimensional matrix obtained by taking the first m columns of $\mathbf{U}$ and $\mathbf{\Lambda}_m$ is the $m \times m$ dimensional diagonal matrix with the first m eigenvalues on the main diagonal. If we define $\mathbf{D}^t = \mathbf{U}_m\mathbf{\Lambda}_m^{1/2}$ and

$$\{\widehat{\beta}_1^t(\mathbf{s}_{\mathrm{pred}}) - \beta_1^t(\mathbf{s}_{\mathrm{pred}})\}\mathbf{D}^{-1} = \mathbf{Q}^t \, ,$$

where $\mathbf{D}^{-1} = \mathbf{U}_m \mathbf{\Lambda}_m^{-1/2}$ is the Moore Penrose right inverse of $\mathbf{D}$ ($\mathbf{D}$ is not square, but is full row rank). Under the null hypothesis, $\beta_1(\mathbf{s}_{\mathrm{pred}}) = 0$, thus we can obtain a p-value of no effect by calculating

$$1 - \mathrm{P}(\max\{|\mathbf{Q}|\} \leq \max\{|\widehat{\beta}_1^t(\mathbf{s}_{\mathrm{pred}})\mathbf{D}^{-1}|\}) ,$$

where $\mathbf{Q}$ is a vector of independent standard normal random variables. This calculation is fairly straightforward, though we use `mvtnorm::pmvnorm` for convenience. The mathematics is clear and nice and it would be perfect if it worked. Below we show how to calculate these quantities and then we explain why this may not be a good idea in general. We leave this as an open problem.

```
beta <- coef(gam_fit)[inx_beta]
#Get Var(hat(beta))
Vbeta <- Bmat %*% Vbeta %*% t(Bmat)
#Get eigenfunctions via svd
eVbeta <- svd(Vbeta, nu = ncol(Bmat), nv = ncol(Bmat))
#Get only positive eigenvalues
inx_evals_pos <- which(eVbeta$d >= 1e-6)
#Get "m" - dimension of lower dimensional RV
m <- length(inx_evals_pos)
#Get associated eigenvectors
U_m <- eVbeta$v[,inx_evals_pos]
#Get D transpose
D_t <- U_m %*% diag(sqrt(eVbeta$d[inx_evals_pos]))
#Get D inverse
D_inv <- MASS::ginv(t(D_t))
#Get Q
Q <- t(Bmat %*% beta) %*% D_inv
#Get max statistic
Zmax <- max(abs(Q))
#Get p-value
p_val <- 1 - pmvnorm(lower = rep(-Zmax, m), upper = rep(Zmax, m),
                     mean = rep(0, m), cor = diag(1, m))
```

Unfortunately, this p-value is not equal to the p-value obtained from simulating from the spline parameters normal distribution. In fact, the p-value obtained using this method is approximately 0, effectively the same p-value reported by `summary.gam` in the `mgcv` package ($p < 2 \times 10^{-16}$). The p-value reported by `mgcv` is based on a very similar approach which modifies slightly the matrix of eigenvalues of the covariance operator [318] and derives a Wald test statistic which very closely matches the sum of the squared $\mathbf{Q}$, a χ^2 random variable under the null hypothesis. In this example, all three approaches would yield the same inference in practice (a statistically significant association), but the discrepancy, which is orders of magnitude, is cause for concern. Indeed, the problem becomes more concerning when considering the results of the multivariate normal approximation, presented below, which provides a p-value that agrees quite closely with the simulations from the spline parameter distribution.

Multivariate Normal Approximation

Another approach may be to ignore the fact that the distribution of $\widehat{\beta}(\mathbf{s}_{\mathrm{pred}})$ in our example is degenerate multivariate normal. To make things work, we slightly jitter the covariance matrix to ensure that the implied correlation function is positive definite. Then we may obtain inference using software for calculating quantiles of a multivariate normal random vector (e.g., `mvtnorm::pmvnorm`). The code to do so is provided below.

```
#Get hat(beta)/SE(hat(beta))
beta_hat_std <- beta_hat / se_beta_hat
#Get Var(hat(beta)/SE(hat(beta)))
Vbeta_hat <- Bmat %*% Vbeta %*% t(Bmat)
#Jitter to make positive definite
Vbeta_hat_PD <- Matrix::nearPD(Vbeta_hat)$mat
#Get correlation function
Cbeta_hat_PD <- cov2cor(matrix(Vbeta_hat_PD, 100, 100, byrow = FALSE))
#Get max statistic
Zmax <- max(abs(beta_hat_std))
#p-value
p_val <- 1 - pmvnorm(lower = rep(-Zmax, 100), upper = rep(Zmax, 100),
                     mean = rep(0, 100), cor = Cbeta_hat_PD)
```

The resulting p-value is 2.9×10^{-6}, which is very close to that obtained from the simulations from the spline parameter distribution. In principle, this method should yield the same result as the exact solution, though we find with some consistency that it does not. In our experience, this approach provides results which align very well with the simulations from the spline parameter distribution, though the theoretical justification for jittering of the covariance function is, as of this writing, not justified. To construct a 95% CMA global confidence interval one may use the `mvtnorm::qmvnorm` function, illustrated below. Unsurprisingly, the resulting global multiplier is almost identical to that obtained from the parametric bootstrap. For that reason we do not plot the results or interpret further.

```
Z_global <- qmvnorm(0.95, mean = rep(0,100), cor = Cbeta_hat_PD, tail =
"both.tails")$quantile
```

5

Function-on-Scalar Regression

We now consider the use of functions as responses in models with scalar predictors. This setting is widespread in applications of functional data analysis, and builds on specific tools and broader intuition developed in previous chapters. We will focus on the linear Function-on-Scalar Regression (FoSR) model, with a brief overview of alternative approaches in later sections.

FoSR is known under different names and was first popularized by [242, 245] who introduced it as a functional linear model with a functional response and scalar covariates; see Chapter 13 in [245]. Here we prefer the more precise FoSR nomenclature introduced by [251, 253], which refers directly to the type of outcome and predictor. It is difficult to pinpoint where these models originated, but it is likely that the life of FoSR models has multiple origins, most likely driven by applications. One of the origins is intertwined with the introduction of linear mixed effects (LME) models for longitudinal data [161]. While mixed effects models have a much longer history, the Laird and Ware paper [161] summarized and introduced the modern formalism of mixed effects models for longitudinal data. Linear mixed models have traditionally focused on a sparsely observed functional outcome and scalar predictors and use specific known structures of random effects (e.g., random slope/random intercept models); for more details see [66, 87, 231, 301]. Another point of origin was the inference for differences between the means of groups of functions [27, 79, 243, 244], as described in Section 13.6 of [245]. These approaches tended to be more focused on the functional aspects of the problem and allowed more flexibility in modeling the random effects as smooth functions. In this book we will unify the data generating mechanisms and methods under the mixed effects models umbrella, though some random effects will be used for traditional modeling of trajectories, while others will be used for nonparametric smoothing of the functional data. We will also emphasize that unifying software can be used for fitting such models and point out the various areas that are still open for research.

FoSR has been applied extensively to areas of scientific research including brain connectivity [251], diffusion brain imaging [103, 109, 280], seismic ground motion [11], CD4 counts in studies of HIV infection studies [77], reproductive behavior of large cohorts of medflies [44, 45], carcinogenesis experiments [207], knee kinematics [6], human vision [218], circadian analysis of cortisol levels [114], mass spectrometry proteomic data [206, 342, 343, 204], eye scleral displacement induced by intraocular pressure [162], electroencephalography during sleep [53], feeding behavior of pigs [100], objective physical activity measured using accelerometers [57, 265, 273, 327], phonetic analysis [8], and continuous glucose monitoring [92, 270], to name a few. Just as discussed in Chapter 4, these papers are referenced here for their specific area of application, but they contain substantial methodological developments that could be explored in detail. Throughout this section, we will focus on methods for the linear FoSR model. Many additional methods exist, including the Functional Linear Array Model (FLAM) [23, 25, 24], Wavelet-based Functional Mixed Models (WFMM) [207], Functional Mixed Effects Modeling (FMEM) [337, 344], longitudinal FoSR using structured penalties [159], and others noted in a recent review of functional regression methods [205].

We will not discuss these approaches here, but they provide alternative model structures and estimation methods that should be considered.

The overall goal of this chapter is not to explore the vast array of published methodological tools for FoSR. Instead, we will focus on a specific group of methods that use penalized splines to model functional effects, connect these models with linear mixed effects models, and show how to implement these methods in software such as `refund` and `mgcv`.

5.1   Motivation and Exploratory Analysis of MIMS Profiles

The code below imports and organizes the processed NHANES data that will be used in this chapter. The code starts with reading the data using the `readRDS` function. For illustration purposes only, a few scalar covariates are retained using the `select()` function. We retain a small number of scalar covariates (`gender` and `age`, as well as the participant ID in `SEQN` which we convert to a factor variable), and rename `MIMS` to `MIMS_mat` to emphasize that this is a matrix. We create a categorical age variable, and then delete data that contains missing covariate information using `drop_na()`. To keep the examples in this chapter computationally feasible, we use the first 250 subjects in this dataset. Data are stored in the `nhanes_df` data frame. As noted elsewhere, we recognize the important distinction between gender and sex assigned at birth. The language and variable names used in this chapter are drawn from NHANES in order to be consistent with the framing of questions in that survey.

```
nhanes_df =
  readRDS(
    here::here("data", "nhanes_fda_with_r.rds")) %>%
  select(SEQN, gender, age, MIMS_mat = MIMS) %>%
  mutate(
    age_cat =
      cut(age, breaks = c(18, 35, 50, 65, 80),
          include.lowest = TRUE),
    SEQN = as.factor(SEQN)) %>%
  drop_na(age, age_cat) %>%
  filter(age >= 25) %>%
  tibble() %>%
  slice(1:250)
```

In the next code chunk, the `MIMS_mat` variable is converted to a `tidyfun` [261] object via the `tfd()` function. As elsewhere, we use the `arg` argument in `tfd()` to define the grid over which functions are observed: $\left[\frac{1}{60}, \frac{2}{60}, \dots \frac{1440}{60}\right]$, so that minutes are in $\frac{1}{60}$ increments and hours of the day fall on integers from 1 to 24. After transformation the `MIMS_tf` variable is a vector that contains all functional observations and makes many operations easier in the `tidyverse` [311]. This transformation is not strictly necessary, and throughout the book we also show how to work directly with the matrix format. However, the `tidyverse` has become increasingly popular in data science and we illustrate here that functional data analysis can easily interface with it.

```
nhanes_df =
  nhanes_df %>%
  mutate(
    MIMS_tf = matrix(MIMS_mat, ncol = 1440),
    MIMS_tf = tfd(MIMS_tf, arg = seq(1/60, 24, length = 1440)))
```

Once data are available in a data frame, one can start to visualize some of their properties. Suppose that one is interested in how the average objectively measured physical activity over a 24-hour interval varies with age and gender. Using the `tidyfun` package, the following code shows how to do that using a `tidy` syntax. Beginning with the NHANES data stored in **nhanes_df**, we group variables by age category and gender using the `group_by` function and obtain the 24-hour means for each subgroup using the `summarize` function. The next component of the code displays the individual mean functions using the `ggplot()` function and the `geom_spaghetti()` function from `tidyfun`.

```
nhanes_df %>%
  group_by(age_cat, gender) %>%
  summarize(mean_mims = mean(MIMS)) %>%
  ggplot(aes(y = mean_mims, color = age_cat)) +
  geom_spaghetti() +
  facet_grid(. ~ gender) +
  scale_x_continuous(breaks = seq(0, 24, length = 5)) +
  labs(x = "Time of day (hours)", y = "Average MIMS")
```

Figure 5.1 displays the results of the preceding code chunk. Individual means are color coded by age category, as indicated by the `color = age_cat` aesthetic mapping and separated into two panels corresponding to males and females, as indicated by the `facet_grid()` function. The means of the objectively measured physical activity exhibit clear circadian rhythms with more physical activity during the day and less during the evening and night. One can also notice a strong effect of age indicating that older individuals tend to be less active than younger individuals on average. This trend is clearer among men, though women in the $(65, 80]$ age range (shown in the right panel in yellow) exhibit, on average, much lower physical activity than younger women. When comparing men to women in the same age category (same color lines in the left and right panels, respectively), results indicate that women are more active. This is consistent with the findings reported in [327] in the Baltimore Longitudinal Study of Aging (BLSA) [265, 284, 327], though in direct contradiction with one of the main findings reported in the extensively cited paper [296] based on the 2003-2005 NHANES accelerometry study.

Exploratory plots like this one are useful for beginning to understand the effects of covariates on physical activity, but are limited in their scope. For example, they cannot be used to adjust for covariates or assess statistical significance, especially in cases when the effects of covariates are not as obvious as those of age and gender. Also, the group-specific mean functions are quite noisy because these are the raw means in each group without accounting for smoothness across time. These are some of the issues that more formal approaches to function-on-scalar regression are intended to address.

5.1.1 Regressions Using Binned Data

To build intuition for interpreting models with functional responses, identify practical challenges, and motivate later developments, we fit a series of regressions to MIMS profiles using binned or aggregate observations. As a first step, we aggregate minute-level data into hour-length bins, so that each MIMS profile consists of 24 observations. To obtain hourly

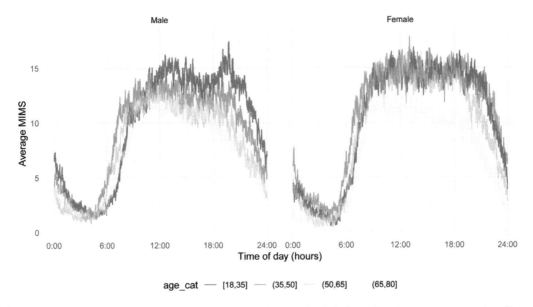

FIGURE 5.1: Pointwise means of MIMS in the NHANES data for four age categories (indicated by different color) and gender (male/female) shown in the left and right panels, respectively.

trajectories, we smooth minute-level data using a moving average approach with a 60-minute bandwidth, and evaluate the resulting functions at the midpoint of each hour.

```
nhanes_df =
  nhanes_df %>%
  mutate(
    MIMS_hour =
      tf_smooth(MIMS, method = "rollmean", k = 60, align = "center"),
    MIMS_hour = tfd(MIMS_hour, arg = seq(.5, 23.5, by = 1)))
```

The resulting `MIMS_hour` data are shown in Figure 5.2, while the code for obtaining the figure is shown in the code chunk below. We pipe the dataset into `ggplot()`, and plotting continues using `geom_spaghetti()` to show functions and adding `geom_meatballs()` to place a dot at every point. This is done both for aesthetic purposes as well as to emphasize the discrete nature of the data.

```
nhanes_df %>%
  ggplot(aes(y = MIMS_hour, color = age)) +
  geom_spaghetti(alpha = .2) +
  geom_meatballs(alpha = .2) +
  facet_grid(. ~ gender)
```

As before, we are interested in the effects of `age`, now treated as a continuous variable, and `gender`. The binning process retains a level of granularity that is informative regarding the diurnal patterns of activity but reflects a substantial reduction in the dimension and detail in the data.

These data can be analyzed using hour-specific linear models that regress bin-average MIMS values on `age` and `gender`. This collection of linear models does not account for the

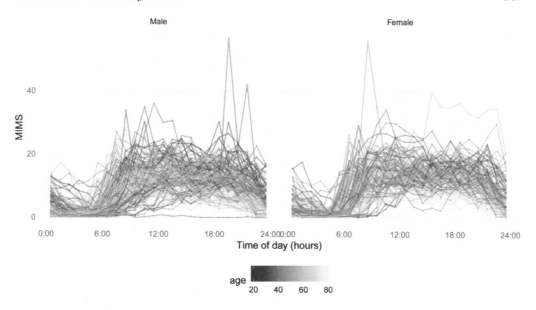

FIGURE 5.2: Subject-specific physical activity profiles averaged in one-hour intervals in NHANES. Each study participant is shown as a different line and a color gradient is used to indicate age of the individual, with darker blue indicating younger individuals and lighter green and yellow indicating older individuals. Left panel: males. Right panel: females.

temporal structure of the diurnal profiles except through the binning that aggregates data to an hour level, but taken together will illustrate the association between the outcome and predictors over the course of the day. As a first example, we will fit a standard linear model with the average MIMS value between 1:00 and 2:00 PM as a response. The model is

$$\text{MIMS}_i = \beta_0 + \beta_1 \text{age}_i + \beta_2 \text{gender}_i + \epsilon_i \,,$$

where MIMS_i represents the average MIMS value between 1:00 and 2:00 PM. For presentation simplicity we avoided adding an index for the time period for the outcome, model parameters and error process.

In the next code chunk, which fits this bin-specific linear model, the first two lines are self-explanatory. The `tf_unnest()` function in the `tidyfun` package transforms the `nhanes_df` dataframe, which includes `MIMS_hour` as a `tidyfun` vector, into a long-format `dataframe` with columns for the functional argument (i.e., hour $0.5, 1.5, \ldots, 23.5$) and the corresponding functional value (i.e., the subject bin-specific average). The `filter()` function subsets the data to include only the mean MIMS between 1:00 and 2:00 PM (the 13.5th hour is the middle of this interval when time is indexed starting at midnight). The resulting dataset is passed as the data argument into `lm()` using the "placeholder," with the appropriate formula specification for the desired model.

```
linear_fit =
  nhanes_df %>%
  select(SEQN, age, gender, MIMS_hour) %>%
  tf_unnest(MIMS_hour) %>%
  filter(MIMS_hour_arg == 13.5) %>%
  lm(MIMS_hour_value ~ age + gender, data = .)
```

TABLE 5.1

Regression results for regressing the mean MIMS between
1:00 and 2:00 PM on age and gender.

term	estimate	std.error	statistic	p.value
(Intercept)	16.442	1.027	16.003	0.000
age	−0.069	0.017	−3.953	0.000
genderFemale	0.963	0.592	1.626	0.105

The results of fitting the linear model are in the following table. Because the age variable is not centered, the intercept is the expected average MIMS between 1:00 and 2:00 PM among males at age 0. The estimated age coefficient implies a decrease in the expected average MIMS in this hour of −0.07 for each one-year increase in age among men, and is strongly statistically significant. Women have, on average, higher MIMS values than men in this time window, although the difference is suggestive rather than statistically significant.

We illustrate this analysis graphically in the next figure, using code we briefly describe but omit. In short, fitted values resulting from the regression were added to the `nhanes_df` data frame. The plot was constructed using `ggplot()`, with data points in the scatterplot shown using `geom_point()` and the fitted values illustrated using `geom_line()`.

The scatterplot and regression results are consistent with our previous observations: the binned-average physical activity decreases with age and is higher for female compared to male participants across the age groups. The benefit of this analysis over visualization-based exploratory techniques is that it provides a formal statistical assessment of these effects and their significance.

So far we have shown regression results for the 1:00 to 2:00 PM interval, but these associations may vary by the hour of the day. The next step in our exploratory analysis is to fit separate regressions at each hour separately. We accomplish this using data nested within each hour. First, we will unnest the subject-specific functional observation and then re-nest

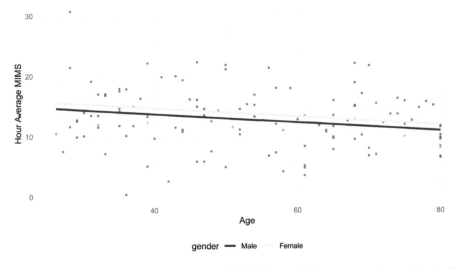

FIGURE 5.3: Age (x-axis) versus average MIMS (y-axis) between 1:00 and 2:00 PM for males (dark purple) and females (yellow). Each dot corresponds to a study participant and regression lines are added for males and females. The analysis is conducted for 250 individuals for didactic purposes.

within an hour; the result is a data frame containing 24 rows, one for each hour, with a column that contains a list of hour-specific data frames containing the MIMS_hour_value, age, and gender. By mapping over the entries in this list, we can fit hour-specific linear models and extract tidied results. The code chunk below implements this analysis.

```
hourly_regressions =
  nhanes_df %>%
  select(SEQN, age, gender, MIMS_hour) %>%
  tf_unnest(MIMS_hour) %>%
  rename(hour = MIMS_hour_arg, MIMS = MIMS_hour_value) %>%
  nest(data = -hour) %>%
  mutate(
    model = map(.x = data, ~ lm(MIMS ~ age + gender, data = .x)),
    result = map(model, broom::tidy)
  ) %>%
  select(hour, result)
```

Before visualizing the results, we do some data processing to obtain hourly confidence intervals and then structure coefficient estimates and confidence bands as `tf` objects. The result is stored in the variable `hour_bin_coefs`, which is a three-row data frame, where each row corresponds to the intercept, age, and gender effects, respectively. Columns include the term name as well as the coefficient estimates and the upper and lower limits of the confidence bands.

```
hour_bin_coefs =
  hourly_regressions %>%
  unnest(cols = result) %>%
  rename(coef = estimate, se = std.error) %>%
  mutate(
    ub = coef + 1.96 * se,
    lb = coef - 1.96 * se
  ) %>%
  select(hour, term, coef, ub, lb) %>%
  tf_nest(coef:lb, .id = term, .arg = hour)
```

We show the analysis results using `ggplot()` and other `tidyfun` functions. Because the coefficient estimates in the variables stored in `coef` are `tf` objects, they can be plotted using `geom_spaghetti()`. To emphasize the estimates at each hour, we add points using the `geom_meatballs()` function. Confidence bands are shown as shaded regions by specifying the upper and lower limits in `geom_errorband()`, and we use the `facet()` function by `term` to plot each coefficient separately.

```
hour_bin_coefs %>%
  ggplot(aes(y)) +
  geom_spaghetti() +
  geom_meatballs() +
  geom_errorband(aes(ymax = ub, ymin = lb, fill = term)) +
  facet_wrap("term", scales = "free")
```

Figure 5.4 displays the point estimators and 95% confidence intervals for the coefficients for regressions of hourly mean MIMS on age and gender. Compared to the regression at a single time point, these results provide detailed temporal information about covariate effects. The left panel corresponds to the intercept and is consistent to a fairly typical

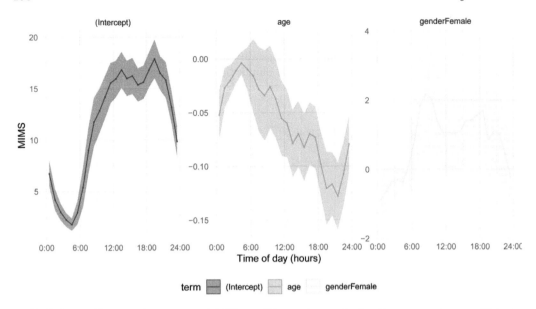

FIGURE 5.4: Point estimators and 95% confidence intervals for the regression coefficients for hourly regressions of mean MIMS on age and gender. Each regression is conducted independent of the other regressions and results are shown for the intercept, age, and gender parameters, respectively.

circadian pattern, with low activity during the night and higher activity during the day. The middle panel displays the age effect adjusted for gender and indicates that older individuals are generally less active at all times (note that point estimators are all negative). Moreover, age has a stronger effect on physical activity in mid-afternoon and evening (note the decreasing pattern of estimated effects as a function of time of the day). This result based on the NHANES data confirms similar results reported by [265, 331] in the Baltimore Longitudinal Study of Aging [284]. The right panel displays the association between gender and objectively measured physical activity after adjusting for age. For this data set results are consistent with less activity for women during the night (potentially associated with less disturbed sleep) and more activity during the day. The associated confidence bands suggest that not all of these effects may be statistically significant – there may not be a significant effect of age in the early morning, for example – but at many times of the day there do seem to be significant effects of age and gender on the binned-average MIMS values. Finally, we note that the confidence bands are narrowest in the nighttime hours and widest during the day, which is consistent with the much higher variability of the physical activity measurements during the day illustrated in Figure 5.2.

The hour-level analysis is an informative exploratory approach, but has several limitations. Most obviously, it aggregates data within prespecified bins, and in doing so loses some of the richness of the underlying data. That aggregation induces some smoothness by relying on the underlying temporal structure, but this smoothness is implicit and dependent on the bins that are chosen – adjacent coefficient estimates are similar only because the underlying data are similar, and not because of any specific model element. To emphasize these points, we can repeat the bin-level analysis using ten-minute and one-minute epochs. These can be implemented using only slight modifications to the previous code, in particular by changing the bin width of the rolling average and the grid of argument values over which functions are observed. We therefore omit this code and focus on the results produced in these settings.

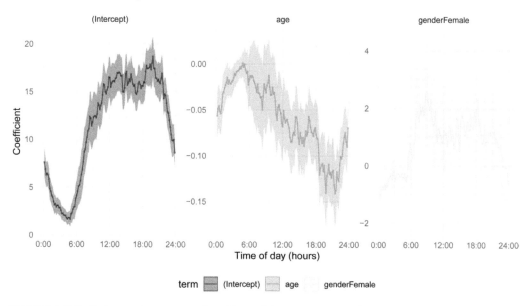

FIGURE 5.5: Point estimators and 95% confidence intervals for the regression coefficients for ten-minute regressions of mean MIMS on age and gender. Each regression is conducted independent of the other regressions and results are shown for the intercept, age, and gender parameters, respectively.

Figures 5.5 and 5.6 display the same results as Figure 5.4, but using ten- and one-minute intervals, respectively, instead of the one-hour intervals. As the bins are getting smaller, the point estimators are becoming more variable, though the overall trends and magnitudes of point estimators remain relatively stable. Such results are reassuring in practice as they show consistency of results and provide a useful sensitivity analysis. The increased variability of point estimators (wigglier curves) and the larger confidence intervals (shaded areas around the curves) make graphs less appealing as the resolution at which analysis is conducted increases.

These graphs indicate some of the difficulties inherent in binning-based analyses because they (1) do not leverage the temporal structure directly in the estimation process; (2) do not induce smoothness of effects across time; and (3) do not account for potentially large within-subject correlations of residuals. Another, more subtle, issue is that we implicitly rely on curves being observed densely over the same grid, or, less stringently, that a rolling mean is a plausible way to generate binned averages. This approach may not work when data are sparse or irregular across subjects. Various approaches to function-on-scalar regression attempt to solve some of these issues, and can work more or less well depending on the characteristics of the data to which they are applied.

5.2 Linear Function-on-Scalar Regression

Let $W_i : S \to \mathbb{R}$ be a functional response of interest for each study participant, $i = 1, \ldots, n$, and Z_{i1} and Z_{i2} be two scalar predictors. In the setting considered in this chapter, the functional response is the MIMS profile for each participant, and the scalar predictors

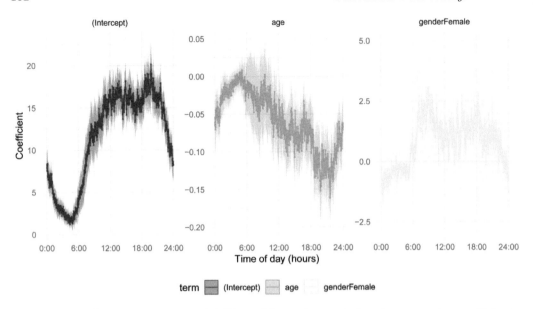

FIGURE 5.6: Point estimators and 95% confidence intervals for the regression coefficients for one-minute regressions of mean MIMS on age and gender. Each regression is conducted independent of the other regressions and results are shown for the intercept, age, and gender parameters, respectively.

of interest are age in years (Z_{i1}) and a variable indicating whether participant i is male $(Z_{i2} = 0)$ or female $(Z_{i2} = 1)$. The linear function-on-scalar regression for this setting is

$$E[W_i(s)] = \beta_0(s) + \beta_1(s)Z_{i1} + \beta_2(s)Z_{i2} \tag{5.1}$$

with coefficients $\beta_q : S \to \mathbb{R}$, $q \in \{0, 1, 2\}$ that are functions measured over the same domain as the functional response $W_i(\cdot)$. Scalar covariates in this model can exhibit the same degree of complexity as in non-functional regressions, allowing any number of continuous and categorical predictors of interest.

Coefficient functions encode the varying association between the response and predictors, and are interpretable in ways that parallel non-functional regression models. In particular, $\beta_0(s)$ is the expected response for $Z_{i1} = Z_{i2} = 0$; $\beta_1(s)$ is the expected change in the response for each one unit change in Z_{i1} while holding Z_{i2} constant; and so on. These are often interpreted at specific values of $s \in S$ to gain intuition for the associations of interest. In the NHANES data considered so far, for example, coefficient functions can be used to compare the effect of increasing age on morning and evening physical activity, keeping gender fixed.

The linear FoSR model addresses the concerns our exploratory analysis raised. Because coefficients are functions observed on S, they can be estimated using techniques that explicitly allow for smoothness across the functional domain. This smoothness, along with appropriate error correlation structures, provides an avenue for statistical inference under a clearly defined set of assumptions. Considering coefficients as functions also opens the possibility for specifying complex data generating mechanisms, such as responses that are observed on grids that are sparse or irregular across subjects.

5.2.1 Estimation of Fixed Effects

Functional responses are observed over a discrete grid $\mathbf{s} = \{s_1, ..., s_p\}$ which, for now, we will assume to be common across subjects so that for subject $i = 1, \ldots, n$ the observed data vector is $\mathbf{W}_i^t = \{W_i(s_1), ..., W_i(s_p)\}$. From the FoSR model (5.1), we have

$$E[\mathbf{W}_i^t] = \mathbf{Z}_i \begin{bmatrix} \beta_0(s_1) & ... & \beta_0(s_p) \\ \beta_1(s_1) & ... & \beta_1(s_p) \\ \beta_2(s_1) & ... & \beta_2(s_p) \end{bmatrix}$$

where $\mathbf{Z}_i = [1, Z_{i1}, Z_{i2}]$ is the row vector containing scalar terms that defines the regression model. This expression is useful because it connects a $1 \times p$ response vector to a recognizable row in a standard regression design matrix and the matrix of functional coefficients.

Estimation of coefficients will rely on approaches that have been used elsewhere with nuances specific to the FoSR setting. As a starting point, we will expand each $\beta_q(s) = \sum_{k=1}^{K} \beta_{qk} B_k(s)$ using the basis $B_1(s), \ldots, B_K(s)$. Here we have used the same basis $B_1(s), \ldots, B_K(s)$ for all three coefficients, though this is not necessary in specific applications. We leave the problem of dealing with different bases and the associated notational complexity as an exercise. While many choices are possible, we will use a spline expansion.

Conveniently, one can concisely combine and rewrite the previous expressions. Let $\mathbf{B}(s_j) = [B_1(s_j), \ldots, B_K(s_j)]$ be the $1 \times K$ row vector containing the basis functions evaluated at s_j and $\mathbf{B}(\mathbf{s})$ be the $p \times K$ matrix with the jth row equal to $\mathbf{B}(s_j)$. Further, let $\boldsymbol{\beta}_q = [\beta_{q1}, \ldots, \beta_{qK}]^t$ be the $K \times 1$ vector of basis coefficients for function q and $\boldsymbol{\beta} = [\boldsymbol{\beta}_0^t, \boldsymbol{\beta}_1^t, \boldsymbol{\beta}_2^t]^t$ be the $(3K) \times 1$ dimensional vector constructed by stacking the vectors of basis coefficients. For the $p \times 1$ response vector $\mathbf{W}_i$, we have

$$E[\mathbf{W}_i] = [\mathbf{B}(\mathbf{s}) \otimes \mathbf{Z}_i]\boldsymbol{\beta}$$

where $\otimes$ is the Kronecker product.

This expression for a single subject can be directly extended to all subjects. Let $\mathbf{Y}$ be the $(np) \times 1$ dimensional vector created by stacking the vectors $\mathbf{W}_i$ and $\mathbf{Z}$ be $n \times 3$ matrix created by stacking the vectors $\mathbf{Z}_i$. Then

$$E[\mathbf{W}] = [\mathbf{Z} \otimes \mathbf{B}(\mathbf{s})]\boldsymbol{\beta} \, .$$

This formulation, while somewhat confusing at first, underlies many implementations of the linear FoSR model. First, we note that the matrix $\mathbf{Z}$ is the design matrix familiar from non-functional regression models, and includes a row for each subject and a column for each predictor. The Kronecker product with the basis $\mathbf{B}(\mathbf{s})$ replicates that basis once for each coefficient function; this product also ensures that each coefficient function, evaluated over $\mathbf{s}$, is multiplied by the subject-specific covariate vector $\mathbf{Z}_i$. Finally, considering $[\mathbf{Z} \otimes \mathbf{B}(\mathbf{s})]$ as the full FoSR design matrix and $\boldsymbol{\beta}$ as a vector of coefficients to be estimated explicitly connects the FoSR model to standard regression techniques.

5.2.1.1 Estimation Using Ordinary Least Squares

Estimation of the spline coefficients, and therefore of the coefficient functions, using (5.1) helps to build intuition for the inner workings of more complex estimation strategies. We will regress the daily MIMS data on age and gender to illustrate this approach. Recall that `nhanes_df` stores MIMS as a `tf` vector; some of the code below relies on extracting information from a `tf` object.

For this model, the code chunk below defines the standard design matrix $\mathbf{Z}$ and the response vector $\mathbf{W}$. The call to `model.matrix()` uses a formula that correctly specifies the predictors, and obtaining the response vector W unnests the `tf` vector and then extracts the observed responses for each subject.

```
Z_des =
  model.matrix(
    SEQN ~ gender + age,
    data = nhanes_df
  )


W =
  nhanes_df %>%
  tf_unnest(MIMS) %>%
  pull(MIMS_value)
```

The next code chunk constructs the spline basis matrix $\mathbf{B}(\mathbf{s})$ and the Kronecker product $[\mathbf{Z} \otimes \mathbf{B}(\mathbf{s})]$. The spline basis is evaluated over the grid $\left[\frac{1}{60}, \frac{2}{60}, \dots, \frac{1440}{60}\right]$, so that minutes are in $\frac{1}{60}$ increments and hours of the day fall on integers from 1 to 24, and uses $K = 30$ basis functions.

```
epoch_arg = seq(1 / 60, 24, length = 1440)


basis =
  splines::bs(epoch_arg, df = 30, intercept = TRUE)


Z_kron_B = kronecker(Z_des, basis)
```

Using these, the direct application of usual formulas provides OLS estimates of the spline coefficients. Reconstructing coefficient functions is possible using model (5.1), and the results can be converted to a `tf` vector for plotting and manipulation. The code chunk below estimates and plots the coefficient functions. Note that we include the `arg` argument when constructing the `tfd` object to ensure the grid on which functions are observed remains consistent throughout this chapter.

```
spline_coefs = solve(t(Z_kron_B) %*% Z_kron_B) %*% t(Z_kron_B) %*% W


OLS_coef_df =
  tibble(
    method = "OLS",
    term = colnames(Z_des),
    coef =
      tfd(
        t(basis %*% matrix(spline_coefs, nrow = 30)),
        arg = epoch_arg)
  )


OLS_coef_df %>%
  ggplot(aes(W = coef, color = term)) +
  geom_spaghetti(size = 1.1, alpha = .75) +
  facet_wrap("term", scales = "free")
```

Figure 5.7 displays the estimators of the time-dependent intercept, age, and gender effects in model (5.1) obtained by expansion of these coefficients in a B-spline basis with 30 degrees of freedom and OLS regression. The outcome is physical activity at every minute.

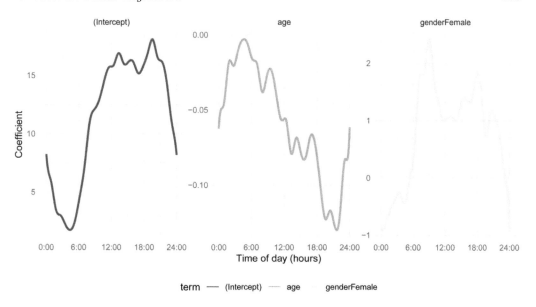

FIGURE 5.7: Coefficient estimates for the intercept, age and gender effects in model (5.1) obtained by expansion of these coefficients in a B-spline basis with 30 degrees of freedom and OLS regression.

These results can be compared to the results in Figures 5.5 and 5.6. Comparing these results to those obtained from epoch-specific regressions begins to indicate the benefits of a functional perspective. Coefficient functions have a smoothness determined by the underlying spline expansion, and estimates borrow information directly across adjacent time points.

Our specification relied on having all observations on the same grid, but this assumption can be relaxed: if data contain subject-specific grids s_i, one can replace the matrix $[\mathbf{Z} \otimes \mathbf{B}(\mathbf{s})]$ with one constructed by row-stacking matrices $[\mathbf{Z}_i \otimes \mathbf{B}(\mathbf{s}_i)]$ (note that it can take some care to ensure the basis is constant across subjects, for example by explicitly defining knot points).

An inspection of results in Figure 5.7, which improves on exploratory analyses by implementing an explicitly functional regression approach, nonetheless indicates that this fit may exhibit undersmoothing of functional coefficients. In the next section, we describe how to induce an appropriate degree of smoothness by starting with a large number of basis functions and then applying smoothing penalties. As we discuss how to incorporate smoothness penalties and account for complex error correlations, it is helpful to view these as variations on and refinements of a regression framework that uses familiar tools like OLS.

5.2.1.2 Estimation Using Smoothness Penalties

Spline expansions for coefficient functions in the linear FoSR model makes the functional nature of the observed data and the model parameters explicit, and distinguishes this approach from multivariate or epoch-based alternatives. However, as we have seen elsewhere, using a spline expansion introduces questions about how best to balance flexibility with goodness-of-fit. Throughout this book, the strategy is to use a large number of basis functions, K, combined with smoothness penalties to prevent overfitting. In this book we use only quadratic penalties. This strategy was shown to be highly successful in nonparametric smoothing [116, 258, 322] because it balances model complexity, computational feasibility, and allows inference via corresponding mixed effects models [258, 322]. A major emphasis of this book is to show how the same ideas can be adapted and extended to functional data regression.

In Section 5.2.1.1 we have used OLS to conduct estimation. These estimators correspond to the maximum likelihood estimation in a model that assumes that residuals $\epsilon_i^t = [\epsilon_i(s_1), ..., \epsilon_i(s_p)]$ are independent and distributed $N(0, \sigma_\epsilon^2)$. Here the variance σ_ϵ^2 is assumed to be constant over time points, $s \in S$, and subjects, i. More precisely, maximizing the (log) likelihood $\mathcal{L}(\beta; \mathbf{W})$ induced by this assumption with respect to spline coefficients is equivalent to the OLS approach. Inducing smoothness on the spline coefficients can be done using penalties on the amount of variation of $\beta_q(s)$. A common measure of this variation is the second derivative penalty $P(\beta_q) = \int_S \{\beta_q''(s)\}^2 ds$, but other penalties can also be used. It can be shown that this penalty has a quadratic form $P(\beta_q) = \beta_q^t \mathbf{D}_q \beta_q$, where $\mathbf{D}_q$ is a known semi-definite positive matrix that provides the structure of the penalty. Many other penalties have a similar quadratic form, but with a different penalty structure matrix $\mathbf{D}_q$. In this book we use only quadratic penalties.

The penalized log likelihood can then be written as

$$\mathcal{L}(\beta, \lambda; \mathbf{W}) = \mathcal{L}(\beta; \mathbf{W}) + \sum_{q=0}^{Q} \lambda_q P(\beta_q) = \mathcal{L}(\beta; \mathbf{W}) + \sum_{q=0}^{Q} \lambda_q \beta_q^t \mathbf{D}_q \beta_q \,, \tag{5.2}$$

where the tuning parameters λ_q, for q between 0 and $Q = 2$, control the balance between goodness-of-fit and complexity of the coefficient functions $\beta_q(s)$. As discussed in Section 2.3.2, the smoothing parameters can be estimated using a variety of criteria, though here we will use the penalized likelihood approach introduced in Section 2.3.3.

For Gaussian outcome data we can reparameterize $\lambda_q = \sigma_\epsilon^2 / \sigma_q^2$, where $\sigma_q^2 \geq 0$ are positive parameters. With this notation and after dividing the criterion in equation (5.2) by σ_ϵ^2 we obtain

$$-\frac{\mathcal{L}(\beta; \mathbf{W})}{2\sigma_\epsilon^2} - \sum_q \frac{\beta_q^t \mathbf{D}_q \beta_q}{2\sigma_q^2} \,. \tag{5.3}$$

Careful inspection of this approach indicates that the model can be viewed as a regression whose structure is defined by the conditional log-likelihood $-\mathcal{L}(\beta; \mathbf{W})/2\sigma_\epsilon^2$, where the coefficients of the splines are treated as random effects with multivariate, possibly rank deficient, multivariate Normal distributions. More precisely, the model can be viewed as the mixed effects model

$$\begin{cases} [\mathbf{W}_i | \beta, \sigma_\epsilon^2] \sim N([\mathbf{B}(\mathbf{s}) \otimes \mathbf{Z}_i]\beta, \sigma_\epsilon^2) \,, \text{for } i = 1, \ldots, n \,, \\ [\beta_q | \sigma_q^2] \quad = \dfrac{\det(\mathbf{D}_q)^{1/2}}{(2\pi)^{K_q/2}\sigma_q} \exp\left(-\dfrac{\beta_q^t \mathbf{D}_q \beta_q}{2\sigma_q^2} \right) \,, \text{for } q = 0, \ldots, Q \,, \end{cases} \tag{5.4}$$

where all conditional distributions are assumed to be mutually independent for $i = 1, \ldots, n$ and $q = 1, \ldots, Q$. The crucial point is to notice that the likelihood of model (5.4) is equivalent to criterion (5.4) up to a constant that depends only on σ_q^2, $q = 0, \ldots, Q$, and σ_ϵ^2. The advantage of model (5.4) is that it provides a natural way for estimating σ_q^2 and σ_ϵ^2 based on an explicit likelihood of a model. Moreover, note that the outcome likelihood $[\mathbf{W}_i | \beta, \sigma_\epsilon^2]$ does not need to be Gaussian. Indeed, it was only in the last step that we made the connection to the Normal likelihood. In general, this is not necessary and the same principles work for exponential family regression. The only difference is that we reparametrize $\lambda_q = 1/\sigma_q^2$ and make sure that the log-likelihood $-\mathcal{L}(\beta; \mathbf{W})$ is correctly specified for the outcome family distribution. This model specification will also allow to include error correlation structures via more complex specifications of the error $\epsilon_i(s)$. Finally, the prior distributions $[\beta_q | \sigma_q^2]$ are normal because we have chosen to work with quadratic penalties. This implies that we work only with Gaussian random effects. This assumption can be further relaxed depending on the requirements of the problem, what is computationally feasible, and how ambitious modeling is.

The idea is remarkably simple and requires only careful inspection of equation (5.4) and a willingness to view spline coefficients as random variables and penalized terms as prior distributions. But once seen, it cannot be unseen. This approach has extraordinary implications because with this structure one can simply use mixed effects model estimation and inference and existing software. This idea is not new in statistics and has been used extensively in non-parametric smoothing. What is powerful is the fact that it extends seamlessly to functional data, which allows us to use and compare powerful software such as `mgcv`, `nlme`, `Rstan` [35, 281, 282], `WinBUGS` [48, 187], or JAGS [232]. The choice of frequentist or Bayesian approaches becomes a matter of personal preference.

We will use the `gam()` function in the well-developed `mgcv` package to fit the FoSR model with smoothness penalties. First, we explicitly create the binary indicator variable for `gender`, and organize data into long format by unnesting the `MIMS` data stored as a `tf` object. The resulting data frame has a row for each subject and epoch, containing the `MIMS` outcome in that epoch as well as the scalar covariates of interest. This is analogous to the creation of the response vector for model fitting using OLS, and begins to organize covariates for inclusion in a design matrix.

```
nhanes_for_gam =
  nhanes_df %>%
  mutate(gender = as.numeric(gender == "Female")) %>%
  tf_unnest(MIMS) %>%
  rename(epoch = MIMS_arg, MIMS = MIMS_value)
```

With data in this form, we can fit the FoSR model using `mgcv::gam()` as follows.

```
gam_fit =
  gam(MIMS ~ s(epoch) + s(epoch, by = gender) + s(epoch, by = age),
      data = nhanes_for_gam)
```

Conceptually, this specification indicates that the expected `MIMS` value is the combination of three smooth functions of time (`epoch`): an intercept function, and the interactions (or products) of coefficient functions and scalar covariates `gender` and `age`. Specifically, the first model term `s(epoch)` indicates the evaluation of a spline expansion over values contained in the `epoch` column of the data frame `nhanes_gam_df`. The second and third terms add `by = gender` and `by = age`, respectively, which also indicate spline expansions over the `epoch` column but multiply the result by the corresponding scalar covariates. This process is analogous to the creation of the design matrix $\mathbf{Z} \otimes \mathbf{B}(\mathbf{s})$, although `gam()`'s function `s()` allows users to flexibly specify additional options for each basis expansion.

In contrast to the OLS estimation in Section 5.2.1.1, smoothness is induced in parameter estimates through explicit penalization. By default, `gam()` uses thin-plate splines with second derivative penalties, and selects tuning parameters for each coefficient using GCV or REML [258, 322]. The results contained in `gam_fit` are not directly comparable to those obtained from OLS, and extracting coefficient functions requires some additional work. The `predict.gam()` function can be used to return each element of the linear predictor for a given data frame. In this case, we show how to return the smooth functions of `epoch` corresponding to the intercept and coefficient functions. We therefore create a data frame containing an `epoch` column consisting of the unique evaluation points of the observed functions (i.e., minutes as stored in `epoch_arg`); a column `gender`, set to 1 for all epochs; and a column `age`, also set to 1 for all epochs.

```
gam_pred_obj =
  tibble(
    epoch = epoch_arg,
    gender = 1,
    age = 1,
  ) %>%
  predict(gam_fit, newdata = ., type = "terms")
```

The result contained in `gam_pred_obj` is a 1440×3 matrix, with columns corresponding to $\beta_0(\mathbf{s})$, $1 \cdot \beta_1(\mathbf{s})$, and $1 \cdot \beta_2(\mathbf{s})$, respectively. We convert these to `tfd` objects in the code below. Note that `gam()` includes the overall intercept as a (scalar) fixed effect, which must be added to the intercept function. With data structured in this way, we can then plot coefficient functions using tools seen previously.

```
gam_coef_df =
  tibble(
    method = "GAM",
    term = c("(Intercept)", "genderFemale", "age"),
    coef =
      tfd(t(gam_pred_obj), arg = epoch_arg)) %>%
  mutate(coef = coef + c(coef(gam_fit)[1], 0, 0))

gam_coef_df %>%
  ggplot(aes(y = coef, color = term)) +
  geom_spaghetti(size = 1.1, alpha = .75) +
  facet_wrap(vars(term), scales = "free")
```

Figure 5.8 provides a comparison of the pointwise, B-spline regression splines, and penalized B-spline smoother estimators of the intercept, age, and gender effects in the FoSR model (5.1). Each approach – separate epoch-level regressions, FoSR using OLS to estimate spline coefficients, and FoSR implemented with smoothness penalties in `mgcv::gam()` – yield qualitatively similar results regarding the effect of age and gender on diurnal MIMS trajectories. This suggests that all approaches can be used at least in exploratory analyses or to understand general patterns. That said, there are also obvious differences. The epoch-level regressions do not borrow information across adjacent time points, and the OLS is sensitive to the dimension of the basis expansion in the model specification; both are wigglier, and perhaps less plausible, than the method that includes smoothness penalties. As a result, methods that explicitly borrow information across time and implement smoothness penalties with data-driven tuning parameters are often preferred for formal analyses.

```
ggplot(mapping = aes(y = coef, color = method)) +
  geom_spaghetti(data = min_regressions, alpha = .5) +
  geom_spaghetti(data = OLS_coef_df, linewidth = 1.2) +
  geom_spaghetti(data = gam_coef_df, linewidth = 1.2) +
  facet_wrap(vars(term), scales = "free")
```

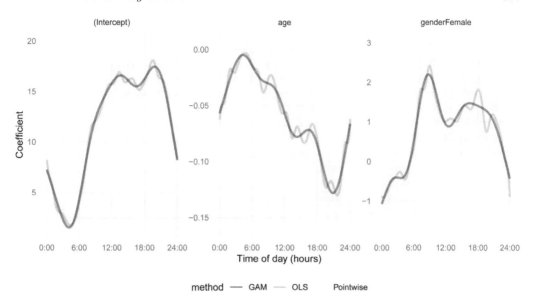

FIGURE 5.8: Comparing the pointwise (yellow), B-spline regression splines with 30 degrees of freedom (green), and penalized B-spline smoothing (purple) of the time-dependent intercept, age, and gender effects in the FoSR model (5.1). The outcome is physical activity (MIMS) measured at the minute level.

5.2.2 Accounting for Error Correlation

To this point, we have developed tools for estimating coefficient functions in model (5.1) while setting aside concerns about residual correlation. Therefore, we have conducted estimation and inference under the "independence of residuals assumption." Of course, in most FoSR settings, residuals are indeed correlated. To show this, Figure 5.9 displays the estimated functional residuals $\widehat{\epsilon}_i(s) = W_i(s) - \sum_{q=1}^{Q} Z_{iq}\widehat{\beta}_q(s)$, where $\widehat{\beta}_q(s)$ are obtained using smoothness penalties and the assumption of independence of residuals. Visual inspection suggests substantial residual correlation, as well as different degrees of residual variance at different times of day.

At this point it is worth taking a step back to understand what the implications of these correlations might be. Indeed, the results in Figures 5.4, 5.5, 5.6 and 5.7, which are based on OLS, are valid even though we have ignored the correlation of residuals. These point estimators are asymptotically consistent and efficient conditional on the information at that particular point, and are very similar to those obtained when smoothness in $\beta_q(\cdot)$ is accounted for through explicit penalization. One could argue that residual correlation should be accounted for to smooth the $\beta_q(\cdot)$ parameters. However, Figure 5.7 indicates that reasonable estimation of the $\beta_q(\cdot)$ parameters can be achieved under the independence of residuals assumption in model (5.1). This has been noted before by [53, 223], who proposed to estimate models pointwise and obtain confidence intervals using the bootstrap of study participants. Therefore, the hope is that by accounting for the correlation of residuals one can construct pointwise and joint confidence intervals that have nominal coverage probability. We will investigate whether this is the case, but for now we focus on conceptual approaches for modeling the residual correlations.

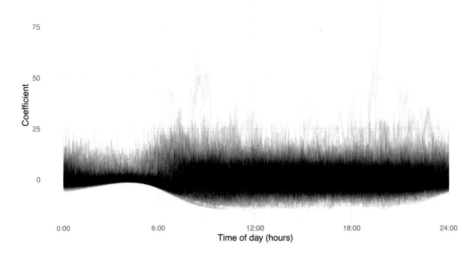

FIGURE 5.9: Estimated functional residuals obtained after fitting model (5.1) using penalized spline smoothing of time-varying coefficients under the independence of residuals assumption.

Most methods would focus on modeling the residuals $\epsilon_i(s) = X_i(s) + e_i(s)$, where $X_i(s)$ follows a mean zero Gaussian Process (GP) with covariance function $\boldsymbol{\Sigma}$ [18, 69] and $e_i(s)$ are independent $N(0, \sigma_e^2)$ random errors. There are many strategies to model jointly the nonparametric mean and error structure of this model including (1) joint modeling based on FPCA decomposition of $X_i(s)$; (2) functional additive mixed models (FAMM) using spline expansions of the error term $X_i(\cdot)$ [263, 260]; and (3) Bayesian posterior simulations for either type of expansion, which is related to Generalized Multilevel Function-on-Scalar Regression and Principal Component Analysis [106]. We first describe these approaches, while in Section 5.3 we introduce a fast scalable alternative based on pointwise regressions for estimation and bootstrap of study participants for inference that follows the philosophy of [57].

Recall that the current model for our example with two scalar covariates Z_{i1} and Z_{i2} can be written as

$$W_i(s) = \beta_0(s) + \beta_1(s)Z_{i1} + \beta_2(s)Z_{i2} + X_i(s) + e_i(s) \,. \tag{5.5}$$

Any joint model is based on the decomposition of the structured error $X_i(s) = \mathbf{B}(s)\boldsymbol{\xi}_i$, where $\mathbf{B}(s)$ is a $1 \times K$ dimensional vector of basis functions evaluated at s and $\boldsymbol{\xi}_i$ is a $K \times 1$ dimensional vector of study participant-specific coefficients. The main difference between the FPCA and FAMM approaches is that the former uses a data-driven FPCA basis and the latter uses a pre-specified spline basis.

We now describe the approaches and their implementation in R. In keeping with the philosophy of this book, models are fit using user-friendly functions that mask to some degree the underlying complexity. However, we emphasize that these model fitting strategies are grounded in familiar regression techniques.

5.2.2.1 Modeling Residuals Using FPCA

Assuming that $\phi_k(\cdot)$ are the eigenfunctions of the covariance operator K_X of $X_i(\cdot)$, one can express $X_i(s) = \sum_{k=1}^{\infty} \xi_{ik}\phi_k(s)$. This leads to the following model

$$W_i(s) = \beta_0(s) + \beta_1(s)Z_{i1} + \beta_2(s)Z_{i2} + \sum_{k=1}^{K} \xi_{ik}\phi_k(s) + e_i(s) , \qquad (5.6)$$

where the scores $\xi_{ik} \sim N(0, \lambda_k)$ and the errors $e_i(s) \sim N(0, \sigma_e^2)$ are mutually independent. Here we used the finite sum $\sum_{k=1}^{K} \xi_{ik}\phi_k(s)$ instead of the infinite sum $\sum_{k=1}^{\infty} \xi_{ik}\phi_k(s)$ for practical purposes. The assumption is that the first K eigenfunctions explain most of the variation in $X_i(s)$ and what is left unexplained is absorbed in the error $e_i(s)$.

From our notation $\epsilon_i(s) = X_i(s) + e_i(s) = W_i(s) - \sum_{q=0}^{Q} Z_{iq}\beta_q(s)$. Therefore, $\epsilon_i(s)$ is a zero mean Gaussian Process with covariance operator equal to $\Sigma + \sigma_e^2 I$, where I is the identity covariance operator. We have already shown that one could estimate the mean structure of model (5.1) using any number of techniques. This allows the estimation of the residuals $\widehat{\epsilon}_i(s) = W_i(s) - \sum_{q=1}^{Q} Z_{iq}\widehat{\beta}_q(s)$. These estimated residuals can then be thought of as functional data and decomposed using the FPCA techniques discussed in Chapter 3.

Thus, our approach is to (1) obtain $\widehat{\beta}_q(s)$, the penalized splines estimators of $\beta_q(s)$ for $q = 0, \ldots, Q$ under the assumption of independence of residuals; (2) obtain $\widehat{\epsilon}_i(s) = W_i(s) - \sum_{q=1}^{Q} Z_{iq}\widehat{\beta}_q(s)$; (3) estimate the eigenfunctions $\widehat{\phi}_k(s)$ using FPCA applied to $\widehat{\epsilon}_i(s)$; and (4) fit the joint model (5.6) where $\widehat{\phi}_k(s)$ are plugged in instead of $\phi_k(s)$ and the $\beta_q(s)$ functions are modeled as penalized splines.

It is important to note that in this model there are two types of random coefficients or random effects. The first type are the spline coefficients, which are treated as random to ensure the smoothness of the $\beta_q(s)$ functions. The second type are the mutually independent scores $\xi_{ik} \sim N(0, \lambda_k)$ corresponding to the orthonormal eigenfunctions describing the residual correlation. These random coefficients play different roles in the model, but, from a purely computational perspective, they can all be treated as random coefficients. This makes model (5.6) a mixed effects model.

Before conducting this analysis, we want to build the intuition first and make connections to well-known concepts in mixed effects modeling. Consider, for example, the case when $K = 1$ and $\phi_1(s) = 1$ for all $s \in S$. In practice we will never have the luxury of this assumption, but let us indulge in simplification. With this assumption the model becomes

$$W_i(s) = \beta_0(s) + \beta_1(s)Z_{i1} + \beta_2(s)Z_{i2} + \xi_{i1} + e_i(s) , \qquad (5.7)$$

where the scores $\xi_{i1} \sim N(0, \lambda_1)$ and the errors $e_i(s) \sim N(0, \sigma_e^2)$ are mutually independent. This model adds a random study participant-specific intercept to the population mean function. It is a particular case of model (5.1), but allows for the dependence of residuals within study participants, i. This should be familiar from traditional regression strategies for longitudinal data. Indeed, one of the first steps to account for the dependence of residuals is to add a random intercept. It also provides a useful contrast between longitudinal data analysis and functional data analysis: the former typically makes assumptions that limit the flexibility of subject-level estimates over the observation interval S, while the latter uses data-driven approaches to add flexibility when appropriate.

To fit the random intercept model, we adapt our previous implementation for penalized spline estimation in a number of ways. Recall that `nhanes_gam_df` contains a long-form data frame with rows for each subject and epoch, and columns containing MIMS and scalar covariates. This data frame also contains a column of subject IDs `SEQN`; importantly, this is encoded as a factor variable that can be used to define subjects in the random effects model.

In the R code, the terms corresponding to fixed effects are unchanged, but we add a term s(SEQN) with the argument bs = "re". This creates a "smooth" term with a random effects "basis" – essentially taking advantage of the noted connection between semiparametric regression and random effects estimation to obtain subject-level random effects estimates and the corresponding variance component. Finally, we note that we use bam instead of gam, and add arguments method = "fREML" and discrete = TRUE. These changes substantially decrease computation times.

```
nhanes_gamm_ranint =
  nhanes_for_gam %>%
  bam(MIMS ~ s(epoch) + s(epoch, by = gender) + s(epoch, by = age) +
          s(SEQN, bs = "re"),
      method = "fREML", discrete = TRUE, data = .)
```

Next, we will use a two-step approach to model correlated residuals using FPCA. The first step is to fit a FoSR model assuming uncorrelated errors to obtain estimates of fixed effects $\beta_p(s)$ and fitted values $\widehat{W}_i(s)$ for each subject. From these, we compute residuals $\widehat{\epsilon}_i(s) = W_i(s) - \widehat{W}_i(s)$ which can be decomposed using standard FPCA methods to produce estimates of FPCs $\hat{\phi}_k(s)$. In the second step, the $\hat{\phi}_k(s)$ are treated as "known" in model (5.6); coefficient functions $\beta_p(s)$ and subject-level scores ξ_{ik} are then estimated simultaneously. The NHANES data considered in this chapter are densely measured over a regular grid that is common to all subjects and we implement the FPCA step using the FACE method [331], but the same two-step approach can be adapted to other data settings.

In the code below, we first construct a data frame that contains fitted values and residuals, which will subsequently be used as the basis for FPCA. We start with nhanes_gam_df, and add the fitted values from our penalized spline estimation assuming independent errors in gam_fit. Residuals are then created by subtracting the fitted values from the MIMS observations. Using only the subject ID SEQN, epoch and the residuals for each subject at each epoch, we nest to create a data frame in which residual curves are tf values. Using this data frame, we conduct FPCA using the rfr_fpca() function in the refundr R package. At the time of writing, refundr is under active development and contains reimplementations of many functions in the refund package using tidyfun for data organization. By default, for data observed over a regular grid, rfr_fpca() uses FACE to conduct FPCA. We specify two additional arguments in this call. First, center = FALSE avoids recomputing the mean function since we are using residuals from a previous fit. Second, we set the smoothing parameter by hand using lambda = 50; this produces smoother FPCs than the default implementation. Of course, the same methodology can be implemented using refund and fpca.face() directly, but here we illustrate a functional tidy-friendly approach.

```
nhanes_fpca_df =
  nhanes_for_gam %>%
  mutate(
    fitted = fitted(gam_fit),
    resid = MIMS - fitted) %>%
  select(SEQN, epoch, resid) %>%
  tf_nest(resid, .id = SEQN, .arg = epoch)


nhanes_resid_fpca =
  rfr_fpca("resid", nhanes_fpca_df, center = FALSE, lambda = 50)
```

The last step in our approach is to treat the resulting eigenfunctions $\widehat{\phi}_k(s)$ as "known" and replace the $\phi_k(s)$ functions in model (5.6). These estimated eigenfunctions are stored in `nhanes_resid_fpca`. We will use a variation on the random intercept implementation to do this. However, in this context ξ_{ik} are uncorrelated random slopes on the "covariates" $\widehat{\phi}_k(s)$, which are the FPCs evaluated over the functional domain. Put differently, we want to scale one component in our model by another; to do this, we will again make use the `by` argument in the `s()` function.

The code chunk below defines the data frame necessary to implement this strategy. It repeats code seen before to convert the `nhanes_df` data frame to a format needed by `mgcv::gam()`, by creating an indicator variable for `gender` and unnesting the `MIMS_tf` data stored as a `tf` object. However, we also add a column `fpc` that contains the first FPC estimated above. The FPCs are the same for each participant and are treated as a `tf` object. We then unnest both `MIMS` and `fpc` to produce a long-format data frame with a row for each subject and epoch. These data contain just the first principal component as the random slope covariate, but one can easily modify the code to add PCs.

```
nhanes_for_gamm =
  nhanes_df %>%
  mutate(
    gender = as.numeric(gender == "Female"),
    fpc = tfd(nhanes_resid_fpca$efunctions[,1])) %>%
  tf_unnest(cols = c(MIMS_tf, fpc)) %>%
  rename(epoch = MIMS_tf_arg, MIMS = MIMS_tf_value, fpc = fpc_value) %>%
  select(-fpc_arg)
```

With these data organized appropriately, we can estimate coefficient functions and subject-level FPC scores using a small modification to our previous random intercept approach. We again estimate subject-level effects using `s(SEQN)` and a random effect "basis" that adds a random intercept for each participant. Including `by = fpc` scales the random effect basis by the FPC value in each epoch, effectively creating the term $\xi_{i1}\phi_1(s)$ by treating $\phi_1(s)$ as known. The code can easily be modified to incorporate additional principal components.

```
nhanes_gamm_fpc =
  nhanes_for_gamm %>%
  bam(MIMS ~ s(epoch, fx = T) + s(epoch, by = gender, fx = T) +
             s(epoch, by = age, fx = T) +
             s(SEQN, bs = "re", by = fpc),
      discrete = TRUE, method = "fREML", data = .)
```

This implementation of the FoSR model using FPCA to account for residual correlation has important strengths. Penalized spline estimation of fixed effects leverages information across adjacent time points and enforces smoothness. The residual correlation is accounted for using FPCA decompositions of the residuals, which is both data-driven and flexible. Some additional refinements are possible, and may be useful in practice. For example, we could incorporate the first K PCs to account for the model structure $\sum_{k=1}^{K} \xi_{ik}\phi_k(s)$. This structure is very similar to the approach we have used in Section 3.2.2 for the decomposition of the linear predictors for conducting generalized functional principal component analysis (GFPCA). The difference is that here we explicitly model fixed effects nonparametrically. In the `mgcv` syntax the random slopes are assumed to not be correlated, which corresponds exactly to our model (5.6). This is different from `lme4` [10], where the default is to assume that random intercepts and slopes are mutually correlated. Another refinement could be to conduct multiple iterations so that FPCs are derived from a model that has accounted for residual correlation in the estimation of fixed effects.

We now show how to extract the quantities necessary to conduct estimation and inference for coefficient function $\beta_q(s)$. Our approach to constructing pointwise confidence intervals is based on the spline-based estimation approaches considered so far, as well as the assumption that spline coefficient estimates have an approximately multivariate Normal distribution. Treating tuning parameters as fixed, for any $s \in S$ the variance of $\widehat{\beta}_q(s)$ is given by

$$\mathrm{Var}\{\widehat{\beta}_q(s)\} = \mathbf{B}(s)\mathrm{Var}(\widehat{\boldsymbol{\beta}}_q)\mathbf{B}(s)^t \ ,$$

where $\widehat{\boldsymbol{\beta}}_q$ is the vector of estimated spline coefficients. From this, one can obtain standard errors and construct a confidence interval for $\beta_q(s)$ using the assumption of normality.

A critical component is the covariance of estimated spline coefficients $\mathrm{Var}(\widehat{\boldsymbol{\beta}}_q)$, which is heavily dependent on the modeling assumptions used to estimate spline coefficients. We have fit three FoSR models using penalized splines for coefficient functions, with different assumptions about residuals, namely that residuals are uncorrelated within a subject; residual correlation can be modeled using a random intercept; and residual correlation can be accounted for using FPCA. We now compare the estimated coefficient functions and pointwise confidence intervals obtained from these methods.

To extract coefficient functions and their standard errors from the objects produced by `mgcv::gam()`, we use of the `predict()` function. This function takes an input data frame that has all covariates used in the model, including the FPC; here we will use the first 1440 rows of the `nhanes_gam_dfm` data frame, which has all epoch-level observations for a single subject. We set `gender` and `age` to 1 as before, so terms produced by `predict()` will correspond to the coefficient functions. When calling `predict()`, we now set the argument `se.fit = TRUE`, so that both the estimated coefficients and their standard errors are returned.

```
nhanes_fpc_pred_obj =
  nhanes_for_gamm[1:1440,] %>%
  mutate(gender = 1, age = 1) %>%
  predict(
    nhanes_gamm_fpc, newdata = .,
    type = "terms", se.fit = TRUE)
```

The output of `predict()` requires some processing before plotting. We construct `coef_df` to contain the relevant output, again using steps that are similar to those seen previously: a `term` variable is created and coefficients are extracted and converted to `tf` objects, and the model's overall intercept is added to the intercept function. There are some important changes, however. Because we set `se.fit = TRUE`, the result contains both coefficients and standard errors, and we extract these from the `fit` and `se.fit` elements of the object returned by `predict`, respectively. The model that uses FPCA to account for residual correlation also includes a coefficient for `SEQN`, and so we use only the first three terms in the model. Finally, the code also includes a step to obtain upper and lower bounds of a 95% pointwise confidence interval, constructed by adding and subtracting 1.96 times the standard error to the estimate for each coefficient.

```
coef_df =
  tibble(
    term = c("(Intercept)", "genderFemale", "age"),
    coef = tfd(t(nhanes_fpc_pred_obj$fit[,1:3]), arg = epoch_arg),
    se = tfd(t(nhanes_fpc_pred_obj$se.fit[,1:3]), arg = epoch_arg)) %>%
  mutate(coef = coef + c(coef(nhanes_gamm_fpc)[1], 0, 0)) %>%
  mutate(
    ub = coef + 1.96 * se,
    lb = coef - 1.96 * se)
```

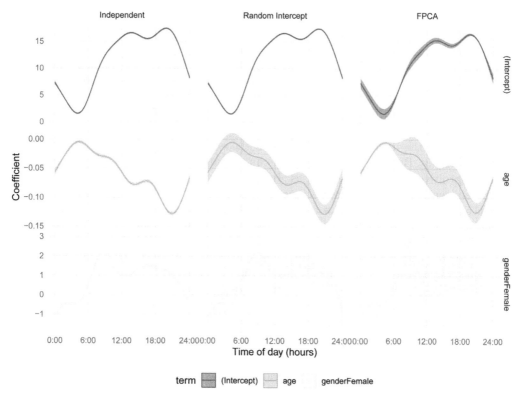

FIGURE 5.10: Point estimators and 95% confidence intervals for the regression coefficients of MIMS on age and gender. The models assume independent errors, parametric random intercepts, and use FPCA to decompose the within-curve error structures.

Although we do not show all steps here, the same approach can be used to extract coefficient functions and confidence intervals for the three approaches to accounting for residual correlation. The results are combined into `comparison_plot_df` and plotted Figure 5.10, again using `geom_errorband()` to show upper and lower confidence limits.

```
comparison_plot_df %>%
  ggplot(aes(y = coef, color = term)) +
  geom_spaghetti() +
  geom_errorband(aes(ymax = ub, ymin = lb, fill = term)) +
  facet_grid(term ~ method, scales = "free")
```

The coefficient functions obtained by all methods are similar to each other and to those based on epoch-level regressions. The confidence bands, meanwhile, differ substantially in a way that is intuitive based on the assumed error structures. Assuming independence fails to capture any of the correlation that exists within subjects, and therefore has overly narrow confidence bands. Using a random intercept accounts for some of the true correlation but makes restrictive parametric assumptions on the correlation structure. Because this approach effectively induces uniform correlation over the domain, the resulting intervals are wider than those obtained under the model assuming independence but have a roughly fixed width over the day. Finally, modeling residual curves using FPCA produces intervals that are narrower in the nighttime and wider in the daytime, which more accurately reflects the variability across subjects in this dataset. This model suggests a significant decrease in MIMS as age increases over much of the day, and a significant increase in MIMS comparing women to men in the morning and afternoon.

In some ways, it is unsurprising that the coefficient function estimates produced under different assumptions are similar. After all, each is a roughly unbiased estimator for the fixed effects in the model, and differences in the error structure is primarily intended to inform inference. But the complexity of the underlying maximization problem can produce counter-intuitive results in some cases. In this analysis, the results of the FPCA model fitting are sensitive to the degree of smoothness in the FPC. When the less-smooth FPCs produced by the default FACE settings were used, the coefficient function estimates were somewhat attenuated. At the same time, the random effects containing FPC scores were dependent on the scalar covariates in a way that exactly offset this attenuation. This does not appear to be an issue with the `gam()` implementation, because "by hand" model fitting showed the same sensitivity to smoothness in the FPCs. Instead, we believe this issue stems from subtle identifiability issues and the underlying complexity of the penalized likelihood.

5.2.2.2   Modeling Residuals Using Splines

A powerful approach to implementing model (5.5) is to model both the fixed effects parameters $\beta_q(s)$, $q = 0, \ldots, 2$ and the structured random residuals $X_i(s)$, $i = 1, \ldots, n$ using penalized splines. This approach has several advantages as it (1) uses a pre-specified basis, which avoids the potential problems associated with the uncertainty of FPCA bases that are estimated from the data; (2) avoids the problem of estimating an FPCA basis, which is especially useful when dealing with non-Gaussian functions; (3) can be implemented in existing software; see the `pffr()` function in the `refund` package; (4) allows joint analysis of the model using a fully specified, explicit likelihood; and (5) is consistent with inferential tools for estimation, testing, and uncertainty quantification.

This approach was developed by [260, 263] and contains a set of inferential tools that extends to many other scenarios including function-on-scalar (FoSR) and function-on-function (FoFR) regression; for more details, see the description of the `pffr()` function in the `refund` package. All methods can also be directly implemented in the `mgcv` package with some modification. The main contribution of the methods was to show how a complex functional regression model can be reduced to a non-parametric smoothing problem with random effects. In turn, this allowed the use of powerful existing software, such as `mgcv`.

We defer the technical details to the papers discussed above. Here we show how to use `pffr()` to fit model (5.5). For computational efficiency, we aggregate minute-level data into hour-length bins and reduce each MIMS profile from 1440 to 24 observations. This step is necessary for the NHANES application, as `pffr()` becomes slow for high-dimensional functional observations. This step is achieved using the same code introduced in Section 5.1.1. The hour-level MIMS is named `MIMS_hour_mat` and stored as a matrix in a column in the data frame `nhanes_df`. We emphasize that this data structure used is different from the data structure used in the `mgcv::bam()` function, where functional responses were organized in long format.

```
nhanes_famm_df =
  nhanes_df %>%
  mutate(
    MIMS_hour_tf =
      tf_smooth(MIMS_tf, method = "rollmean", k = 60, align = "center"),
    MIMS_hour_tf = tfd(MIMS_hour_tf, arg = seq(.5, 23.5, by = 1))) %>%
  tf_unnest(MIMS_hour_tf) %>%
  rename(epoch = MIMS_hour_tf_arg, MIMS_hour = MIMS_hour_tf_value) %>%
  pivot_wider(names_from = epoch, values_from = MIMS_hour)

MIMS_hour_mat =
  nhanes_famm_df %>%
  select(as.character(seq(0.5, 23.5, by = 1))) %>%
  as.matrix()

nhanes_famm_df =
  nhanes_famm_df %>%
  select(SEQN, gender, age) %>%
  mutate(MIMS_hour_mat = I(MIMS_hour_mat))
```

We now fit the function-on-scalar model using the `pffr()` function to analyze the association between age and gender on physical activity, where each residual $X_i(s)$ is modeled using penalized splines. The syntax is shown below.

```
nhanes_famm =
  nhanes_famm_df %>%
  pffr(MIMS_hour_mat ~ age + gender + s(SEQN, bs = "re"),
    data = ., algorithm = "bam", discrete = TRUE,
    bs.yindex = list(bs = "ps", k = 15, m = c(2, 1)))
```

Notice that the fixed effects are now represented differently. The functional intercept $\beta_0(s)$ is specified automatically in the `pffr()` syntax. In addition, the functional fixed effects $\beta_q(s)Z_{iq}$ are specified by indicating the covariate Z_{iq} only. For example, recall that in `mgcv::bam` this term was specified as `s(SEQN, by = age)`. In `refund::pffr` the same term is specified as `age`. To specify the functional random effects $X_i(s)$, the syntax in `refund::pffr()` is `s(SEQN,bs="re")`, which indicates that a subject-specific penalized spline is used. Although this syntax is similar to the one used in the function `mgcv::bam()`, in `refund::pffr()` it actually specifies a *functional* random intercept instead of a *scalar* random intercept for each subject. The two approaches have a very similar syntax, which could lead to confusion, but they represent different models for the residual correlation.

The characteristics of the penalized splines used for fitting the functional residuals $X_i(s)$ are indicated in the `bs.yindex` variable. This is a list that indicates the penalized spline structures. In our case, `bs="ps"` indicates the P-splines proposed by [71], `k=15` indicates the number of basis functions, and `m=c(2,1)` specifies a second-order penalized spline basis (quadratic spline), with a first-order difference penalty. These were modified from the default number of knots `k=5`, which is insufficient in our application. In this example, increasing the number of basis functions to 15 results in significantly longer computation times due to the high time complexity of `pffr()`. Specifically, it takes about 70 seconds to complete on a local laptop using 5 basis functions, while it takes about 27 minutes using the same laptop when specifying 15 basis functions. Once specified, the spline penalty and basis functions are the same for all $X_i(s)$. The spline structures for the fixed functional intercept were left unchanged to their defaults. They could be changed using the `bs.int` variable. As

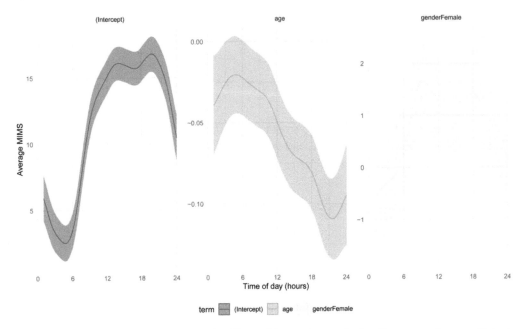

FIGURE 5.11: Point estimators and 95% confidence intervals for the regression coefficients for function-on-scalar regressions of hour-level MIMS on age and gender using `pffr()`, where each residual is modeled using penalized splines. Results are shown for the intercept, age, and gender parameters, respectively.

elsewhere, we used `algorithm="bam"` and `discrete = TRUE`, which substantially accelerates the method.

The fitted model is stored as the object `nhanes_famm`. To visualize the fixed effects estimates and confidence bands, we use a similar approach to the one that extracted estimates and SE from `mgcv` objects earlier in this chapter. An alternative simple solution is to use the base R `plot()` function, which has generic methods to show smooth functions. Figure 5.11 shows the fixed effects estimates obtained using `pffr()`. The first panel shows the estimated functional intercept $\beta_0(s)$. By default, the intercept is centered at 0 in `pffr()`, and the actual coefficient of the intercept can be obtained from `nhanes_famm$coefficients[1]`. The second panel shows the fixed functional coefficient $\beta_1(s)$, which is the effect of age on physical activity. The third panel shows the fixed functional coefficient $\beta_2(s)$, which is the effect of gender on physical activity. Notice that these estimates are similar to the one using separate regressions shown in Section 5.1.1, and in particular those using similar hour-level data in Figure 5.4. An advantage of this joint modeling approach is that it gives smoother estimates and accounts for error correlations.

5.2.2.3   A Bayesian Perspective on Model Fitting

There is an unmistakable appeal to fitting (5.5) using Bayesian modeling. Indeed, it is well known that individual components can be expressed using splines or other approaches. The full likelihood of the model becomes complicated, though the full conditionals are quite tractable. For nonparametric penalized spline regression, [51] showed how to use WinBUGS, while [48] showed how to extend ideas to functional nonparametric methods. Ideas can easily be adapted to other Bayesian inferential platforms. See for example, [106] for using Stan [35, 281, 282], and [59] for using JAGS [232] in complex functional regression contexts.

Here we will not provide the details of Bayesian implementations, but we will show the general ideas for simulating from the complex posterior distribution. Assume that we have prior distributions for all parameters in model (5.5). Then the general structure of the full conditionals of interest is

1. $[\beta_q(s)|\text{others}]$ for $q = 0, \ldots, Q$;

2. $[X_i(s)|\text{others}]$ for $i = 1, \ldots, n$;

3. $[\sigma_e^2|\text{others}]$.

The notation here is focused on the concepts, as the detailed description of each conditional distribution is notationally burdensome. For example, the full conditional $[\beta_q(s)|\text{others}]$ refers to the full conditional distribution of the spline basis coefficients used for the expansion of $\beta_q(s)$ and of the smoothing parameter associated with this function. Recall that when sampling from $[\beta_q(s)|\text{others}]$, all other parameters, including $X_i(s)$ and σ_e^2 are fixed. Therefore, these full conditionals are relatively simple and well understood. For example, with standard choices of prior distributions, the full conditional of the spline coefficients of $\beta_q(s)$ is a multivariate Normal distribution, while that of the smoothing parameter is an inverse Gamma prior [51, 258]. Thus, this step can be conducted using direct sampling from the full conditional distribution without the need for a Rosenbluth-Metropolis-Hastings (RMH) step. The full conditionals for $\beta_q(s)$ contain information from all study participants, as they appear in the likelihood for every subject in the study.

At every step of the simulation, the full conditionals $[X_i(s)|\text{others}]$ depend only on the likelihood for study participant i. The information from the other study participants is encoded in the population level parameters $\beta_q(s)$ and σ_e^2, which are fixed because they are conditioned on at this step. If $X_i(s)$ are expanded into a basis (e.g., FPCA or splines), the basis coefficients have multivariate Normal distributions if standard Normal priors are used. The smoothing parameters are updated depending on the structure of the penalties. For example, for the FPCA basis the scores on the kth component are assumed to follow a $N(0, \lambda_k)$ distribution. It can be shown that with an inverse-Gamma prior on $\sigma_k^2 = \lambda_k$, the full conditional for $[\sigma_k^2|\text{others}]$ is an inverse Gamma. Similarly, if we use a spline basis expansion for $X_i(s)$, the variance parameter that controls the amount of smoothing has an inverse-Gamma full conditional distribution if an inverse-Gamma prior is used. The specific derivation looks different if we use one smoothing parameter per function $X_i(s)$ or one smoothing parameter for all functions. We leave the details to the reader. Irrespective of the modeling structure (FPCA or splines), the number of full conditionals $[X_i(s)|\text{others}]$ increases with the number of subjects, though this increase is linear in the number of subjects.

Finally, with an inverse-Gamma prior on σ_e^2 it can be shown that $[\sigma_e^2|\text{others}]$ is an inverse-Gamma, which makes the last step of the algorithm relatively straightforward.

Therefore, in the case of Gaussian FoSR with functional residuals modeled as a basis expansion, all full conditionals are either multivariate normals or inverse-Gamma. This allows the use of Gibbs sampling [37, 95] without the RMH [42, 113, 200, 256] step, which tends to be more stable and easier to implement.

When the functional data are not Gaussian, many of the full conditionals will require an RMH step, which substantially increases computational times and requires tuning of the proposal distributions. This can be done, but requires extra care when implementing the software.

The Bayesian perspective is very useful and provides extraordinary flexibility. It has several advantages including (1) provides a joint-modeling approach that more fully accounts for the uncertainty in model parameters; (2) introduces a more unified approach to inference, where all parameters are random variables and the difference between "random" and "fixed" effects is modeled via distributional assumptions; (3) simulates the full joint distribution of all model parameters given the data; (4) can produce predictions and uncertainty quantification for missing data within and without the observed domain of the functions; and (5) it provides an inferential framework for more complex analyses that cannot be currently handled by existing non-Bayesian software. However, it also has several limitations, including (a) deciding what priors to choose and what priors are non-informative is very difficult in highly complex models; (b) some priors, such as inverse Gamma priors on variance components, do not allow the variance to be zero, which would correspond to simpler parametric models; (c) computations tend to take longer, though they may still be scalable with enough care; (d) small changes in models still require substantial changes in implementations; and (e) as implementations are slow, realistic simulation analysis of software performance and accuracy is often not conducted.

5.3   A Scalable Approach Based on Epoch-Level Regressions

FoSR modeling often involves high-dimensional data with resulting computational pressures. In regressing MIMS values on age and gender, we have largely avoided a discussion of these issues by focusing on a subset comprised of 250 subjects. However, it is worth noting the scale of required matrix computations. For 250 subjects observed over 1440 epochs each, design matrices will have 360,000 rows; if 30 spline basis functions are used in the estimation of coefficient functions, this design matrix will have 90 columns. When using FAMM there are an additional 30 columns for each subject-specific spline function. One can see how these matrices can quickly explode both in terms of number of rows and columns, which can bring any software to a standstill. Increasing the number of subjects or predictors will exacerbate problems of scale. The tools we have used so far, including state-of-the-art software such as mgcv and pffr(), can nonetheless struggle to meet the demands of some modern datasets.

A very simple alternative strategy revisits the epoch-level regressions that previously motivated a switch to functional techniques. Recall that the epoch-level regression models did not account for the temporal structure, but simply fit standard regression models at each epoch separately. The functional approach explicitly accounted for temporal structure by expanding coefficient functions of interest and shifting focus to the estimation of spline coefficients. Instead, one can smooth epoch-level regression coefficients to obtain estimates of coefficient functions. Computationally, this requires fitting many simple models rather than one large model, and can scale easily as the number of subjects or covariates increases. This idea was motivated by the work on Fast Univariate Inference (FUI, [57]) in the context of more complex structured functional data. Here we investigate whether the same family of methods can be applied in the simpler context of function-on-scalar (FoSR) regression.

In Section 5.1.1, we saw results of minute-level regression. The code below implements that analysis, and is similar to the approach seen for one-hour epochs. In particular, we unnest the MIMS column containing a tf vector for functional observations, which produces a long-format data frame containing all subject- and epoch-level observations. We then re-nest by epoch, so that data across all subjects within an epoch are consolidated; this step allows the epoch-by-epoch regressions of MIMS on age and gender by mapping across epoch-level datasets. The results of the regressions are turned into data frames by mapping the broom::tidy() function across model results. Finally, epoch-level regression results are unnested so that the intercept and two regression coefficients are available for each epoch.

```
min_regressions =
  nhanes_df %>%
  select(SEQN, age, gender, MIMS_tf) %>%
  tf_unnest(MIMS_tf) %>%
  rename(epoch = MIMS_tf_arg, MIMS = MIMS_tf_value) %>%
  nest(data = -epoch) %>%
  mutate(
    model = map(.x = data, ~ lm(MIMS ~ age + gender, data = .x)),
    result = map(model, broom::tidy)
  ) %>%
  select(epoch, result) %>%
  unnest(result)
```

The next step in this analysis is to smooth the regression coefficients across epochs.
Many techniques are available for this; the approach implemented below organizes epoch-
level regression coefficients as `tf` objects in a term called `coef`, and then smooths the
results using a lowess smoother in the `tf_smooth()` function. The resulting data frame has
three columns: `term` (taking values `Intercept`, `age` and `genderFemale`), `coef` (containing
the unsmoothed results of epoch-level regressions), `smooth_coef` (containing the smoothed
versions of values in `coef`). Figure 5.12 below contains a panel for each term, and shows
epoch-level and smooth coefficients. Note the similarity between the smoothed coefficients
and those obtained by "functional" approaches, including penalized splines; this suggests
that this technique provides plausible results, even though smoothing is conducted by ig-
noring the correlation of functional residuals.

```
ui_coef_df =
  min_regressions %>%
  select(epoch, term, coef = estimate) %>%
  tf_nest(coef, .id = term, .arg = epoch) %>%
  mutate(smooth_coef = tf_smooth(coef, method = "lowess"))

fui_coef_df %>%
  ggplot(aes(y = smooth_coef, color = term)) +
  geom_spaghetti(linewidth = 1.2) +
  geom_spaghetti(aes(y = coef), alpha = .2) +
  facet_wrap(~ term, scales = "free")
```

A main focus of Section 5.2.2 was to model error structures and thereby obtain accurate
inference. The scalable approach we suggest in this section models each epoch separately,
but the residual correlation is implicit: regression coefficients across epochs are related
through the residual covariance. This fact, and the scalability of the estimation algorithm,
suggests that bootstrapping is a plausible inferential strategy in this setting. In particular,
we suggest the following resample participants, including full response functions, with replacement to create bootstrap samples; fit epoch-level regressions for each bootstrap sample
and smooth the results; and construct confidence intervals based on the results. This resampling strategy preserves the within-subject correlation structure of the full data without
making additional assumptions on the form of that structure and account for missing data
at the subject level using standard mixed effects approaches. From a computational perspective, bootstrap always increases computation time because one needs to refit the same
model multiple times. However, pointwise regression and smoothing is a simple and relatively fast procedure, which makes the entire process much more computationally scalable

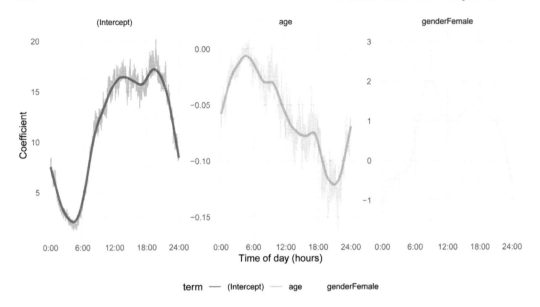

FIGURE 5.12: Epoch-level regression coefficients and smoothed coefficient functions.

than the joint modeling approaches described in Section 5.2.2. Moreover, the approach is easy to streamline and parallelize, which can further improve computational times.

Our implementation of this analysis relies on a helper function `nhanes_boot_fui`, which has arguments `seed` and `df`. This function contains the following steps: first, it sets the seed to ensure reproducibility and then creates a bootstrap sample from the provided data frame; second, it creates the hat matrix that is shared across all epoch-level regressions; third, it estimates epoch-level coefficients by multiplying the hat matrix by the response vector at each epoch; and fourth, it smooths these coefficients and returns the results. Because these steps are relatively straightforward, we defer the function to an online supplement. The code chunk below uses `map` and `nhanes_boot_fui` to obtain results across 250 bootstrap samples. In practice one may need to run more bootstrap iterations, but this suffices for illustration. After unnesting, we have smooth coefficients for each iteration and can plot the full-sample estimates with the results for each bootstrap resample in the background.

```
fui_bootstrap_results =
  tibble(iteration = 1:250) %>%
  mutate(
    boot_res = map(iteration, nhanes_boot_fui, df = nhanes_df)
  ) %>%
  unnest(boot_res)
```

Figure 5.13 shows results of this analysis. We include the full-sample estimates and estimates obtained in 25 bootstrap samples in the background. We also show pointwise confidence intervals by adding and subtracting 1.96 times the pointwise standard errors obtained via the bootstrap to the full-sample estimates. Constructing correlation and multiplicity adjusted (CMA) confidence intervals requires one to account for the complete joint distribution of the functions. As we have already described how to do that in Chapters 2 and 4, we leave this as an exercise.

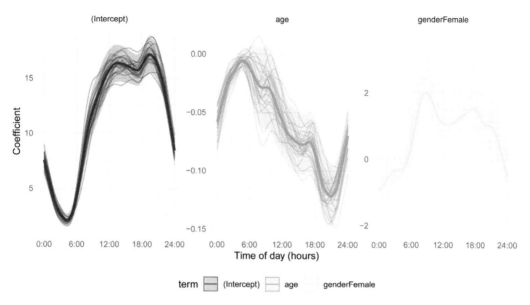

FIGURE 5.13: Effects of age and gender on physical activity in the NHANES data using a fast, scalable approach for estimation and bootstrapping for inference.

6

Function-on-Function Regression

We now consider the case when the response is a function and at least one predictor is a function. This is called function-on-function regression (FoFR), which generalizes the scalar-on-function regression (SoFR) in Chapter 4 and the function-on-scalar regression (FoSR) in Chapter 5. This generalization provides the flexibility to model the association between functions observed on the same study participant. We will focus on the linear FoFR and provide details about possible extensions. FoFR is known under different names and was first popularized by [242, 245] who introduced it as a functional linear model with a functional response and functional covariates; see Chapter 16 in [245]. Here we prefer the more precise FoFR nomenclature introduced by [251, 253], which refers directly to the type of outcome and predictor. It is likely that this area of research was formalized by the paper [241], where the model was referred to as the "linear model" and was applied to a study of association between precipitation and temperature at 35 Canadian weather stations. FoFR may have had other points of origin, as well, but we were not able to find them. The history of the origins of ideas is fascinating and additional information may become available. There is always the second edition, if we remain healthy and interested in FDA.

FoFR has been under intense methodological research, at least in statistics. In particular, it has been applied to multiple areas of scientific research including biliary cirrhosis and association between systolic blood pressure and body mass index [149, 335], medfly mortality [122, 326], forecasting pollen concentrations [300], weather data [16, 195], online virtual stock markets [78], evolutionary biology [308], traffic prediction [46], jet engine temperature [125], daily stock prices [249], bike sharing [148], lip movement [190], environmental exposures [115], and electroencephalography [201], to name a few. Just as discussed in Chapters 4 and 5, these papers are referenced here for their specific area of application, but each contains substantial methodological developments that could be explored in detail.

Here we will focus on methods closely related to the penalized function-on-function regression framework introduced and extended by [110, 128, 263, 260, 262] and implemented in the rich and flexible function `refund::pffr`. The software and methods allow multiple functional predictors, smooth effects of scalar predictors, functional responses and/or covariates observed on possibly different non-equidistant or sparse grids, and provide inference for model parameters. Regression methods are based on the general philosophy described in this book: (1) expand all model parameters in a rich spline basis and use a quadratic penalty to smooth these parameters; (2) use the connection to linear mixed effects models (LME) for estimation and inference; (3) use existing powerful software designed for nonparametric smoothing to fit nonparametric functional regression models; and (4) produce and maintain user friendly software. Developing such software is not straightforward and simplicity of use should be viewed as a feature of the approach. This simplicity is the result of serious methodological development as well as trial and error of a variety of approaches. This philosophy and general approach can be traced to [48, 102], who developed the methods for SoFR, but the same ideas can be generalized and extended to many other models. It is this philosophy that allows for generalization of knowledge and seamless implementation of methods.

In this chapter we will focus on the intuition and methods behind FoFR as implemented in `refund:pffr`. As we mentioned, there are many other powerful methods that could be considered, but, for practical purposes, we focused on `pffr` and the mixed effects representation of nonparametric functional regression. To start, we first describe two motivating examples: (1) quantifying the association between weekly excess mortality in the US before and after a particular date (e.g., week 20); and (2) predicting future growth measurements in children from measurements up to a particular time point (e.g., day 100 after birth).

6.1   Examples

6.1.1   Association between Patterns of Excess Mortality

Consider the problem of studying the association between patterns of US weekly excess mortality in 52 US states and territories before and after a particular time during 2020. Figure 6.1 provides an illustration of this problem where the "future" trajectories are regressed on the "past" trajectories. The boundary between "past" and "present" is indicated by the blue vertical line and separates data before May 23, 2020 and after, but including, May 23, 2020. Data for all states and territories are displayed as light gray lines, while several states are highlighted using color: New Jersey (green), Louisiana (red), and California (plum). For these states, data from the "past" is shown as solid lines, while data from the "future" is shown as dots. The general problem is to identify patterns of associations across US states between the trajectories before and after May 23, 2020. Of course, this particular separation between "past" and "future" is arbitrary and other cutoffs could be considered. In this chapter we will discuss several ways to conceptualize and quantify such associations.

6.1.2   Predicting Future Growth of Children from Past Observations

Here we revisit the CONTENT data introduced in Chapter 1. In Chapter 3 we discussed how to use the entire data set to conduct sparse FPCA, estimate the underlying dynamics (eigenfunctions), and predict the unobserved data within the range of the observations and in the future. Here we are interested in a different problem. More precisely, at every time point, s^*, we would like to predict the future growth trajectory of an individual based only on the data from that individual up to s^*.

Figure 6.2 provides an illustration of the problem using data for one study participant. The left panel illustrates the z-score for weight (zwei) and the right panel illustrates the z-score for length (zlen) for the same child. In this case the gray vertical line separates the "past" from the "future," the blue dots represent the past information and the red dots represent the points that we would like to predict based on the "past." The difference between this data structure and the one of weekly excess mortality in the US described in the previous section is that these data are observed irregularly (unequal sampling times) and sparsely (some children have only few observations).

6.2   Linear Function-on-Function Regression

In both of these examples, we have an outcome function of interest $W_i(s)$, $s \in S$ and one or multiple functional predictors $X_{ir}(u_r)$, $u_r \in U_r$. For simplicity of presentation we assume

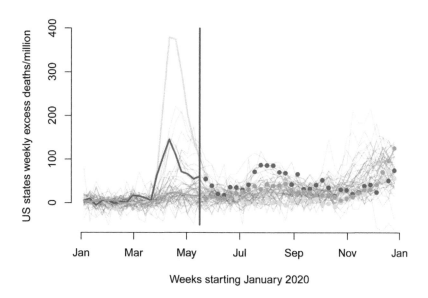

FIGURE 6.1: Illustration of the concept of function-on-function regression (FoFR) as a regression of the future trajectory conditional on the past trajectory. Data are weekly all-cause mortality data per one million people in 50 US states, Puerto Rico and Washington DC. The blue vertical line separates "past" and "future." Each gray line corresponds to one state or territory. Three states are emphasized using color: New Jersey (green), Louisiana (red) and California (plum). For these states, "past" is shown as lines and "future" is shown as dots.

that $r = 1$ and $S = U_1 = \ldots = U_R = [0, 1]$. However, methods apply more generally with more complex notation and integral operations. For this case, the linear function-on-function regression (FoFR) model was first proposed by [241] and has the following form

$$W_i(s) = f_0(s) + \int_U X_i(u)\beta(s, u)du + \epsilon_i(s) \,, \tag{6.1}$$

where $\epsilon_i(t) \sim N(0, \sigma_\epsilon^2)$ are independent random noise variables. While this assumption is often made implicitly or explicitly in FoFR, it is a very strong assumption as the residuals from such a regression often have substantial residual correlations. However, model (6.1) is a good starting point for estimation, while acknowledging that inference requires additional considerations.

There is a large literature addressing this problem; see, for example, [1, 16, 122, 123, 125, 234, 237, 249, 326, 335]. While some of these approaches have been implemented, their use in applications should be considered on a case-by-case basis.

For presentation purposes we assume that $f_0(s) = 0$, even though $f_0(s)$ and $\beta(s, u)$ can be estimated simultaneously. Before diving deeper into methods for estimation, it is worth taking a step back to build some intuition about the interpretation of the model. To start, let us fix the location of the outcome function, s. Then model (6.1) is simply a scalar-on-function regression (SoFR), as discussed in Chapter 4. Therefore, one approach

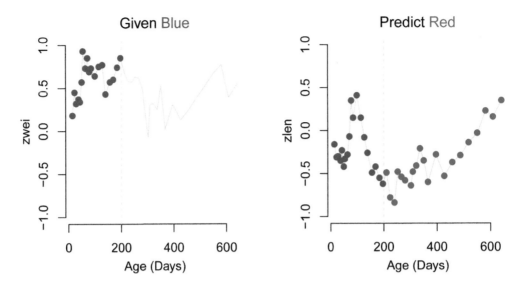

FIGURE 6.2: Illustration of the concept of function-on-function regression (FoFR) in the CONTENT study. In this case FoFR is a regression of future trajectory (z-score for length, shown in red) conditional on two past trajectories (z-score for length and weight, shown in blue).

to estimation could be to set a grid of points $s_1, \ldots, s_p$, run a SoFR regression at every time point, and obtain the estimators $\widehat{\beta}(s_j, u)$ for $j = 1, \ldots, p$ and $u \in U$. An estimator of $\beta(s, u)$ can then be obtained by simply "stitching together" the $\widehat{\beta}(s_j, u)$ estimators. One could argue that such an approach will not account for smoothness of $\beta(s, u)$ over s and/or u. This can be addressed in various ways, but it is useful to have the option of using FoSR to provide an estimator for the FoFR model parameters. Imperfect alternatives are better than no alternatives.

6.2.1   Penalized Spline Estimation of FoFR

Assume for now that $X_i(u)$ is measured without noise and it is observed at every sampling point, u, on a grid. The problem is to estimate the bivariate function $\beta(\cdot, \cdot)$ in model (6.1) and we will use the strategy described in Section 2.3.1.3 to expand the function in a bivariate spline basis. Recall that bivariate spline models are of the type

$$\beta(s, u) = \sum_{k=1}^{K} \beta_k B_k(s, u) ,$$

where $B_k(s, u)$ is a basis in $\mathbb{R}^2$. For presentation purposes we use the tensor product of splines, where k is an indexing of the pair (k_1, k_2) and $B_k(s, u) = B_{k_1, k_2}(s, u) = B_{k_1, 1}(s)B_{k_2, 2}(u)$ for $k_1 = 1, \ldots, K_1$ and $k_2 = 1, \ldots, K_2$, where K_1 and K_2 are the number of bases in the first and second dimension, respectively. With this notation, the total number of basis functions is $K = K_1 K_2$ and

$$\beta(s, u) = \sum_{k_1=1}^{K_1} \sum_{k_2=1}^{K_2} \beta_{k_1 k_2} B_{k_1, 1}(s) B_{k_2, 2}(u) .$$

The functional predictor in (6.1) becomes

$$\int_U X_i(u)\beta(s,u)du = \sum_{k_1=1}^{K_1} \sum_{k_2=1}^{K_2} \beta_{k_1 k_2} B_{k_1,1}(s) \int_U X_i(u)B_{k_2,2}(u)du \ ,$$

where $D_{i,k_2} = \int_U X_i(u)B_{k_2,2}(u)du$ is a random variable that depends on i (the study participant) and k_2 (the index of the basis function in the second dimension), but not on u (the argument of the $X_i(u)$ function). In practice, the function $X_i(u)$ is not observed at every u and the Riemann sum approximations to the integrals $\int_U X_i(u)B_{k_2,2}(u)du$ will be used instead. Similarly, the function $W_i(s)$ is not observed at every point s and is observed on a grid $s_1, \ldots, s_p$ instead. Denote by $C_{j,k_1} = B_{k_1,1}(s_j)$. With this notation, model (6.1) becomes the following standard regression model

$$W_i(s_j) = \sum_{k_1=1}^{K_1} \sum_{k_2=1}^{K_2} \beta_{k_1 k_2} C_{j,k_1} D_{i,k_2} + \epsilon_i(s_j) \ , \tag{6.2}$$

where $A_{ji,k_1,k_2} = C_{j,k_1} D_{i,k_2}$ are the predictors of $W_i(s_j)$ and the regression parameters are $\boldsymbol{\beta} = (\beta_{11}, \beta_{12}, \ldots, \beta_{K_1 K_2})^t$. Denote by

$$\mathbf{W} = \{W_1(s_1), W_2(s_1), \ldots, W_n(s_1), W_1(s_2), \ldots, W_n(s_p)\}^t$$

the $(np) \times 1$ dimensional vector of outcomes, by $\mathbf{C}$ the $p \times K_1$ dimensional matrix with entry (j, k_1) equal to C_{j,k_1}, by $\mathbf{D}$ the $n \times K_2$ dimensional matrix with entry (i, k_2) equal to D_{i,k_2}, and by $\mathbf{X} = \mathbf{C} \otimes \mathbf{D}$ the $(np) \times (K_1 K_2)$ dimensional matrix obtained by taking the Kronecker product of the $\mathbf{C}$ and $\mathbf{D}$ matrices. With this notation the functional model (6.1) and its parametric form (6.2) can be rewritten as the standard linear regression model

$$\mathbf{W} = \mathbf{X}\boldsymbol{\beta} + \boldsymbol{\epsilon} \ , \tag{6.3}$$

where $\boldsymbol{\epsilon}$ is the $(np) \times 1$ dimensional with entries $\epsilon_i(s_j)$ ordered the same way as $W_i(s_j)$ in the vector $\mathbf{W}$. Therefore, fitting a parametric FoFR model is equivalent to fitting a standard regression model. Fitting a nonparametric smoothing model adds quadratic penalties on the $\boldsymbol{\beta}$ parameters, as described in Section 2.3.2. If $\mathbf{D}_{\boldsymbol{\lambda}}$ is a penalty matrix that depends on the vector of smoothing parameters $\boldsymbol{\lambda}$ then a penalized spline criterion would be of the type

$$||\mathbf{W} - \mathbf{X}\boldsymbol{\beta}||^2 + \boldsymbol{\beta}^t \mathbf{D}_{\boldsymbol{\lambda}} \boldsymbol{\beta} \ . \tag{6.4}$$

As discussed in Section 2.3.3, this criterion is equivalent to fitting a linear mixed effects model, where $\boldsymbol{\lambda}$ are ratios of variance components. Therefore, the nonparametric model can be estimated using mixed effects methods, which also produce inference for all model parameters. By now we have repeated this familiar tune: (1) functional regression can be viewed as a standard regression; (2) smooth FoFR can be viewed as a bivariate penalized regression with specific design matrices; and (3) smooth FoFR estimation and inference can be conducted using a specific mixed effects model that can be fit using existent software. Each step is relatively easy to understand, but when considered together they provide a recipe for using existing software directly for estimation and inference for FoFR.

In this case we have considered the tensor product of splines, but the algebra works similarly for any type of bivariate splines, including thin-plate splines. Each choice of basis can be used with its standard quadratic penalties; see, for example the construction of tensor product bases using the `te`, `ti`, and `t2` options in the `mgcv` package. The main message here is that the FoFR regression can be reduced to a nonparametric bivariate regression with a specific regression design matrix and appropriate penalties. Once this is done, software implementation requires only careful accounting of parameters and model structure.

While we have considered the case when the outcome functions $W_i(s)$ are continuous and Gaussian, this assumption is not necessary. The same exact approach works with non-Gaussian data including binary and count observations. The method could also be extended to any other bases, as long as they involve a quadratic penalty on the model coefficients. While non-quadratic penalties are possible, they are not covered in this book.

In Section 6.3 we show how to use `pffr` to fit such models directly and how to expand models to multiple functional predictors as well as nonparametric time-varying effects of covariates. We also show how to change the approach to account for sparse FoFR in Section 6.3 using `mgcv`. Recall that `pffr` is based on `mgcv`, though making the connection and showing exactly how to do that was a crucial contribution [128, 260, 263]. Indeed, other packages for FoFR exist, including the `linmod` and `fRegress` in the `fda` R package [246] and PACE [334] in MATLAB. However, `linmod` and PACE cannot currently handle multiple functional predictors or linear or non-linear effects of scalar covariates. The `fRegress` function is restricted to concurrent associations with the predictor and outcome functions being required to be observed on the same domain. Another package that could be considered is the `fda.usc` [80]. Here we focus on the `refund::pffr` and `mgcv::gam` functions, which provide a highly flexible estimation and inferential approach for a wide variety of FoFR models. We encourage the use of multiple approaches and deciding for oneself what and when works for a particular application.

6.2.2   Model Fit and Prediction Using FoFR

We have shown that fitting model (6.1) is equivalent to the linear model $\mathbf{W} = \mathbf{X}\boldsymbol{\beta} + \boldsymbol{\epsilon}$ (6.3), which is fit using penalized splines (6.4), which, in turn, can be fit using a particular mixed effects model. From this model we can obtain both the estimator $\widehat{\boldsymbol{\beta}}$ of $\boldsymbol{\beta}$ and its joint estimated covariance $\widehat{\mathbf{V}}(\widehat{\boldsymbol{\beta}})$. Under the assumption of multivariate normality of $\widehat{\boldsymbol{\beta}}$ confidence intervals can be obtained for any linear combination of these parameters. In general, the parameter $\boldsymbol{\beta}$ can be partitioned into a subset of parameters $\boldsymbol{\beta}^t = (\boldsymbol{\beta}_1^t, \ldots, \boldsymbol{\beta}_A^t)$, where each subset of parameters corresponds to a particular component of the FoFR model. For example, for model 6.1 $A = 2$ and $\boldsymbol{\beta}_1^t$ are the coefficients of the spline expansion of the univariate function $f_0(s)$, while $\boldsymbol{\beta}_2^t$ are the coefficients of the spline expansion of the bivariate surface $\beta(s, t)$.

Thus, the model can provide predicted values at all points $s_j \in S$ as $\widehat{W}_i(s_j) = \widehat{f}_0(s_j) + \int_U X_i(u)\widehat{\beta}(s_j, u)du$ using the simple matrix multiplication $\widehat{\mathbf{W}}_i = \mathbf{X}_i\widehat{\boldsymbol{\beta}}$, where $\mathbf{X}_i$ is the submatrix of $\mathbf{X}$ obtained by selecting the p rows corresponding to study participant i. Moreover, the covariance matrix of these predictors is $\widehat{\mathbf{V}}(\widehat{\mathbf{W}}_i) = \mathbf{X}_i^t\widehat{\mathbf{V}}(\widehat{\boldsymbol{\beta}})\mathbf{X}_i$. A pointwise 95% confidence band for the predictor $f_0(s_j) + \int_U X_i(u)\beta(s_j, u)du$ can be calculated as

$$\widehat{\mathbf{W}}_i \pm 1.96\sqrt{\text{diag}\{\widehat{\mathbf{V}}(\widehat{\mathbf{W}}_i)\}}\,,$$

where the square root operator is applied separately for each vector and the diag operator is the diagonal operator for a matrix. As discussed in Section 2.4, correlation and multiplicity adjusted (CMA) confidence intervals can be obtained under the normality assumption of $\widehat{\boldsymbol{\beta}}$ or using simulations from the max deviation statistic under the assumption of spherically symmetric distribution of $\widehat{\mathbf{W}}$. Pointwise and global CMA p-values can then be obtained by searching the supremum value of α where the corresponding $1 - \alpha$ level confidence intervals do not cross zero. We will address this in depth in Section 6.5.2.

To construct 95% prediction intervals for individual observations one needs to account for the residual variance, σ_ϵ^2. Thus, the pointwise 95% confidence intervals have the following structure

$$\widehat{\mathbf{W}}_i \pm 1.96 \sqrt{\text{diag}\{\widehat{\mathbf{V}}(\widehat{\mathbf{W}}_i) + \widehat{\sigma}_\epsilon^2 \mathbf{I}_p\}} \,,$$

where $\widehat{\sigma}_\epsilon^2$ is an estimator of the residual variance, σ_ϵ^2. Here $\widehat{\mathbf{V}}(\widehat{\mathbf{W}}_i)$ characterizes the uncertainty of the estimated prediction and $\widehat{\sigma}_\epsilon^2 \mathbf{I}_p$ characterizes the uncertainty of the observations around this prediction. The prediction intervals for individual observations tend to be much larger than for the predictor. The problem of building correlation and multiplicity adjusted (CMA) prediction intervals is as of yet, an open methodological problem.

Confidence and prediction confidence intervals can be obtained at any set of points s_0, not just at $s_j \in S$. This requires recalculating the design matrix that corresponds to these points, say $\mathbf{X}_0$, and conducting the same inference for $\widehat{\mathbf{W}}_0 = \mathbf{X}_0 \widehat{\boldsymbol{\beta}}$ instead of $\widehat{\mathbf{W}}_i = \mathbf{X}_i \widehat{\boldsymbol{\beta}}$. The index 0 here is a a generic notation for data at a new point, while the index i is the notation for data at an observed point i within the sample.

Once predictions are calculated it is easy to compute the estimated residuals $\widehat{\epsilon}_i(s_j) = W_i(s_j) - \widehat{W}_i(s_j)$ for every i and j. These residuals can be used to investigate whether the assumptions about these residuals hold in a particular model and data set. Simple tests for normality and zero serial correlations can be applied to these estimated residuals to point out potential problems with the model and suggest modeling alternatives. Large residuals can also be investigated to identify portions of functions that are particularly difficult to fit. See Section 6.3.1 for an example of residual analysis using the `refund::pffr` function.

If model assumptions are violated, questions can be raised about the validity and performance of confidence intervals. In such situations, we recommend to use the point estimators and supplement the confidence intervals obtained from the model with confidence intervals obtained from a nonparametric bootstrap of study participants.

6.2.3 Missing and Sparse Data

So far, we have focused on the case when all data points are observed, both among the functional predictors and outcomes. In many applications this may not always be the case. If data are densely observed and some observations are missing, one could use smoothing within each study participant to produce complete data. Alternatively, one could use functional PCA as described in Chapter 3, which borrows strength from the other study participants to reconstruct individual trajectories. This approach is especially useful when portions of the functions are missing, in which case we recommend the `refund::fpca.face` function in R. If data are sparsely observed then sparse functional PCA could be used to predict functional data and we recommend `face::face.sparse` function in R. Once missing data are imputed, `pffr` can be used on the complete data. More research is needed to address the effects of the imputation process on the inference about the model parameters.

6.3 Fitting FoFR Using `pffr` in `refund`

We now show how to use `pffr` in the `refund` package to fit some of the FoFR models described in Section 6.1. We start by studying the association between excess mortality in each US state and two territories in the weeks starting with and including May 23, 2020 (last 32 weeks of the year) given the same data starting from January 1, 2020 up to the week starting on May 23, 2020 (first 20 weeks of the year). This problem was first described in Section 6.1.1 and illustrated in Figure 6.1.

In this case, the outcome functions are $W_i : \{21, \ldots, 52\} \to \mathbb{R}$, where $W_i(s_j)$ is the excess mortality for state or territory i in week $s_j \in \{21, \ldots, 52\}$. The predictor is $X_i : \{1, \ldots, 20\} \to \mathbb{R}$, where $X_i(u)$ is the excess mortality for state or territory i in week $u \in \{1, \ldots, 20\}$. Here 21 corresponds to the week of May 23, 2020.

We now show how to implement this approach using `pffr`. Assume that the data are stored in the matrix `Wd`, which is 52×52 dimensional because there are 52 states and territories in the data and 52 weeks in 2020. Each row corresponds to one state and each column corresponds to one week in 2020. The code below shows how data are separated in functional outcomes, $W_i(\cdot)$, and predictors, $X_i(\cdot)$.

```
#Define where prediction is done
cutoff <- 20
#Create the predictor and outcome matrices
X <- Wd[,1:cutoff]
W <- Wd[,(cutoff + 1):dim(Wd)[2]]
#Create the arguments of the predictor and outcome functions
s <- 1:cutoff
t <- (cutoff + 1):dim(Wd)[2]
#Fit pffr with one functional predictor
m1 <- pffr(W ~ ff(X,xind = s), yind = t)
```

This model is the FoFR model (6.1) with a domain-varying intercept, $f_0(\cdot)$, and a bi-variate smooth functional effect, $\beta(\cdot, \cdot)$. Both of these effects are modeled nonparametrically using penalized splines. We show now how to extract the estimators of these functions from the `pffr-object` `m1`, which is a gam-object with some additional information.

```
#Extract all the coefficient information
allcoef <- coef(m1)
#Extract domain-invariant intercept
intercept_fixed <- allcoef$pterms[1]
#Extract the domain-varying intercept
intercoef <- allcoef$smterms$Intercept$coef
#Obtain an estimator of f_0()
intersm <- intercept_fixed + intercoef$value
#Obtain the standard error of the domain-varying intercept
interse <- intercoef$se
```

With this code, an estimator of the function $f_0(\cdot)$ is obtained at every point $s \in [20, 52]$ and it is stored in the vector `intersm`. The standard error of the $f_0(\cdot)$ function without taking into account the domain-invariant intercept is stored in the vector `interse`. Figure 6.3 displays the estimator of $f_0(\cdot)$ (solid blue line) together with the pointwise confidence interval for the domain-varying component (shaded blue area). Here we added the domain-invariant intercept stored in `intercept_fixed` to the domain-varying component `intercoef$value` to make results more interpretable. The estimated function $\widehat{f_0}(s)$ is positive for every week starting with the week of May 23, 2020. This indicates that for every week there is a non-zero excess mortality in the US, though these results do not say anything about individual states. The estimators of excess mortality vary from around 6 the week of May 23, 2020 to around 80 the week of December 26, 2020 for every one million individuals in the US. The trend is increasing, with the exception of the summer period when the excess mortality is estimated to be around 30 for every one million individuals in the US. A direct comparison with the data shown in Figure 6.1 indicates that this estimator is visually compatible with the real data. Indeed, the excess mortality is lower around May

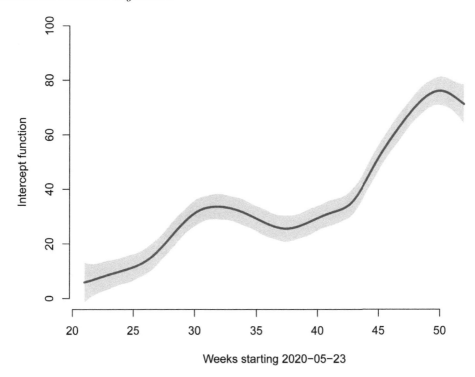

FIGURE 6.3: Intercept and 95% confidence interval for the FoFR regression implemented using `pffr`. The regression is of the excess mortality in 50 US states and 2 territories in the last 32 weeks of 2020 on the excess mortality in the first 20 weeks of 2020.

23, 2020 with some states having negative excess mortality. However, towards the end of the period almost all states have positive excess mortality with the mean around 80 for every one million individuals in the US. In fact, the nonparametric estimator of the mean displayed in Figure 6.3 is similar to the observed data for the state of California (plum) displayed in Figure 6.1.

We now show how to extract the surface $\beta(s, u) : [0, 20] \times [20, 52] \to \mathbb{R}$. The code below extracts the surface as well as the s and u coordinates of where the surface is estimated.

```
#Extract the smooth coefficients, stored as a vector
smcoef <- allcoef$smterms$`ff(Wpred,s)`$value
#Extract the predictor functional arguments
xsm <- allcoef$smterms$`ff(Wpred,s)`$x
#Extract the outcome functional arguments
ysm <- allcoef$smterms$`ff(Wpred,s)`$y
#Transform the smooth coefficients into a matrix
smcoef <- matrix(smcoef, nrow = length(xsm))
```

The estimated coefficient is now stored in the variable `smcoef`, which is then plotted in Figure 6.4. The x-axis corresponds to the first 20 weeks of the year, while the second axis corresponds to the last 32 weeks of the year. For interpretation purposes, recall that the excess mortality in the first 10-12 weeks of the year was probably not affected by COVID-19. Strong effects start to be noticed somewhere between week 13 and week 20 of the year.

A close inspection of Figure 6.4 indicates that the strongest effects appear in the right bottom corner of the graph (note the darker shades of red). This is to be expected, as these are the associations between the observations in week 20 and observations immediately after week 20. This means that states that had a high excess mortality in weeks 18-20 had a high

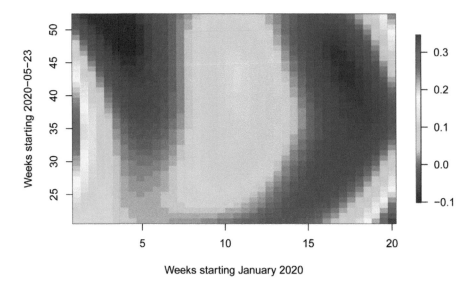

Weeks starting January 2020

FIGURE 6.4: Smooth estimator of the association between US excess mortality in the first 20 weeks and last 32 weeks of 2020 using FoFR implemented via `pffr`.

excess mortality in weeks 21-24. Similarly, states that had a low excess mortality in weeks 18-20 had a low excess mortality in weeks 21-24. The upper-right corner indicates that states with a high/low excess mortality in weeks 18-20 had a high/low excess mortality at the end of the year. There are also two vertical bands of darker blue (negative coefficients) corresponding to weeks 2-7 and weeks 12-17. The darker blue band for weeks 12-17 indicates that states that had a higher than average initial excess mortality after COVID-19 started to affect excess mortality, tended to have lower than average excess mortality for the rest of the year. Similarly, states that had a lower initial excess mortality between weeks 12-17 tended to have a higher than average excess mortality for the rest of the year.

The darker blue in the left-upper corner of the graph indicates that states that had a higher than average mortality between weeks 3-7 had a lower excess mortality towards the end of the year. This finding is a little surprising and may require additional investigation. However, it is unlikely that excess mortality in the first 3 to 5 weeks was affected by COVID-19. This indicates that these associations may be due to other factors, such the demography or geography of different states or territories. A light blue vertical band in the middle of the plot corresponds to very small or zero effects indicating that excess mortality between weeks 8 to 12 is not strongly associated with excess mortality after week 20.

Our next steps are to extract the fitted values $\widehat{W}_i(s_j) = \widehat{f}_0(s_j) + \int X_i(u)\widehat{\beta}(s_j, u)du$ and the residuals $\widehat{\epsilon}_i(s_j) = W_i(s_j) - \widehat{W}_i(s_j)$. Both of these results are stored as 52×32 dimensional matrices, where every row corresponds to a state, i, and every column corresponds to a week, j, between 21 and 52.

```
#Extract the fitted values
fitted_values <- fitted(m1)
#Extract the residuals
residual_values <- residuals(m1)
```

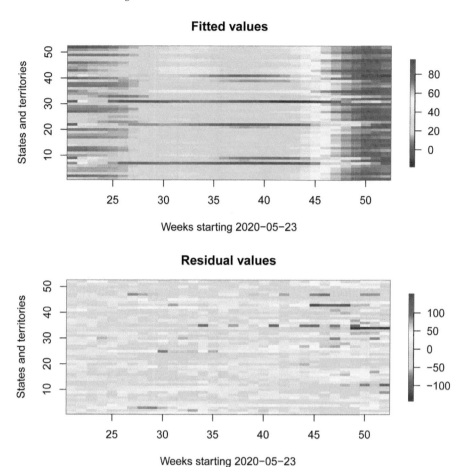

FIGURE 6.5: Predicted values and residuals for the FoFR regression of US excess mortality in the last 32 weeks on the first 20 weeks of 2020 using `pffr`.

Figure 6.5 displays the fitted values in the top panel and the residuals in the bottom panel. The x-axis in both plots represents a week number starting from May 23, 2020 (indicated as week 21 of the year). The y-axis corresponds to a state or territory number.

The fitted values (top plot) indicate that for most states there is a substantial increase from week 21 to week 52 of the year; note the shades of blue changing to shades of red as time passes (moving from left to right in the fitted values panel). This is consistent with the estimated intercept $\widehat{f}_0(s_j)$ displayed in Figure 6.3. The residual plot is also useful to check whether the assumptions about residuals $\epsilon_i(s_j)$ are reasonable. First, it seems that residuals are reasonably well centered around 0, though some residual correlations seem to persist after model fitting. This can be seen as persistent shades of a color for individual states (along the rows of the image). Also, some of the residuals seem to be particularly large in the range of 100 and -100, which are very large values for weekly excess mortality per one million individuals in the US.

6.3.1 Model Fit

We now investigate whether model (6.1) is a reasonable model for the excess mortality data. The idea is to investigate whether the model predictions make sense and assumptions hold.

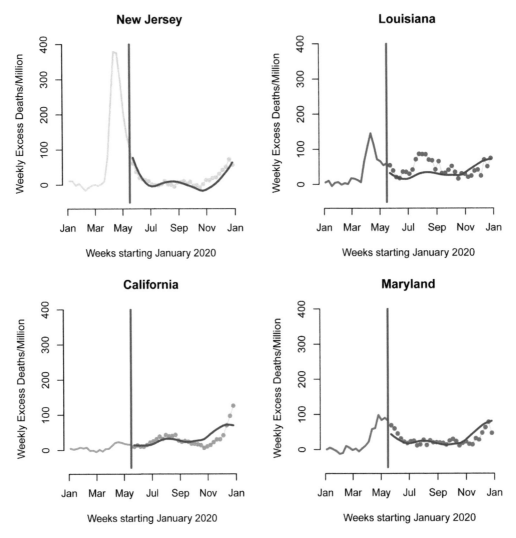

FIGURE 6.6: Observed and predicted weekly excess mortality for four states: New Jersey, Louisiana, California, and Maryland. Observed data are shown as lines for the first 20 weeks and as dots for the last 32 weeks of 2020. The blue vertical line indicates the threshold between "past" and "future." The dark red line indicates the prediction of the data based on model (6.1) implemented in `pffr`.

Figure 6.6 displays prediction and actual data for four states: New Jersey, Louisiana, California, and Maryland. Data used for prediction is shown as a continuous line before May 23, 2020 while the predicted data are shown as dots of the same color after May 23, 2020. The model prediction is indicated as a dark red solid line starting from May 23, 2020. For these four states the predictions and the real data look reasonably close and are on the same scale. For New Jersey the predictions immediately after May 23, 2020 tend to be pretty close to the actual data and capture the down-trend in excess mortality. The model correctly captures the increase in mortality towards the end of the year, though the observed excess mortality tends to be consistently higher than the predicted excess mortality. For Louisiana the prediction model considerably underestimates the observed data during July and August and slightly overestimates it during October, November and December. In general, when predictions do not match the observations they tend to under or overshoot for at least a month or so. This raises questions about the potential residual correlations.

Moreover, the second panel in Figure 6.5 indicated that some of the residuals are very large with differences as large as 100 weekly excess deaths per one million individuals. Here we take a closer look at the estimated residuals $\widehat{\epsilon}_i(s_j) = W_i(s_j) - \widehat{W}_i(s_j)$, which are stored in the `residual_values` variable. We are interested both in investigating the normality of these residuals as well as some of the larger residuals irrespective of whether or not their distribution is normal.

Figure 6.7 provides a more in-depth analysis of the residuals. The top-left panel provides a QQ-plot of the residuals relative to a normal distribution. This indicates a reasonably symmetric distribution with much heavier tails than a normal. We now investigate where the large residuals originate and identify that the large negative residuals can be traced to the state of North Carolina. The top-right panel displays the data for North Carolina and provides an insight into exactly why these residuals occurred. Note that the weekly excess mortality in the state of North Carolina remained close to zero in the last part of the year and even dipped down substantially under zero in the last month of the year. Based on the data from the other states, the predicted values went steadily up during the last part of the year. Many of the large positive residuals can be traced to the states of North and South Dakota (the two panels in the second row). In contrast to North Carolina, both these states had a strong surge in the weekly excess mortality in the last part of the year. These surges were much larger than what was anticipated by the model.

The results displayed both in Figures 6.6 and 6.7 show that under- and over-prediction tend to happen for several weeks and even months, which raises questions about whether residuals can be assumed to be independent. Figure 6.8 displays the estimated residual correlations as a function of week number from May 23, 2020 (first week when model predictions are conducted). This is obtained by using the `cor` function applied to the variable `residual_values`, which has 52 rows corresponding to states and territories and 32 columns corresponding to weeks after May 23, 2020. As anticipated from the visual inspection of results, strong positive and negative correlations can be observed. Residuals for weeks 25 to 35 tend to be positively associated with residuals for weeks 25 to 35 and negatively associated with residuals for weeks 40 to 50. This indicates that if the model under/over predicts for one week between weeks 25 and 35, then it will tend to under/over predict for multiple weeks in this period and it will tend to over/under predict for weeks 40 to 50.

So, we have found that residuals have a relatively symmetric distribution, with much heavier tails than the normal distribution, large absolute deviations that correspond to particular states, and very strong local correlations that persist for a few months. Essentially, almost all model assumptions are rejected, which raises questions about the validity of inference for these models. This is an open problem in functional data analysis that is rarely acknowledged. One potential solution could be to consider the bootstrap of sampling units (in our case states and territories), but it is not clear whether this is a good approach in general. Another potential solution could be to model the residuals, but this would substantially increase computational complexity.

6.3.2 Additional Features of `pffr`

While we have discussed at length how to fit model (6.1), the `pffr` function allows for a much richer class of models. Suppose, for example, that we are interested in the

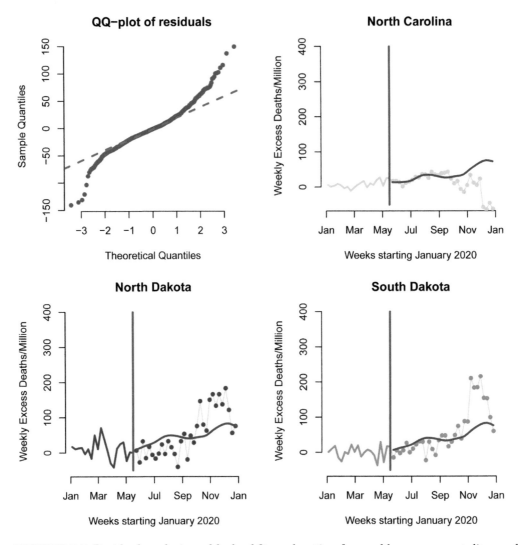

FIGURE 6.7: Residual analysis and lack of fit exploration for weekly excess mortality analysis using `pffr`. Top-left panel: QQ-plot analysis of the estimated residuals for model (6.1) applied to weekly excess mortality in the US. Other panels: observed and predicted data for three states (North Carolina, North Dakota, South Dakota) with the largest residuals.

time-varying effect of state population size on the observed weekly excess mortality. Model (6.1) can easily be extended to the model

$$W_i(s) = f_0(s) + P_i f_1(s) + \int_U X_i(u)\beta(s, u)du + \epsilon_i(s) , \qquad (6.5)$$

where P_i is the population size of state or territory i. Here $f_1(\cdot)$ is the time-varying effect of the population size. This effect is modeled using penalized splines, like all other effects, using a similar basis expansion and corresponding penalties.

The function `pffr` seamlessly incorporates such effects using the following code structure (note the minimal change to code):

Residual correlations

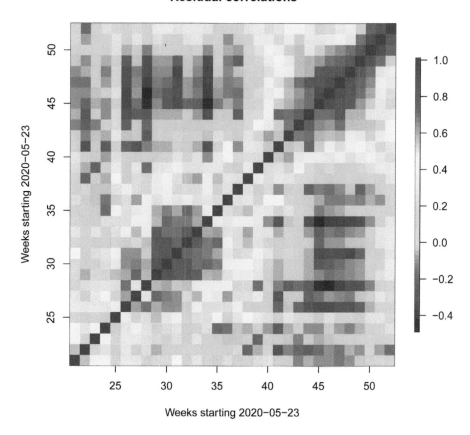

FIGURE 6.8: Residual correlations for the model (6.1) applied to weekly US excess mortality data using `pffr`.

```
#Fit pffr with one functional predictor and one time-varying effect
m2 <- pffr(W ~ ff(X, xind = s) + pop_state_n, yind = t)
```

This code can now be used to extract estimators of interest from the model fit `m2` using similar approaches indicated for model fit `m1`. Below we show how to extract the time-varying effect and standard error for the population size variable.

```
#Extract the time varying population effect
pop_size_effect <- allcoeff$smterms$'pop_state_n(t)'$value
#Extract the standard error of the time varying population effect
pop_size_se <- allcoeff$smterms$'pop_state_n(t)'$se
```

The point estimator for $f_1(s)$ is contained in the variable `pop_size_effect` and its standard error is contained in the variable `pop_size_se`. Figure 6.9 displays the point estimators and 95% pointwise confidence intervals for $f_1(s)$ as a function of time from the week of May 23, 2020 (week 21 of the year). The effect corresponds to one million individuals. The shape of the effect indicates that states with a larger population had larger excess mortality in July

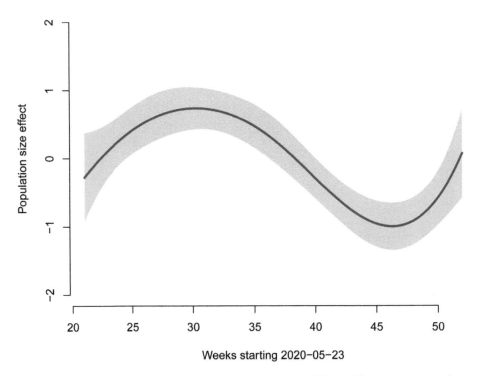

FIGURE 6.9: Time-dependent effect of population size on US weekly excess mortality after May 23, 2020 using `pffr`. This is the effect for every one million additional individuals.

and August and lower excess mortality in November and December. However, the effects for both periods are relatively small compared to some of the other effects. Of course, one could explore other characteristics of states including geography, implementation of mitigation policies, or population density. Such variables can be added either as time varying or fixed predictors in `pffr`; see the help file for the `pffr function` for additional details.

Note that we have conducted all analyses for a particular time point, May 23 2020. However, this choice is arbitrary and the same analyses can be conducted at any other time point during the year. An analysis conducted at all time points provides a dynamic FoFR regression [43, 46, 101, 126, 127].

6.3.3 An Example of `pffr` in the CONTENT Study

We now show how to use `pffr` in the CONTENT study, where both the outcome and the predictor are sparsely observed functions. In Section 6.1.2, we discussed the problem of predicting the future growth trajectory of an individual at a particular time point based on their data up to that point. Specifically, we now consider the association between the z-score of length in the 101 days or later and the z-scores of length and weight in the first 100 days. The choice of 100 days as the threshold is arbitrary, and other thresholds could be used instead.

The code below shows how to upload the CONTENT data directly using the `refund` package. The package `face` is also used for smoothing the sparse growth data. The first step is to split the data based on time 100. The variable `content_old` contains all data observed before 100 days, while variable `content_new` contains all data observed after 100 days.

```
#The refund package contains the refund data CONTENT data
library(refund)

#Function used for smoothing
library(face)

#Load CONTENT data
content <- data(content)

#Split the data at 100 days
content_old <- content[which(content$agedays < 100),]
content_new <- content[which(content$agedays >= 100),]
```

There are two key differences between the CONTENT data and the US Covid-19 excess mortality. First, the CONTENT data collected for each individual are sparse both before and after the time when prediction is conducted. Second, instead of only including one functional predictor, we are now including the z-score of length and weight in the first 100 days, both of which are functional predictors. There are many ways to handle sparse functional data. Here we follow a simple technique that combines smoothing the data before and after the time when prediction is conducted using `face::face.sparse` and then using `refund::pffr` to conduct regression of these smooth estimates evaluated on a regular grid. This approach is computationally fast, though more research may be necessary to address estimation accuracy especially in areas with very sparse observations.

Conceptually, the outcome functions are $W_i : \{101, 103, \ldots, 701\} \to \mathbb{R}$, where $W_i(s_j)$ is the z-score for length (`zlen`) for child i on day s_j. The predictors are $X_{i1} : \{1, \ldots, 100\} \to \mathbb{R}$ and $X_{i2} : \{1, \ldots, 100\} \to \mathbb{R}$, where $X_{i1}(u)$ and $X_{i2}(u)$ are the z-score for length (`zlen`) and z-score for weight (`zwei`) for child i on day u, respectively. We are interested in conducting the following function-on-function regression (FoFR)

$$W_i(s_j) = f_0(s_j) + \int_U X_{i1}(u)\beta_1(s_j, u)du + \int_U X_{i2}(u)\beta_2(s_j, u)du + \epsilon_i(s_j) , \qquad (6.6)$$

where $\epsilon_i(s_j) \sim N(0, \sigma_\epsilon^2)$ are independent random noise variables. Of course, we do not observe $W_i(s_j)$, $X_{i1}(u)$, $X_{i2}(u)$ and we will use estimates of these functions obtained from sparse noisy data. An important point in our example is that only information from the past (observations before day 100 from birth) is used to reconstruct the smooth functions in the past ($X_{i1}(u)$ and $X_{i2}(u)$ for $u < 100$) on an equally spaced grid. Similarly, only information from the future (observations after day 100 after birth) is used to reconstruct the smooth functions in the future ($W_i(s_j)$, $s_j \geq 100$) on an equally spaced grid. Once the functions are estimated on equally spaced grids, the same fitting approach described in Section 6.2 can be used to fit model (6.6).

Below we show how to reorganize the CONTENT data into the required format for conducting smoothing using `face::face.sparse`. The variables `content_zlen_old`, `content_zwei_old`, `content_zlen_new` contain the CONTENT data corresponding to `zlen` and `zwei` data before day 100 and `zlen` data after day 100, respectively. Each one of these sparse data sets is then smoothed using sparse functional smoothing implemented in `face.sparse` and the results are stored in `fpca_zlen_old`, `fpca_zwei_old`, and `fpca_zlen_new`. These variables contain the scores and the estimated eigenvalues necessary to reconstruct the smooth trajectories $X_{i1}(u)$, $X_{i2}(u)$, and $W_i(s)$ in this order.

After performing sparse FPCA, the estimated z-scores of length and weight for each day of the first 100 days are stored in `zlen_old_it` and `zwei_old_it` respectively, both of which are matrices with 100 columns (one column for each day from birth). For the outcome, we obtain the estimated z-scores of length after day 100 every other day, that is, days 101, day 103, ..., day 701. Therefore, the outcome is stored in `zlen_new_it`, a matrix with 301 columns. All these matrices have 197 rows, each row corresponding to one child. At this point we have everything we need to run a standard FoFR model with outcome and predictor functions observed on an equally spaced grid of points.

```
#Reorganize the data into required format for face.sparse
content_zlen_old <- data.frame(argvals = content_old$agedays,
                               subj = content_old$id,
                               y = content_old$zlen)
content_zwei_old <- data.frame(argvals = content_old$agedays,
                               subj = content_old$id,
                               y = content_old$zwei)
content_zlen_new <- data.frame(argvals = content_new$agedays,
                               subj = content_new$id,
                               y = content_new$zlen)


#Fit sparse FPCA
fpca_zlen_old <- face.sparse(data = content_zlen_old,
                             calculate.scores = TRUE,
                             argvals.new = 1:100)
fpca_zwei_old <- face.sparse(data = content_zwei_old,
                             calculate.scores = TRUE,
                             argvals.new = 1:100)
fpca_zlen_new <- face.sparse(data = content_zlen_new,
                             calculate.scores = TRUE,
                             argvals.new = seq(101,
                                 max(content_new$agedays), 2))

#Obtain functional predictors and responses on a regular grid
id <- fpca_zlen_old$rand_eff$subj
xind <- fpca_zlen_old$argvals.new
yind <- fpca_zlen_new$argvals.new

zlen_old_it <- fpca_zlen_old$rand_eff$scores %*%
               t(fpca_zlen_old$eigenfunctions)
zwei_old_it <- fpca_zwei_old$rand_eff$scores %*%
               t(fpca_zwei_old$eigenfunctions)
zlen_new_it <- fpca_zlen_new$rand_eff$scores %*%
               t(fpca_zlen_new$eigenfunctions)
```

The syntax to implement an FoFR with two functional predictors is similar to the one used for a single predictor, as introduced at the beginning of Section 6.3. Specifically, the `ff` function is used to specify an FoFR term. Note that the `pffr` function includes a functional intercept by default. After fitting the model, the coefficients can be extracted from the fitted object using the `coef()` function. Notice that here we only show the code to fit FoFR and extract coefficients, since the remaining steps are similar to those described in the COVID-19 example.

```
#Fit PFFR
m_content <- pffr(zlen_new_it ~ ff(zlen_old_it, xind = xind) +
                  ff(zwei_old_it, xind = xind), yind = yind)

#Extract coefficients
allcoef <- coef(m_content)
```

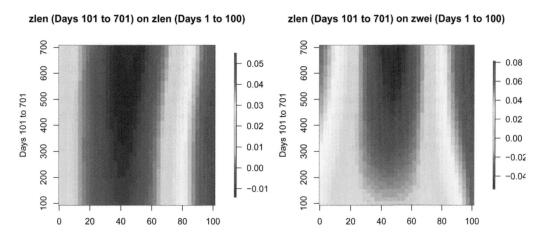

FIGURE 6.10: Smooth estimator of the association between the z-score of length in the 101 days or later and the z-score of length (left panel) and weight (right panel) in the first 100 days in the CONTENT study using FoFR implemented via `pffr`.

The estimated coefficients are plotted using the `image.plot` function in the `fields` package [217] and are shown in Figure 6.10. The left panel shows the estimated association between the z-score for length after day 101 (y-axis) and the z-score of length before day 100 day (x-axis). The right panel displays the estimated association between the z-score of length after day 100 (y-axis) and the z-score of weight before day 100 (x-axis).

We first focus on the left panel, which displays the estimated coefficient $\beta_1(s, u)$, which quantifies the association between the z-score for length before and after day 100. The z-score of length at around day 100 has the strongest association with the z-score of length after 100 days, and this association is consistent across days after day 100. This is indicated by the red vertical band between day 90 and 100 after birth. The shades of red get darker closer to day 100 indicating that the length of the baby closer to the time when prediction is made is a strong predictor of future length. The fact that the red vertical band extends to day 90 seems to indicate that sustained high values of length before day 100 may improve prediction above and beyond the length of the baby at day 100. This is not surprising, as day 100 is the closest day to the days after day 100 and the estimated trajectories are quite smooth in the neighborhood of an observation. It may be surprising that there is a large vertical band of negative estimated coefficients around day 40. The coefficients are smaller (-0.01 compared to 0.05) and some may not be statistically significant. But results seem to imply that a baby who is tall at day 100 and tall at day 40 is predicted to be shorter than a baby who is as tall at day 100 but shorter at day 40. This seems to also make sense as this would capture faster growth trajectories. This type of analysis suggests the possibility of a more in-depth analysis of individual trajectories.

The right panel displays the estimated coefficient $\beta_2(s, u)$, which quantifies the association between the z-score for weight before day 100 and the z-score for length after day 100. This surface has many similarities to the surface for $\beta_1(s, u)$. This indicates that individuals who are heavier around day 100 tend to be longer in the future. Some of these associations, may be due to the correlation between height and weight and indicate that babies who are heavier at day 100 will tend to be taller later on even if they are as long at day 100. Just as in the case of the $\beta_1(s, u)$ surface, there is a blue vertical band around day 40 from birth, which corresponds to negative coefficients. One possible interpretation of these results is that of two babies of the same height at time 100 and 40, the baby who is lighter at day 40 will tend to be taller after day 100. While this is counter intuitive, it may correspond to the fact that the lighter baby is on faster growth trajectory in terms of weight between days 40 and 100, which is enough to increase their likelihood of being taller later on. These plots also illustrate the timeline on which the growth processes are associated with future growth. More precisely, the current height and weight are the most important predictors of future height and weight, but changes in either height or weight over two months may also be indicative of future growth.

6.4 Fitting FoFR Using `mgcv`

The `refund::pffr` function provides a convenient wrapper for functions from the `mgcv` package. While the `mgcv` package was developed for semiparametric smoothing, throughout this book we show that nonparametric functional regression models have a mixed effects model structure that is closely related to that of semiparametric models. Translating the functional nonparametric models into structures that can be used by `mgcv` is a major methodological development and the exact steps on how to do that are not obvious. The first papers that illustrated how to use `mgcv` to fit a large class of functional regression models were [110, 128, 263]. Making the connection between penalized functional regression and mixed effects models completed with software implementation can be traced to [102, 103].

We now show how to estimate the PFFR models in Section 6.2 using the `mgcv` package directly, as well as how to extract the relevant estimators. Note that here we consider regularly observed functional predictors and responses. The general approach to irregular/sparse data is similar in principle, but requires modification of the code to set up the data for model fitting (estimation syntax remains the same once the data frame supplied to the relevant estimation functions is appropriately constructed).

We consider again the problem of regressing the weekly all-cause excess mortality for US states and two territories after week 20 of year 2020 on the data before week 20. Recall the FoFR model:

$$W_i(s) = f_0(s) + \int_U X_i(u)\beta(s,u)du + \epsilon_i(s) \,,$$

where $\epsilon_i(s)$ are independent identically distributed $N(0,\sigma_\epsilon^2)$ random variables with $s \in S$. In our example, the functional predictor, $X_i(u)$, $u \in U = \{1,\ldots,20\}$ and the response $W_i(s)$, $s \in S = \{21,\ldots,52\}$ have time in weeks as argument, though their domains do not overlap.

Recall that the `refund::pffr` approach to fitting FoFR models involves applying a bivariate spline basis to $\beta(s,u)$, then approximating the term $\int_S X_i(u)\beta(s,u)du$ numerically. By default `refund::pffr` uses a tensor product smooth of marginal spline bases to estimate $\beta(s,u)$, though other bivariate bases could be used (e.g., thin plate regression splines). Below we illustrate how this approximation is implemented.

$$W_i(s) = f_0(s) + \int_U X_i(u)\beta(s,u)ds + \epsilon_i(t)$$
$$\approx f_0(s) + \sum_l q_l X_i(u_l)\beta(s,u_l) + \epsilon_i(s) \quad \text{Numeric approximation}$$
$$= f_0(s) + \sum_l q_l X_i(u_l) \sum_{k_1=1}^{K_1}\sum_{k_2=1}^{K_2} \beta_{k_1 k_2} B_{k_1}(s) B_{k_2}(u_l) + \epsilon_i(s) \,. \quad \text{Spline basis}$$

The second line is the numeric approximation to the integral, where q_l are the Riemann sum weights (length of the interval where the integral is approximated) and u_l are the points where the components of the Riemann sum are evaluated. The third line is obtained by expanding $\beta(s, u_l)$ in a tensor product of marginal spline bases.

To estimate this model using `mgcv::gam` we first create a long format data matrix with the elements needed to fit the model. Specifically, each row will contain a single functional response $\{W_i(s), s = 21, \ldots, 52\}$, the entire functional predictor $\{X_i(u) : u = 1, \ldots, 20\}$, as well as matrices associated with the domain of the functional predictor $u \in U = \{1, \ldots, 20\}$, of the functional response, $s \in S$, and the quadrature weights, q_l, multiplied element-wise by the functional predictor $X_i(u_l)$ for numeric approximation.

An important insight is that the component $\sum_{k_1=1}^{K_1} \sum_{k_2=1}^{K_2} \beta_{k_1 k_2} B_{k_1}(s) B_{k_2}(u_l)$ of the model is a standard bivariate spline and the quantities $q_l X_i(u_l)$ can be viewed as additional linear terms that multiply the linear spline decomposition. We will show how to construct this structure in the `mgcv` package using the `by=` option. This is a crucial technical detail that is used throughout this book to conduct functional regression using software originally designed for semiparametric smoothing.

First, let $\boldsymbol{W}$, $\boldsymbol{X}$ be $n \times |S|$ and $n \times |U|$ dimensional matrices containing the functional response and predictor, respectively, with each row corresponding to a subject $i = 1, \ldots, n$. In the R code below the matrix $\boldsymbol{W}$ is identified as `W` and the matrix $\boldsymbol{X}$ is identified as `X`. Next, let $\text{vec}(A)$ denote the vectorization of a matrix A (stacking columns of the matrix). We construct the $n|S| \times 1$ dimensional vector $\boldsymbol{w}^v = \text{vec}(\boldsymbol{W}^t)$, which contains the stacked outcome vectors in the order: study participants and then observations within study participant. This vector is identified as `wv`.

We also construct $\mathbf{1}_{|S|}$, which is a column vector of ones with length equal to the number of observations of the functional outcome, $|S|$. Recall that in our example $|S| = 32$. The next step is to construct the functional predictor matrix $\boldsymbol{X}^v = \boldsymbol{X} \otimes \mathbf{1}_{|S|}$ where $\otimes$ denotes the Kronecker product. The matrix $\boldsymbol{X}$ is $n \times |U|$ dimensional and the matrix $\boldsymbol{X}^v$ is $n|S| \times |U|$ dimensional. The matrix $\boldsymbol{X}^v$ is referred to as `Xv` in the code below. Next, let $\boldsymbol{q} = [q_1, \ldots, q_{|U|}]^t$ be the $|U| \times 1$ dimensional column vector of quadrature weights (q_l in the model above) and $\boldsymbol{L} = \mathbf{1}_{n|S|} \otimes \boldsymbol{q}^t$ be the $n|S| \times |U|$ dimensional matrix where each row is equal with the vector $\boldsymbol{q}$, which contains the quadrature weights. Here $\mathbf{1}_{n|S|}$ is the column vector of ones of length $n|S|$. This matrix is referred to as `L` in the code below. Then we construct the $n|S| \times |U|$ dimensional matrix $\boldsymbol{X}_L^v = \boldsymbol{X}^v \odot \boldsymbol{L}$ with $\odot$ denoting the element-wise product and is referred to as `XvL` in the code. This matrix will be supplied to the `by=` argument, as mentioned in the paragraph above. That is, the matrix $\boldsymbol{X}_L^v$ corresponds to the $q_l X_i(u_l)$ components of the FoFR model.

To complete the data set up we need two additional matrices for the definition of the bivariate spline basis and one vector of arguments for defining the population mean effect $f_0(s)$. Specifically, we construct the $n|S| \times |U|$ dimensional matrix $\boldsymbol{U}^v = \mathbf{1}_{n|S|} \otimes \mathbf{U}^t$, where $\mathbf{U}^t = (u_1, \ldots, u_{|U|})$ is the vector of arguments for the predictor function. The matrix $\boldsymbol{U}^v$ has $n|S|$ rows, where each row consists of $\mathbf{U}^t$, the $1 \times |U|$ dimensional row vector containing the domain of the functional predictor. It is referred to as `Uv` in the code below. The next step is to build the $n|S| \times 1$ dimensional vector $\boldsymbol{s}^v = \mathbf{1}_n \otimes \mathbf{S}$, where $\mathbf{1}_n$ is the $n \times 1$ dimensional column vector of ones and $\mathbf{S} = (s_1, \ldots, s_{|S|})^t$ is the $|S| \times 1$ column vector of arguments for the outcome function. This vector is obtained by repeating the vector $\mathbf{S}$ denoted by `sv` in the code below n times, where n is the number of functions. The vector is labeled `"sv"` in the data frame. We also construct the $n|S| \times |U|$ dimensional matrix $\boldsymbol{S}^v = \boldsymbol{s}^v \otimes \mathbf{1}_{|U|}^t$ with $|U|$ identical columns, and each column being equal to $\boldsymbol{s}^v$, the $n|S| \times 1$ dimensional vector that contains the domain of the functional response repeated n times. This matrix is referred to as `Sv` and is recorded as `"Sv"=I(Sv)` in the data frame.

The code below constructs these matrices and puts them in a data frame. To be consistent with `refund::pffr` defaults we use Simpson's rule for creating quadrature weights.

```r
#Number of observed functions
n <- nrow(W)
#Vectorize the response
wv <- as.vector(t(W))
#Create matrix of functional predictor associated with each response
Xv <- kronecker(X, matrix(1, length(S), 1))
#Matrix of the domain of functional predictor
Uv <- kronecker(matrix(1, length(wv), 1), t(U))
#Quadrature weight vector (Simpson's rule)
q <- (Uv[length(Uv)] - Uv[1]) / length(Uv) / 3 *
        c(1, rep(c(4, 2), length = length(Uv) - 2), 1)
#Matrix of quadrature weights
L <- kronecker(matrix(1, length(wv), 1), t(q))
#Quadrature weights multiplied elementwise by the functional predictor
XvL <- Xv * L
#Vector and matrix associated with domain of the functional response
sv <- kronecker(matrix(1, n, 1), S)
Sv <- kronecker(sv, matrix(1, 1, length(U)))
#Combine into a dataframe
df_mgcv <- data.frame("wv" = wv, "XvL" = I(XvL),
                      "sv" = sv, "Sv" = I(Sv), "Uv" = I(Uv))
```

The data frame for fitting the model directly using `mgcv` then contains

- w^v: an $n|S| \times 1$ dimensional vectorized functional response, denoted as `wv`

- X_L^v: an $n|S| \times |U|$ matrix corresponding to the functional predictor, repeated within the repeated measures of the same functional response and multiplied by the quadrature weights for numeric integration, denoted as `XvL`

- U^v: an $n|S| \times |U|$ dimensional matrix, where each row consists of U^t, the $1 \times |U|$ dimensional row vector containing the domain of the functional predictor, denoted as `Uv`

- s^v: an $n|S| \times 1$ dimensional vector obtained by repeating the vector S n times, denoted below as `sv`

- S^v: an $n|S| \times |U|$ dimensional matrix with $|U|$ identical columns, and each column being equal to s^v, denoted as `Sv`

These objects are all stored in a data frame, with the matrices stored as objects of the class `AsIs` via the `I()` function. Having set up the required data frame for model fitting, estimation is done using the following call to `mgcv::gam()`.

```r
#Fit the model
fit_mgcv <- gam(wv ~ s(sv, bs = "ps", k = 20, m = c(2,1)) +
                  te(Uv, Sv, by = XvL, bs = "ps",
                  k = c(5, 5), m = list(c(2, 1), c(2, 1))),
               method = "REML", data = df_mgcv)
```

We now describe the syntax and how it relates to the elements of our FoFR model. Note that specific arguments, which are not the `mgcv::gam()` defaults were chosen so that the results are identical to the `refund::pffr()` fit. The syntax above specifies that the functional response, $\boldsymbol{w}^v$ (wv) is the outcome variable. The two components of the linear predictor, $f_0(s)$, $\int_U X_i(u)\beta(u,s)du$ are specified by the calls to the s() and te() functions, respectively.

First consider the specification of $f_0(t)$. The syntax s(sv, bs = "ps", k = 20, m = c(2, 1)) adds to the linear predictor a smooth function of sv, $f_0(s)$, which is modelled using penalized B-splines (bs = "ps") with 20 (k = 20) knots. The B-splines are of order 2 and have a first-order difference penalty indicated by (m=c(2, 1)). This part of the function is standard in nonparametric smoothing using mgcv.

Next, consider the specification of $\int_U X_i(u)\beta(u,s)du$. The te() function specifies a tensor product smooth of marginal bases, with the marginal bases defined by the first unnamed arguments to the function call. In this case, we specify a bivariate smooth of Uv, the functional domain of the predictor ($\boldsymbol{U}^v$), and Sv, the functional domain of the outcome ($\boldsymbol{S}^v$). The marginal bases are penalized B-splines (bs="ps", argument is recycled for the second basis), each with 5 knots (k=c(5, 5)). Of course, these arguments could be changed and one could specify a different number of knots and type of spline for each direction in the bivariate space. Again, the B-splines are of order 2 and have a first-order difference penalty applied indicated by (m=c(2, 1)). The summation over the domain of the functional predictor, multiplying the coefficient surface $\beta(u,s)$ by the covariate and quadrature weights, $\sum_l q_l X_i(u_l)$, is specified by the option by = XLmat. Smoothing parameter selection is done using residual marginal likelihood (method = "REML") [121, 228, 317]. As this was implemented in mgcv, one can easily use any of the other criteria available for estimation of the smoothing parameters. Together these components add to the desired term, the linear predictor $\sum_l q_l X_i(u_l) \sum_{k_1=1}^{K_1} \sum_{k_2=1}^{K_2} \beta_{k_1 k_2} B_{k_1}(s) B_{k_2}(u)$.

Extracting the estimated coefficients, $\widehat{f}_0(s)$ and $\widehat{\beta}(u,s)$, requires a bit more work than if one used refund::pffr(), but the exercise is useful for understanding how to extract estimated quantities from mgcv::gam(). We extract the estimated coefficients with estimated standard errors separately. Recall that the **predict** method associated with objects obtained from mgcv::gam() allows for predictions evaluated at: the link scale (type = "link", sum of the linear predictor), the response scale (type = "response", inverse link function), the design matrix (type = "lpmatrix"), or the individual components of the linear predictor (type = "terms"). Here we will use primarily the output from type = "terms", but we will also illustrate a use of type = "lpmatrix".

First consider $\widehat{f}_0(s)$. We must first construct a dataframe which contains the domain of the functional response at each point where we wish to evaluate. In our case, it makes sense to evaluate at each week $21, \ldots, 52$, though due to the smoothness assumption we could evaluate the function on a denser or coarser grid if desired. Accordingly, we must create a dataframe where each row contains the domain of the response, s, at each point we wish to make predictions on. Denote this vector as $s_{\text{pred}} = [21, \ldots, 52]^t$, which is denoted as S in the code below. Because we only need the argument tvec to evaluate $\widehat{f}_0(s)$, the other arguments which must be supplied to predict.gam() (tmat, smat, XLmat) are arbitrary. The code below does this

```
#Set up data frame for predictions
df_pred_f0 <- data.frame("sv" = S, "Uv" = 21, "Sv" = 1, "XvL" = 1)
#Get predictions on for each smooth term
f0_hat <- predict(fit_mgcv, type = "terms", newdata = df_pred_f0,
                  se.fit = TRUE)
```

The object f0_hat is then a list with the first element containing a matrix with two columns, corresponding to a set of predictions $\widehat{f}_0$ and $\widehat{\beta}$ evaluated at the values contained in df_pred_f0. In this case, the matrix will have 32 rows. The second element of f0_hat is a matrix with the same structure, but contains the estimated standard errors associated with the predictions in the first element.

Next consider how to extract the $\widehat{\beta}(u, s)$. To get estimates for this bivariate surface, we need to create a dataframe which has all pairwise combinations of s and t that we wish to obtain predictions on. We use an equally spaced grid on the domain of both the functional predictor and response of length 100. Specifically, consider the equally spaced grid of points $\boldsymbol{u}_{\text{pred}} = [1, \ldots, 20]^t$ with $|\boldsymbol{u}_{\text{pred}}| = 100$ and $\boldsymbol{s}_{\text{pred}} = [21, \ldots, 52]^t$ with $|\boldsymbol{s}_{\text{pred}}| = 100$. The matrix of all pairwise combinations is what we need to create. Specifically, we want to create the $|\boldsymbol{u}_{\text{pred}}||\boldsymbol{s}_{\text{pred}}| \times 2$ dimensional matrix $[\boldsymbol{u}_{\text{pred}} \otimes \mathbf{1}_{|\boldsymbol{s}_{\text{pred}}|}, \boldsymbol{s}_{\text{pred}} \otimes \mathbf{1}_{|\boldsymbol{u}_{\text{pred}}|}]$, where $\mathbf{1}_{|\boldsymbol{u}_{\text{pred}}|}$ and $\mathbf{1}_{|\boldsymbol{s}_{\text{pred}}|}$ are both columns of ones of length $|\boldsymbol{u}_{\text{pred}}|$ and $|\boldsymbol{u}_{\text{pred}}|$, respectively. In our case, $|\boldsymbol{u}_{\text{pred}}| = |\boldsymbol{u}_{\text{pred}}| = 100$, but these choices can vary by application. This can be conveniently created using the `expand.grid` function, as we show in the code below when we build the `df_pred_beta` data frame. In addition, we need to consider the fact that the linear predictor component `predict.gam()` evaluates the entire term $\sum_l q_l X_i(u_l) \sum_{k_1=1}^{K_1} \sum_{k_2=1}^{K_2} \beta_{k_1 k_2} B_{k_1}(u_l) B_{k_2}(s) = \sum_l q_l X_i(u_l)\beta(u_l, s)$. As such, to obtain $\widehat{\beta}(u, s)$ we need $q_l X_i(u_l) = 1$ for $u_l \in \boldsymbol{u}_{\text{pred}}$. This is handled by setting `XvL=1` in the call to `predict.gam()`. *What may look like a "computational trick" is actually a sine-qua-non procedure for extracting and conducting inference on the functional parameter.* Since we do not care here what values $\widehat{f}_0(s)$ is evaluated at, the choice is arbitrary. The code to obtain the predictors as described above is presented in the code chunk below.

```
#Set up grid for predictions on domain of predictor, response
uv_pred <- seq(1, max(U), len = 100)
sv_pred <- seq(min(S), max(S), len = 100)
#Set up data frame for predictions
df_pred_beta <-
    expand.grid(Uv = uv_pred, Sv = sv_pred) %>%
    mutate(sv = 21, XvL = 1)
#Get predictions on for each smooth term
coef_mgcv <- predict(fit_mgcv, type = "terms", se.fit = TRUE,
                     newdata = df_pred_beta)

#Put estimates in matrix format for beta(u,s)
beta_hat <- matrix(coef_mgcv$fit[,2],
                   length(uv_pred), length(sv_pred), byrow = FALSE)
```

The predictions obtained from `predict.gam` contain the estimated coefficient in long format. Figure 6.11 presents the results obtained from calling `mgcv::gam` directly (left panel) versus the results from `refund::pffr()` (right panel). The resulting coefficient surfaces are identical, as expected given that we used the default settings in `refund::pffr`. Note that there is a visual difference in the estimated surfaces due to the fact that the `predict` method associated with `refund::pffr()` by default evaluates on a coarser grid than we used for the `mgcv::gam` predictions (40×40 versus 100×100). All parameters estimated by the models are identical.

6.5   Inference for FoFR

Inference for the FoFR model can be done using either `refund::pffr` or `mgcv::gam`. When specified correctly, the results are identical. The difference in implementation is ease of use. For `refund::pffr`, the point estimates and pointwise standard errors are obtained

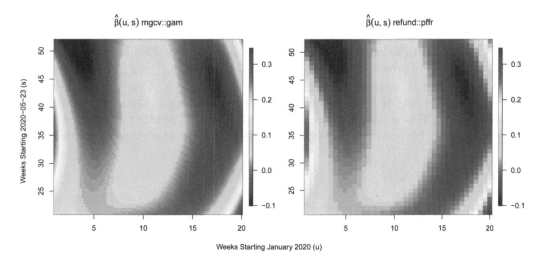

FIGURE 6.11: Estimated $\widehat{\beta}(u,s)$ in the FoFR model (6.1) for the regression of weekly all-cause excess mortality in the US in the last 32 weeks on the first 20 weeks of 2020. Results obtained using `mgcv::gam()` directly (left panel) and `refund::pffr()` (right panel).

using a simple call to the `coef` method associated with `refund::pffr`. Using the code provided in Section 6.4 we can obtain pointwise standard errors for $\widehat{\beta}(s,t)$ which can be used for inference. Here we first discuss unadjusted pointwise inference and then discuss correlation and multiplicity adjusted (CMA) inference for the coefficient surface. We will follow closely the methodological approaches described in Section 2.4 and we will point out specific methods and software adjustments required for FoFR inference.

6.5.1 Unadjusted Pointwise Inference for FoFR

Pointwise inference in the context of the FoFR model can be done by constructing pointwise confidence intervals for a fixed type I error rate α (e.g., 0.05). This can be done in many ways, but one straightforward approach is to construct pointwise 95% Wald confidence intervals for the surface of the form $\widehat{\beta}(u,s) \pm 1.96 \times \text{SE}\{\widehat{\beta}(u,s)\}$ for $u \in [0,20]$, $s \in [21,52]$. Correspondingly, one can obtain pointwise p-values associated with the null hypothesis $H_0 : \beta(u,s) = 0$ for $u \in U$, $s \in S$ using the fact that, under the null, $\widehat{\beta}(u,s)/\text{SE}\{\widehat{\beta}(u,s)\}$ is approximately distributed $N(0,1)$ for large N. We refer readers to [316] for a discussion on the implementation of standard error calculation in `mgcv`.

Let $\boldsymbol{u}_{\text{pred}}$, $\boldsymbol{s}_{\text{pred}}$ denote the vectors that define the grid over which we wish to obtain the standard error for $\widehat{\beta}(u,s)$. When using a tensor product smoother for fixed u,s we have $\beta(u,s) = [\boldsymbol{B}_1(u) \otimes \boldsymbol{B}_2(s)]\boldsymbol{\beta}$, where $\otimes$ denotes the Kronecker product and $\boldsymbol{B}_1(u)$, $\boldsymbol{B}_2(s)$ are the row vectors containing the marginal bases used to model $\beta(u,s)$ evaluated at u and s, respectively. When we evaluate $\beta(u,s)$ over the grid defined by $\boldsymbol{u}_{\text{pred}}, \boldsymbol{s}_{\text{pred}}$ we have $\beta(\boldsymbol{u}_{\text{pred}}, \boldsymbol{s}_{\text{pred}}) = [\boldsymbol{B}_1(u) \otimes \boldsymbol{B}_2(s)]_{(u,s)\in \boldsymbol{u}_{\text{pred}} \times \boldsymbol{s}_{\text{pred}}}\boldsymbol{\beta}$. Here we abused notation for defining $\beta(\boldsymbol{u}_{\text{pred}}, \boldsymbol{s}_{\text{pred}})$, but we will be specific about what we mean. Note that for a fixed (s,u) $\boldsymbol{B}_1(u) \otimes \boldsymbol{B}_2(s)$ is a $1 \times (K_1K_2)$ dimensional vector, while $\boldsymbol{\beta}$ is a $(K_1K_2) \times 1$ dimensional vector, where K_1 and K_2 are the number of bases in the first and second dimension, respectively. The matrix $[\boldsymbol{B}_1(u) \otimes \boldsymbol{B}_2(s)]_{(u,s)\in \boldsymbol{u}_{\text{pred}} \times \boldsymbol{s}_{\text{pred}}}$ is obtained by stacking the $1 \times (K_1K_2)$ dimensional rows $[\boldsymbol{B}_1(u)\otimes\boldsymbol{B}_2(s)]$ for each $u \in \boldsymbol{u}_{\text{pred}}$ and each $s \in \boldsymbol{s}_{\text{pred}}$. Thus, this matrix is

$|\boldsymbol{u}_{\text{pred}}||\boldsymbol{u}_{\text{pred}}| \times (K_1 K_2)$ dimensional, where each row corresponds to a point where prediction is conducted. Therefore, $\beta(\boldsymbol{u}_{\text{pred}}, \boldsymbol{s}_{\text{pred}})$ is a $|\boldsymbol{u}_{\text{pred}}||\boldsymbol{s}_{\text{pred}}| \times 1$ dimensional vector, where each entry corresponds to an estimated functional parameter at a location $(u, s) \in \boldsymbol{u}_{\text{pred}} \times \boldsymbol{s}_{\text{pred}}$.

For simplicity of presentation, let $\boldsymbol{A} = [\boldsymbol{B}_1(u) \otimes \boldsymbol{B}_2(s)]_{(u,s) \in \boldsymbol{u}_{\text{pred}} \times \boldsymbol{s}_{\text{pred}}}$. It follows that for the bivariate grid of points $\boldsymbol{u}_{\text{pred}} \times \boldsymbol{s}_{\text{pred}}$, $\text{Var}\{\widehat{\beta}(\boldsymbol{u}_{\text{pred}}, \boldsymbol{s}_{\text{pred}}\} = \boldsymbol{A}\text{Var}(\widehat{\boldsymbol{\beta}})\boldsymbol{A}^t$. The diagonal of this covariance matrix is all that is needed for constructing unadjusted pointwise confidence intervals. This is provided directly by `coef.pffr` and `predict.gam` (when selecting `type="terms"`). We provide code below showing the implementation using `refund::pffr`.

```
#Get coefficient estimate and pointwise SE
fofr_pffr <- coef(m_2)
#Extract the relevant term from the fitted object
beta_hat_pffr <- fofr_pffr$smterms$`ff(X,s)`$coef
#Create 95% CI and get p-values
beta_hat_pffr <-
    beta_hat_pffr %>%
    mutate(LB = value - qnorm(0.975) * se,
           UB = value + qnorm(0.975) * se,
           p_val = 2 * pnorm(abs(value / se)))
```

The data frame `beta_hat_pffr` created by the code above can then be passed to `ggplot2::ggplot` for plotting, or transformed to matrix format and plotted using, for example, `fields::image.plot` [217]. Figure 6.12 plots the unadjusted pointwise p-values (left panel) along with lower and upper bounds for the 95% confidence intervals (middle and left panels, respectively). First consider the plot of p-values in the left panel of Figure 6.12. Regions that are blue represent areas where the pointwise p-value is less than 0.05, with white regions indicating no statistical significance at the level $\alpha = 0.05$. The regions which are statistically significant indicate that weeks $\approx$ 18-20 are associated with future death rates in the subsequent weeks ($\approx$ 18-25) and weeks toward the end of the year ($\approx$ 50-52). In addition, deaths between weeks 8-13 and 15-18 are significantly associated with future deaths over the majority of the follow-up period. Interestingly, if we look at the estimated coefficient surface presented in Figure 6.11, we can see that the direction of the effect is reversed (positive for weeks 8-13, negative for weeks 15-18). This suggests a cyclic nature to the death rate, consistent with our exploratory findings.

We urge caution in interpreting statistical significance in this data application due to the autocorrelation of residuals highlighted in Section 6.3.1. This autocorrelation violates key assumptions of the model and may impact inference.

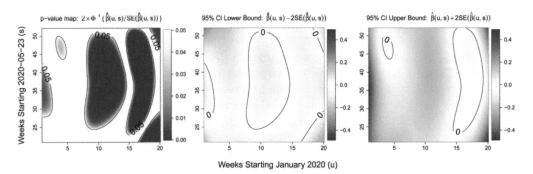

FIGURE 6.12: Unadjusted pointwise inference of FoFR: estimated p-values (left panel) along with lower (middle panel) and upper (right panel) bounds for the 95% pointwise confidence intervals for $\widehat{\beta}(u, s)$ from the FoFR model

We can obtain the same results using the `mgcv` approach discussed in Section 6.4. Specifically, the `predict.gam()` function returns pointwise standard errors which can be used to construct pointwise 95% Wald confidence intervals. The code below constructs both pointwise confidence intervals and p-values using `mgcv`.

```
#Place estimates in matrix format for SE(beta(u,s))
beta_hat_SE <- matrix(coef_mgcv$se.fit[,2],
                      length(uv_pred), length(sv_pred), byrow = FALSE)

#Get lower and upper bounds for pointwise 95% CI
beta_hat_LB <- beta_hat - qnorm(0.975) * beta_hat_SE
beta_hat_UB <- beta_hat + qnorm(0.975) * beta_hat_SE
#Calculate pointwise p-values 95% CI
beta_hat_p <- 2 * pnorm(abs(beta_hat / beta_hat_SE), lower.tail = FALSE)
```

6.5.2   Correlation and Multiplicity Adjusted Inference for FoFR

The unadjusted pointwise inference discussed in Section 6.5.1 is useful for interpretation and building intuition about the signal strength at various points on the surface. However, this approach does not account for the correlation between $\widehat{\beta}(u, s)$ across (u, s) or the multiplicity problem associated with the number of tests conducted.

To address these problems we focus on the correlation and multiplicity adjusted (CMA) inference described in Section 2.4. We consider testing the null hypotheses

$$H_0 : \beta(u, s) = 0, \quad u \in \mathbf{u}_{\text{pred}}, s \in \mathbf{s}_{\text{pred}} .$$

Note that there are $|\mathbf{u}_{\text{pred}}||\mathbf{s}_{\text{pred}}|$ tests that are conducted. This number of tests is arbitrary because it depends on our choice of the bivariate grid considered for estimation. In our case there are 10,000 tests because we have chosen a 100 by 100 grid, but this could have easily been 100 or 10 billion tests. This illustrates the limitation of the "simple" approach of using a Bonferonni correction to control the family-wise error rate (FWER) or the Benjamini-Hochberg false discovery rate correction (FDR) [15]. Indeed, even defining the number of tests that are conducted is meaningless and the width of the confidence intervals depends on this choice. At one extreme, applying the Bonferonni correction for an infinite number of tests is impractical and incorrect, as this would result in confidence intervals that cover the entire real line. Given this discussion, one starts to build the intuition that "there is probably a limit to the practical choice of the density of the grid" and "maybe something should be done about the number of tests". The first part of the intuition is related to the correlation between the tests and the second part is related to the multiplicity of testing.

Some methods exist for addressing this problem. Notably, the `summary` method for `mgcv::gam` fitted objects contains a p-value associated with this null hypothesis [318], though we urge caution in using these p-values in the context of functional regression, as they do not seem to be consistent with p-values obtained from alternative approaches. We will discuss this in more details, but we consider that this area of research is definitely still open.

To address the problem of correlation and multiplicity adjusted (CMA) confidence intervals, three methods were introduced in Section 2.4. The first method discussed in Section 2.4.1 introduced the joint confidence intervals based on multivariate normality. This method requires the calculation of the exact quantiles of a multivariate normal distribution. Unfortunately, in our context this distribution is 10,000 dimensional as we have chosen a 100×100 grid of points. The function `mvtnorm::qmvnorm` required for this calculation does not work for such dimensions. Instead we focus on the other two approaches, joint confidence intervals based on parameter simulations introduced in Section 2.4.2 and joint confidence intervals based on the max absolute statistics introduced in Section 2.4.3. We also discuss some potential pitfalls associated with using the PCA or SVD decomposition of the covariance operator.

Simulations From the Spline Parameter Distribution

Under certain conditions, conditional on the estimated smoothing parameter, the joint distribution of the spline coefficients is approximately normal with mean $\widehat{\boldsymbol{\beta}}$ and variance $\mathrm{Var}(\widehat{\boldsymbol{\beta}})$. Noting that $\widehat{\beta}(\boldsymbol{s},\boldsymbol{t}) = \boldsymbol{A}\widehat{\boldsymbol{\beta}}$, where $\boldsymbol{A}$ is the row-wise concatenated tensor product of marginal bases associated with each element of $\boldsymbol{s}_{\mathrm{pred}} \times \boldsymbol{t}_{\mathrm{pred}}$ (see Section 6.5.1). The parametric bootstrap proceeds quite simply, is presented in Algorithm 2, and follows directly from the approach presented in [57].

Algorithm 2 Algorithm for Simulations from the Spline Parameter Distribution

---

    Input $\boldsymbol{A}, \widehat{\boldsymbol{\beta}}, \widehat{\beta}(\boldsymbol{s},\boldsymbol{t}), \mathrm{Var}(\widehat{\boldsymbol{\beta}}), B$
    Output
  for $b = 1,\ldots,B$ **do**
    $\beta^b \sim N\{\widehat{\boldsymbol{\beta}}, \mathrm{Var}(\widehat{\boldsymbol{\beta}})\}$
    $\beta^b(\boldsymbol{s},\boldsymbol{t}) = \boldsymbol{A}\beta^b$
    $d^b = \max_{s\in\boldsymbol{s}_{\mathrm{pred}}, t\in\boldsymbol{t}_{\mathrm{pred}}} |\beta^b(s,t) - \widehat{\beta}(s,t)|/\mathrm{SE}\{\widehat{\beta}(s,t)\}$
  end for

The simulated test statistic d^b, $b = 1,\ldots,B$ is a sample from the distribution of the maximum of the point-wise Wald statistic over $\boldsymbol{s}_{\mathrm{pred}}, \boldsymbol{t}_{\mathrm{pred}}$. The $1 - \alpha$ quantile of $\{d^b : 1 \leq b \leq B\}$, denoted as $q(\mathbf{C}_\beta, 1 - \alpha)$, then represents the value such that $\mathrm{Pr}_{H_0:\beta(s,t)=0}[\{|\widehat{\beta}(s,t)|/\mathrm{SE}\{\widehat{\beta}(s,t)\} \geq q(\mathbf{C}_\beta, 1 - \alpha) : s \in \boldsymbol{s}_{\mathrm{pred}}, t \in \boldsymbol{t}_{\mathrm{pred}}\}] \approx \alpha$, with the approximation due to the variability associated with the simulation procedure and the normal approximation of the distribution of $\widehat{\boldsymbol{\beta}}$. Here we have used the notation $q(\mathbf{C}_\beta, 1 - \alpha)$ to indicate the dependence of the quantile on the correlation matrix $\mathbf{C}_\beta$ of $\widehat{\beta}(\mathbf{u}_{\mathrm{pred}}, \mathbf{s}_{\mathrm{pred}})$ and to keep the notation consistent with the one introduced in Section 2.4.

Thus, the $1 - \alpha$ level correlation and multiplicity adjusted (CMA) confidence interval is

$$\widehat{\beta}(u, s) \pm q(\mathbf{C}_\beta, 1 - \alpha)\mathrm{SE}\{\widehat{\beta}(u, s)\}$$

for all values $u \in \mathbf{u}_{\mathrm{pred}}$ and $s \in \mathbf{s}_{\mathrm{pred}}$. This confidence intervals can be inverted to form both pointwise and global CMA p-values. Consider first pointwise confidence intervals and fix (u, s). We can simply find the smallest value of α for which the above interval does not contain zero (the null hypothesis is rejected). We denote this probability by $p_{\mathrm{pCMA}}(u, s)$ and refer to it as the pointwise correlation and multiplicity adjusted (pointwise CMA) p-value. To calculate the global pointwise correlation and multiplicity adjusted (global CMA) p-value, we define the minimum α level at which at least one confidence interval $\widehat{\beta}(u, s) \pm q(\mathbf{C}_\beta, 1 - \alpha)\mathrm{SE}\{\widehat{\beta}(u, s)\}$ for $u \in \mathbf{u}_{\mathrm{pred}}$ and $s \in \mathbf{s}_{\mathrm{pred}}$ does not contain zero. We denote this p-value by $p_{\mathrm{gCMA}}(\mathbf{u}_{\mathrm{pred}}, \mathbf{s}_{\mathrm{pred}})$. As discussed in Section 2.4.4, it can be shown that

$$p_{\mathrm{gCMA}}(\mathbf{u}_{\mathrm{pred}}, \mathbf{s}_{\mathrm{pred}}) = \min\{p_{\mathrm{pCMA}}(u, s) : u \in \mathbf{u}_{\mathrm{pred}}, s \in \mathbf{s}_{\mathrm{pred}}\}.$$

Calculating both the pointwise and global CMA adjusted p-values requires only one iteration of the simulation $b = 1,\ldots,B$, though a large number of simulations, B, may be required to estimate extreme p-values. The entire procedure is fairly fast as it does not involve model refitting, simply simulating from a multivariate normal of reasonable dimension, in our case $K_1 K_2$ dimensional. Below we show how to conduct these simulations and calculate the CMA adjusted confidence intervals and global p-values. An essential step is to extract the covariance matrix of $\widehat{\beta}$ from the `mgcv` fit. This is accomplished in the expression `Vtheta <- vcov(fit_mgcv)[inx_beta,inx_beta]`. Recall that a similar approach was used in Chapter 4, which also depended intrinsically on extracting the covariance of the spline coefficients, $\mathrm{Var}(\widehat{\beta})$. The first code chunk focuses on obtaining the CMA confidence intervals at a particular confidence level, in this case $\alpha = 0.05$.

```
#Get the design matrix columns associated with s,t
lpmat <- predict(fit_mgcv, newdata = df_pred_beta, type = 'lpmatrix')
#Get the column indices associated with the functional term
inx_beta <- which(grepl("te\\(Uv,Sv\\):XvL\\.[0-9]+", dimnames(lpmat)[[2]]))
#Get the design matrix A associated with beta(s,t)
Amat <- lpmat[,inx_beta]
#Get Var spline coefficients (beta_sp) associated with beta(u,s)
beta_sp <- coef(fit_mgcv)[inx_beta]
Vbeta_sp <- vcov(fit_mgcv)[inx_beta, inx_beta]
#Number of bootstrap samples
nboot <- 10000
#Set up container for bootstrap
beta_mat_boot <-
        array(NA, dim = c(nboot, length(svec_pred), length(tvec_pred)),
              dimnames = list("boot" = 1:nboot, "Uv" = uv_pred, "Sv" = sv_pred))
#Do the bootstrap
for(i in 1:nboot){
        beta_sp_i <- MASS::mvrnorm(n = 1, mu = beta_sp, Sigma = Vbeta_sp)
        beta_mat_boot[i,,] <- matrix(Amat %*% beta_sp_i, length(uv_pred),
                                     length(sv_pred), byrow = FALSE)
}

#Find the max statistic
dvec <- apply(beta_mat_boot, 1, function(x) max(abs(x - beta_hat) / beta_hat_SE))
#Get 95% global confidence band
Z_global <- quantile(dvec, 0.95)
beta_hat_LB_global <- beta_hat - Z_global * beta_hat_SE
beta_hat_UB_global <- beta_hat + Z_global * beta_hat_SE
```

We now show how to invert these confidence intervals and obtain the global CMA adjusted p-value $p_{\mathrm{gCMA}}(\mathbf{u}_{\mathrm{pred}}, \mathbf{s}_{\mathrm{pred}})$. In practice the p-value we estimate is limited by the number of bootstrap samples we use.

```
#Set up a vector of quantiles to search over
qs_vec <- seq(0.99, 1, len = 10000)
#Get the corresponding quantiles of the max statistic
qs_vec_d <- quantile(dvec, qs_vec)
#Loop over quantiles of max statistic, evaluate which values result
# in global confidence bands covering 0 over all regions of interest
cover_0 <-
   vapply(qs_vec_d, function(x){
            all((beta_hat - x * beta_hat_SE) < 0 & (beta_hat + x * beta_hat_SE > 0))
   }, numeric(1))
#Get the p-value
inx_sup_p <- which(cover_0 == 1)
p_val_g <- ifelse(length(inx_sup_p) == 0, 1 / nboot, 1 - min(qs_vec[inx_sup_p]) )
```

In this example, the p-value $p_{\mathrm{gCMA}}(\mathbf{u}_{\mathrm{pred}}, \mathbf{s}_{\mathrm{pred}})$ is extremely small and we estimate that it is < 0.001. We cannot say more unless we consider an increased number of bootstrap samples, but this resolution of p-value is sufficient for most purposes. The conclusion being that there is a statistically significant association between historical excess mortality and future excess mortality, even after adjusting for the correlation between tests and multiple comparisons.

We now show how to calculate the pointwise correlation and multiplicity adjusted p-value $p_{\mathrm{pCMA}}(\mathbf{u}_{\mathrm{pred}}, \mathbf{s}_{\mathrm{pred}})$. P-values obtained in this fashion are also constrained by the number of bootstrap samples. The R code below calculates this quantity using the results from the parametric bootstrap above.

```
#Quantiles to search over
qs_vec_lg <- seq(0, 1, len = 1000)
#Corresponding quantiles of the max statistic
qs_vec_d_lg <- quantile(dvec, qs_vec_lg)
#Loop over quantiles, search for coverage of 0 pointwise
cover_lg <-
    sapply(qs_vec_d_lg, function(x){
            cover_x <- (beta_hat-x*beta_hat_SE) < 0 & (beta_hat+x*beta_hat_SE > 0)
            matrix(cover_x, length(svec_pred),
            length(tvec_pred), byrow = FALSE)
    }, simplify = "array")
#Loop over grid search, find the max p-value
p_val_lg <-
    apply(cover_lg, c(1, 2), function(x,...){
            inx_gl <- which(x == TRUE)
            ifelse(length(inx_gl) == 0, 1 / nboot, 1 - qs_vec_lg[min(inx_gl)])
    }, qs_vec, nboot)
```

The $p_{\mathrm{pCMA}}(\mathbf{u}_{\mathrm{pred}}, \mathbf{s}_{\mathrm{pred}})$ p-values are stored in the matrix (p_val_lg). Figure 6.13 displays the CMA inference results using the same structure as Figure 6.12 using simulations from the spline parameter distribution. The left panel of Figure 6.13 now presents the pointwise CMA adjusted p-values $p_{\mathrm{pCMA}}(\mathbf{u}_{\mathrm{pred}}, \mathbf{s}_{\mathrm{pred}})$. We find that after adjusting for correlations and multiple comparisons, the regions which remain statistically significant are generally the same, with the exception that the early portion of the history (weeks 0-10) is no longer statistically significant, and the late history (weeks 18-20) is no longer associated with excess mortality at the end of the follow up (weeks 50-52). The middle and right panels in Figure 6.13 display the CMA 95% confidence intervals for $\beta(\mathbf{u}_{\mathrm{pred}}, \mathbf{s}_{\mathrm{pred}})$.

Non-parametric Bootstrap of the Max Absolute Statistic

The non-parametric bootstrap proceeds similarly to the parametric bootstrap, but involves re-estimating the model within each sample. Specifically, the non-parametric bootstrap: (1) re-samples units (in this case states) with replacement; (2) re-estimates the model within each re-sampled data set; (3) extracts the estimated coefficient on the grid of interest; (4) calculates the quantiles of the distribution of the maximum statistic to obtain global inference. This procedure does not require the normality of $\widehat{\beta}$, but implicitly requires an assumption of spherical symmetry of the distribution of $\widehat{\beta}(\mathbf{u}_{\mathrm{pred}}, \mathbf{s}_{\mathrm{pred}})$. The approach was proposed by [258] in the context of nonparametric smoothing and was adapted for functional inference by [53] and used in multiple subsequent papers.

The non-parametric bootstrap of the max statistic can be be used to construct pointwise CMA 95% confidence intervals. To be precise, suppose that we obtain $b = 1, \ldots, B$ non-parametric bootstrap estimates of $\widehat{\beta}(s, t)$, denoted as $\widehat{\beta}^b(s, t)$. Then $\{\widehat{\beta}^b(s, t) : 1 \leq b \leq$

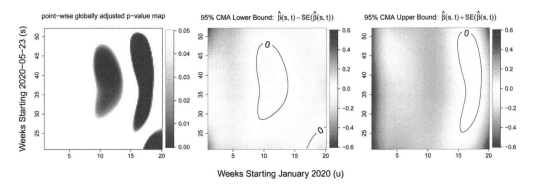

FIGURE 6.13: Estimated pointwise CMA p-values denoted by $p_{\mathrm{pCMA}}(u,s)$ (left panel) along with CMA lower (middle panel) and upper (right panel) bounds for the 95% confidence intervals for $\widehat{\beta}(s,t)$ for FoFR based on simulations from the distribution of spline coefficients

$B, s \in \mathbf{s}_{\mathrm{pred}}, t \in \mathbf{t}_{\mathrm{pred}}\}$ is the collection of bootstrap estimates over the grid $\{\mathbf{s}_{\mathrm{pred}}, t \in \mathbf{t}_{\mathrm{pred}}\}$. As with the parametric bootstrap, we then obtain $d^b = \max_{s \in \mathbf{s}, t \in \mathbf{t}} |\beta^b(s,t) - \widehat{\beta}(s,t)|/\mathrm{SE}\{\widehat{\beta}(s,t)\}$. This procedure requires to extract the variances of $\widehat{\beta}(\mathbf{u}_{\mathrm{pred}}, \mathbf{s}_{\mathrm{pred}})$, but not the entire covariance matrix. However, it requires to refit the model B times, which increases the computational complexity.

The code below illustrates how to perform the non-parametric bootstrap with repeated model estimation done using `refund::pffr`. Using `pffr` makes the non-parametric bootstrap easy to implement as the data are stored in wide format (as opposed to long format for `mgcv::gam` estimation). However, to extract the coefficient predictions on the grid we need to call the function `predict.gam`. To do so, we need to modify the variable names in our data frame which we pass to `predict.gam` to mirror the naming conventions `refund::pffr` uses when it constructs the long format data frame that is passed to `mgcv::gam` for model estimation. The code below shows how to do this. Note that internally `pffr` refers to the predictor domain as S and the response domain as T, where we use U and S to denote the response and predictor domains, respectively.

```
#Set up container for bootstrap
beta_mat_boot <-
        array(NA, dim = c(nboot, length(uv_pred), length(sv_pred)),
              dimnames = list("boot" = 1:nboot, "Uv" = uv_pred, "Sv" = sv_pred))
#Modifiy data frame for calling predict.gam
df_pred_beta_pffr <-
        df_pred_beta %>%
        mutate(t.vec = sv,
               Wpred_i.smat = Uv,
               Wpred_i.tmat = Sv,
               L.Wpred_i = XvL)
#Conduct the bootstrap
for(i in 1:nboot){
   inx_i <- sample(1:nrow(Wpred), size = nrow(Wpred), replace = TRUE)
   Wpred_i <- Wpred[inx_i,]
   Wout_i <- Wout[inx_i,]
   fit_pffr_i <- pffr(Wout_i ~ ff(Wpred_i, xind = s), yind = t)
   beta_i <- predict.gam(fit_pffr_i, newdata = df_pred_beta_pffr, type = "terms")
   beta_mat_boot[i,,] <- matrix(beta_i[,2], length(uv_pred),
                                length(sv_pred), byrow = FALSE)
}
```

Once the bootstrap simulations are obtained, calculations proceed similar to the ones shown for the case when we simulate from the distribution of the spline coefficients. For this reason, we omit this code chunk. Figure 6.14a and 6.14b display the unadjusted and CMA inference using non-parametric bootstrap of the max absolute statistic, respectively. Results are markedly different from the ones obtained using simulations from the spline parameter distribution. Consider first the unadjusted inference shown in Figure 6.14a. We find that fewer regions are statistically significant in Figure 6.14a compared to Figure 6.12, which is based on simulations from the spline parameter distribution. Moreover, Figure 6.14a displays the pointwise CMA inference based on the nonparametric bootstrap of the max statistic and should be compared to Figure 6.13, which is based on simulations from the spline coefficients distribution. Results are remarkably different, indicating that, effectively, no region of the surface is statistically significant. This is due to increased variability of the estimated surface when using a non-parametric bootstrap. Indeed, the p-value associated with the global test using the non-parametric bootstrap is very large, which explains why there are some small regions which are blue in the left panel of Figure 6.14b. This discrepancy between the parametric and non-parametric bootstrap results can occur due to non identifiability of the surface, high variability in smoothing parameter estimation, or heavier than expected tails of the max statistic distribution compared to what would be obtained from a multivariate normal distribution. We note that when the distribution of $\widehat{\boldsymbol{\beta}}$ is reasonably close to a multivariate normal distribution, the non-parametric bootstrap of the max statistic and the simulation from the spline coefficients distribution yield similar estimates. When the smoothing parameter is close to the boundary of 0, then these approaches may give very different results.

Analytic Solution to the Parametric Bootstrap Approximation

We now describe an approach based on the PCA decomposition of the covariance operator that is often used in practice. *However, we will point out problems that can occur in this context and we do not recommend this approach at this time.*

We consider the case when the distribution of $\widehat{\boldsymbol{\beta}}$ is well approximated by a multivariate normal distribution; this can be justified either based on the distribution of the data or by asymptotic considerations. We have already seen that in this application, this assumption is probably incorrect. Regardless, we would like to find an analytic solution to this problem and compare the results with those obtained via simulations from the spline coefficients distribution. Under the assumption of normality of $\widehat{\boldsymbol{\beta}}$ it follows that $\widehat{\beta}(\boldsymbol{s}_{\mathrm{pred}}, \boldsymbol{t}_{\mathrm{pred}}) = \boldsymbol{A}\widehat{\boldsymbol{\beta}}$ has a multivariate normal distribution with covariance matrix $\mathrm{Var}\{\widehat{\beta}(\boldsymbol{s}_{\mathrm{pred}}, \boldsymbol{t}_{\mathrm{pred}})\} = \boldsymbol{A}\mathrm{Var}(\widehat{\boldsymbol{\beta}})\boldsymbol{A}^t$. Using this fact, we can construct the CMA confidence intervals as well as pointwise and global p-values for testing the null hypothesis. H_0 . $\beta(u, s) - 0, \forall s \in \boldsymbol{s}_{\mathrm{pred}}, t \in \boldsymbol{t}_{\mathrm{pred}}$.

Using the same argument discussed in Section 2.4.1

$$[\{\widehat{\beta}(u, s) - \beta(u, s)\}/\mathrm{SE}\{\widehat{\beta}(u, s)\} : s \in \boldsymbol{u}_{\mathrm{pred}}, s \in \boldsymbol{s}_{\mathrm{pred}}] \sim N(0, \mathbf{C}_\beta) \ ,$$

where $\mathbf{C}_\beta$ is the correlation matrix corresponding to the covariance matrix $\boldsymbol{A}\mathrm{Var}(\widehat{\boldsymbol{\beta}})\boldsymbol{A}^t$. As we discussed in Section 2.4.1, we need to find a value $q(\mathbf{C}_\beta, 1 - \alpha)$ such that

$$P\{q(\mathbf{C}_\beta, 1 - \alpha) \times \mathbf{e} \leq \mathbf{X} \leq q(\mathbf{C}_\beta, 1 - \alpha) \times \mathbf{e}\} = 1 - \alpha \ ,$$

where $\mathbf{X} \sim N(0, \mathbf{C}_\beta)$ and $e = (1, \ldots, 1)^t$ is the $|\mathbf{u}_{\mathrm{pred}}||\mathbf{s}_{\mathrm{pred}}| \times 1$ dimensional vector of ones. Once $q(\mathbf{C}_\beta, 1 - \alpha)$ is available, we can obtain a CMA $(1 - \alpha)$ level confidence interval for $\beta(u, s)$ as

$$\widehat{\beta}(u, s) \pm q(\mathbf{C}_\beta, 1 - \alpha)\mathrm{SE}\{\widehat{\beta}(u, s)\} : \forall \ s \in \boldsymbol{u}_{\mathrm{pred}}, s \in \boldsymbol{s}_{\mathrm{pred}} \ .$$

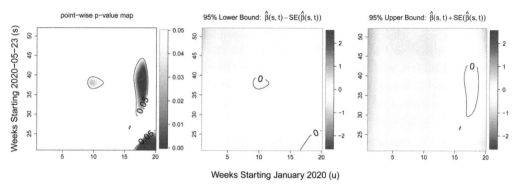

(a) Estimated pointwise unadjusted p-values (left panel) along with unadjusted lower (middle panel) and upper (right panel) bounds for the 95% confidence intervals for $\widehat{\beta}(s,t)$ for FoFR based on non-parametric bootstrap

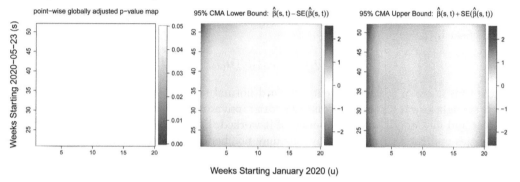

(b) Estimated pointwise CMA p-values denoted by $p_{\mathrm{pCMA}}(u,s)$ (left panel) along with CMA lower (middle panel) and upper (right panel) bounds for the 95% confidence intervals for $\widehat{\beta}(s,t)$ for FoFR based on the non-parametric bootstrap of the max statistic

FIGURE 6.14: FoFR Inference using the non-parametric bootstrap

Luckily, the function `qmvnorm` in the R package `mvtnorm` [96, 97] is designed to extract such quantiles. Unluckily, the function does not work for matrices $\mathbf{C}_\beta$ that are singular and very high dimensional. Therefore we need to find a theoretical way around the problem.

Indeed, $\widehat{\beta}(\mathbf{u}_{\mathrm{pred}}, \mathbf{s}_{\mathrm{pred}}) = \boldsymbol{A}\widehat{\boldsymbol{\beta}}$ has a degenerate normal distribution, because its rank is at most $K_1 K_2$, where K_1 and K_2 are the number of basis functions used in each dimension, respectively. Moreover, we have evaluated $\widehat{\beta}(s,t)$ on a grid of $|\boldsymbol{u}_{\mathrm{pred}}| \times \boldsymbol{s}_{\mathrm{pred}}| = 10{,}000$, which implies that the covariance and correlation matrices of $\widehat{\beta}(\mathbf{u}_{\mathrm{pred}}, \mathbf{s}_{\mathrm{pred}})$ is $10{,}000 \times 10{,}000$ dimensional. To better understand this, consider the case where $\widehat{\beta}(s,t) = s + t + st$. Then $\mathrm{rank}(\mathbf{C}_\beta) \leq 4$ for any choice of $\boldsymbol{u}_{\mathrm{pred}}$ and $\boldsymbol{s}_{\mathrm{pred}}$.

We will use a statistical trick based on the eigendecomposition of the covariance matrix. Recall that if $\widehat{\beta}(\mathbf{u}_{\mathrm{pred}}, \mathbf{s}_{\mathrm{pred}})$ has a degenerate multivariate normal distribution of rank $m \leq K_1 K_2$, then there exists some random vector of independent standard normal random variables $\boldsymbol{Q} \in \mathbb{R}^m$ such that

$$\widehat{\beta}^t(\boldsymbol{s}, \boldsymbol{t}) = \boldsymbol{Q}^t \boldsymbol{D} \,,$$

with $\boldsymbol{D}^t \boldsymbol{D} = \mathbf{A}\mathrm{Var}(\widehat{\boldsymbol{\beta}})\mathbf{A}^t$. If we find m and a matrix $\mathbf{D}$ with these properties, the problem is solved, at least theoretically. Consider the eigendecomposition $\mathbf{A}\mathrm{Var}(\widehat{\boldsymbol{\beta}})\mathbf{A}^t = \boldsymbol{U}\boldsymbol{\Lambda}\boldsymbol{U}^t$, where $\boldsymbol{\Lambda}$ is a diagonal matrix of eigenvalues and the matrix $\mathbf{UU}^t = \mathbf{I}_{|\boldsymbol{u}_{\mathrm{pred}}| \times |\boldsymbol{s}_{\mathrm{pred}}|}$ is an orthonormal matrix with the kth column being the eigenvector corresponding to the kth eigenvalue. Note that all eigenvalues $\lambda_k = 0$ for $k > m$ and

$$\mathbf{A}\mathrm{Var}(\widehat{\boldsymbol{\beta}})\mathbf{A}^t = \boldsymbol{U}_m \boldsymbol{\Lambda}_m \boldsymbol{U}_m^t \, ,$$

where $\boldsymbol{U}_m$ is the $|\boldsymbol{u}_{\mathrm{pred}}| \times |\boldsymbol{s}_{\mathrm{pred}}| \times m$ dimensional matrix obtained by taking the first m columns of $\mathbf{U}$ and $\boldsymbol{\Lambda}_m$ is the $m \times m$ dimensional diagonal matrix with the first m eigenvalues on the main diagonal. If we define $\boldsymbol{D}^t = \boldsymbol{U}_m \boldsymbol{\Lambda}_m^{1/2}$ and

$$\{\widehat{\beta}^t(\boldsymbol{u}_{\mathrm{pred}}, \boldsymbol{s}_{\mathrm{pred}}) - \beta^t(\boldsymbol{u}_{\mathrm{pred}}, \boldsymbol{s}_{\mathrm{pred}})\}\boldsymbol{D}^{-1} = \boldsymbol{Q}^t \, ,$$

where $\boldsymbol{D}^{-1} = \boldsymbol{U}_m \boldsymbol{\Lambda}_m^{-1/2}$ is the Moore Penrose right inverse of $\boldsymbol{D}$ ($\boldsymbol{D}$ is not square, but is full row rank). Under the null hypothesis, $\beta(\boldsymbol{u}_{\mathrm{pred}}, \boldsymbol{s}_{\mathrm{pred}}) = 0$, thus we can obtain a p-value of no effect by calculating

$$1 - \mathrm{P}(\max\{|\boldsymbol{Q}|\} \leq \max\{|\widehat{\beta}^t(\boldsymbol{u}_{\mathrm{pred}}, \boldsymbol{s}_{\mathrm{pred}})\boldsymbol{D}^{-1}|\}) \, ,$$

where $\boldsymbol{Q}$ is a vector of independent standard normal random variables. This calculation is fairly straightforward, though we use `mvtnorm::pmvnorm` for convenience. The mathematics is clear and nice and it would be perfect if it worked. Below we show how to calculate these quantities and then we explain why this may not be a good idea in general. We leave this as an open problem.

Unfortunately, this p-value is not equal to the p-value obtained from simulating from the spline parameters normal distribution. In fact, the p-value obtained using this method is approximately 0 (specifically, $p \approx 6.2 \times 10^{-62}$), which is consistent with the p-value reported by `summary.gam` in the `mgcv` package ($p < 2 \times 10^{-16}$). The p-value reported by `mgcv` is based on a very similar approach which modifies slightly the matrix of eigenvalues of the covariance operator [318] and derives a Wald test statistic which very closely matches the sum of the squared $\boldsymbol{Q}$, a χ^2 random variable under the null hypothesis. In this example, all three approaches approximately agree, though we note that it is unrealistic to perform the number of parametric bootstraps required to estimate a p-value this close to zero, so a discrepancy may exist.

However, we have found that, in general these approaches are not guaranteed to give the same result. We have investigated this problem further and we identified two potential problems. First, the number of non-zero eigenvalues, m, is unknown. One potential solution could be to fix $m = K_1 K_2$, though if some zero or close to zero eigenvalues "sneak in", then the observed statistic will be artificially inflated through the division by these zero eigenvalues (note the multiplication by $\boldsymbol{D}^{-1} = \boldsymbol{U}_m \boldsymbol{\Lambda}_m^{-1/2}$). An alternative would be to stop with $m \ll K_1 K_2$, though it is not clear at what value would be reasonable to stop. Indeed, if $m = 1$, we are back to the case of not adjusting for correlation and multiplicity.

```r
#Get Var(hat(beta))
Vbeta <- Amat %*% Vtheta %*% t(Amat)
#Get eigenfunctions via svd
eVbeta <- svd(Vbeta, nu = ncol(Amat), nv = ncol(Amat))
#Get only positive eigenvalues
inx_evals_pos <- which(eVbeta$d >= 1e-6)
#Get "m" - dimension of lower dimensional RV
m <- length(inx_evals_pos)
#Get associated eigenvectors
U_m <- eVbeta$v[,inx_evals_pos]
#Get D transpose
D_t <- U_m %*% diag(sqrt(eVbeta$d[inx_evals_pos]))
#Get D inverse
D_inv <- MASS::ginv(t(D_t))
#Get Q
Q <- t(Amat %*% theta) %*% D_inv
#Get max statistic
Zmax <- max(abs(Q))
#Get p-value
p_val <- 1 - pmvnorm(lower = rep(-Zmax, m), upper = rep(Zmax, m),
                     mean = rep(0, m), cor = diag(1, m))
```

7

Survival Analysis with Functional Predictors

Survival analysis, also known as time-to-event analysis, is an important area of research which is focused on quantifying the association between predictors and the time to a particular event. The most common type of survival data is observed when studying individuals with a known enrollment time. For some individuals the event of interest and time to event is observed, while for others the event is not observed by the end of the study. The outcome information for individuals who did not experience the event is considered "censored." This is not strictly true for general time-to-event models, as, for example, in cure models, a subset of individuals never experience the event. Here we consider the simplest example of a right censoring without cure subgroup. For example, if the event of interest is mortality, the time from entering the study to mortality is observed for participants who die during the study follow-up period. For participants who did not experience the event (mortality), how long they were in the study is known, which is a lower bound on their true survival time. This type of right censoring is referred to as "administrative censoring." While other types of censoring are possible, here we focus on non-recurrent survival outcomes with non-informative right censoring. Analyzing functional data in the context of other types of survival outcomes (e.g., recurrent events) and censoring (e.g., left censoring, informative right censoring, interval censoring) is an area of open and active research.

Surprisingly, there are very few published methods for analyzing time-to-event data with functional predictors. In particular, [94, 155, 238] proposed different versions of the "linear functional Cox model," which included a linear functional term of the form $\int X_i(s)\beta(s)ds$ in the log-hazard expression to capture the effect of the functional covariate $\{X_i(s), s \in S\}$. Recently [56] introduced the Additive Functional Cox Model, which contains a term of the type $\int F\{s, X_i(s)\}ds$, where $F(\cdot, \cdot)$ is not specified and is estimated from the data. The paper also discusses the identifiability and estimability of such complex models and was inspired by the work of [94] on Cox regression with functional covariates and of [198] on functional generalized additive models for scalar-on-function regression (SoFR).

In this chapter we focus on the same philosophy suggested by the paper of Goldsmith and co-authors [102]. In particular, we expand the functional parameter into a rich spline basis, use quadratic penalties on the smoothness, and employ pseudo-likelihood of the induced mixed effects structures. The original paper of [94] was published in 2015 and developed much earlier. It contained all these connections, but was based on a software package developed from scratch. Through a discussion with Fabien Scheipl, we have become aware of the deployment of the Cox model option in `mgcv` in 2016 [322]. In combination with earlier ideas developed for functional regression, this allowed us to quickly expand the scope and versatility of survival models with functional predictors. Some of these approaches are developed in `refund` and `mgcv`, while others are under active development. Therefore, here we focus on simplicity of implementation and on providing simple code that can be used in a variety of survival data analysis context. Far from being an exhausted field of inference, we believe that this is an area that requires much additional research.

We start by introducing the traditional notation and methods for survival analysis. We then provide examples and extensions to the case when some predictors are high-dimensional functions.

7.1   Introduction to Survival Analysis

The observed outcomes in survival studies are $Y_i = \min(T_i, C_i)$ and the event indicator $\delta_i = I(T_i \leq C_i)$ for each study participant $i = 1, \ldots, n$, where T_i is the survival time, C_i is the censoring time, and $I(\cdot)$ denotes the indicator function. The censoring time, C_i, is often assumed to be independent of the event time, T_i. For example, if the ith study participant dies 3.2 years after entering the study and the study is still ongoing at their time of death, then $T_i = 3.2$ years, $T_i \leq C_i$, and $\delta_i = 1$. The outcome in this case is the pair $(Y_i, \delta_i) = (3.2, 1)$. If the ith study participant is still alive 6.5 years after entering the study and no additional information is available after this time, then $C_i = 6.5$ years, $T_i > C_i$, and $\delta_i = 0$. The outcome in this case is the pair $(Y_i, \delta_i) = (6.5, 0)$.

Given this data structure, we are interested in how the survival time, T_i, is associated with a set of predictors, $\mathbf{Z}_i = (Z_{i1}, \ldots, Z_{iQ})^t$. What makes survival regression different from standard regression is that survival time is positive, often skewed, and incompletely observed because of censoring. For this reason, specialized regression models, such as the Cox proportional hazards model [47], are used instead. The Cox model imposes a structure on the hazard function, $\lambda_i(t|\mathbf{x}_i)$, which is defined as

$$\lambda_i(t|\mathbf{Z}_i) = \frac{f_i(t|\mathbf{Z}_i)}{S_i(t|\mathbf{Z}_i)} \ , \tag{7.1}$$

where $f_i(t|\mathbf{Z}_i)$ is the conditional probability density function of the event time and $S_i(t|\mathbf{Z}_i) = \int_t^\infty f_i(u|\mathbf{Z}_i)du$ is the probability of surviving at least t. Modeling the hazard function is different from standard regression approaches, which focus on imposing structure on the mean of the survival time distribution.

The Cox proportional hazards model assumes that

$$\lambda_i(t|\mathbf{Z}_i) = \lambda_0(t) \exp(Z_{i1}\gamma_1 + \ldots + Z_{iQ}\gamma_Q) = \lambda_0(t) \exp(\mathbf{Z}_i^t \boldsymbol{\gamma}) \ , \tag{7.2}$$

where $\lambda_0(t)$ is the baseline hazard, $\boldsymbol{\gamma} = (\gamma_1, \ldots, \gamma_Q)^t$, and $\exp(\gamma_q)$, $q = 1, \ldots, Q$, are the hazard ratios. A value of $\gamma_q > 0$, or $\exp(\gamma_q) > 1$, indicates that larger values of Z_{iq} corresponds to shorter survival time. Cox models focus primarily on estimating the $\boldsymbol{\gamma}$ parameters using the log partial likelihood

$$l(\boldsymbol{\gamma}) = \sum_{i:\delta_i=1} [\mathbf{Z}_i^t \boldsymbol{\gamma} - \log\{ \sum_{j:Y_j \geq Y_i} \exp(\mathbf{Z}_j^t \boldsymbol{\gamma})\}] \ , \tag{7.3}$$

which does not depend on the baseline hazard, $\lambda_0(t)$. The consistency and asymptotic normality of the maximum log partial likelihood estimator were established in [297].

There are many ways to estimate the baseline hazard, but the best known approach is due to Breslow [22]. Assume that an estimator, $\widehat{\boldsymbol{\gamma}}$, of $\boldsymbol{\gamma}$ is available, say from maximizing the log partial likelihood (7.3). By treating $\lambda_0(t)$ as a piecewise constant function between uncensored failure times, the Breslow estimator of the baseline cumulative hazard function $\Lambda_0(t) = \int_0^t \lambda_0(u)du$ is

$$\widehat{\Lambda}_0(t) = \sum_{i=1}^n \frac{I(Y_i \leq t)}{\sum_{j:Y_j \geq Y_i} \exp(\mathbf{Z}_j^t \widehat{\boldsymbol{\gamma}})} \ . \tag{7.4}$$

The hazard function (7.1) can be rewritten as

$$\lambda_i(t|\mathbf{Z}_i) = -\frac{\partial}{\partial t} \log\{S_i(t|\mathbf{Z}_i)\} \ ,$$

which allows us to obtain the conditional survival function for the Cox model as

$$S_i(t|\mathbf{Z}_i) = \exp\{-\int_0^t \lambda_i(u|\mathbf{Z}_i)du\} = \exp\{-\exp(\mathbf{Z}_i^t\gamma)\Lambda_0(t)\} .$$

This indicates that the conditional cumulative distribution function can be estimated by

$$\widehat{F}_i(t|\mathbf{Z}_i) = 1 - \exp\{-\exp(\mathbf{Z}_i^t\widehat{\gamma})\widehat{\Lambda}_0(t)\} . \tag{7.5}$$

This is a result that can be used for predicting survival time given a set of covariates. Indeed, we can use either the mean or the median of the distribution $\widehat{F}_i(t|\mathbf{Z}_i)$ to predict survival time. There is no closed-form solution for either of these predictors, but they can be obtained easily using numerical approaches. We will also use the survival functions (7.5) to conduct realistic simulations that mimic complex observed data.

The Cox proportional hazards model can be implemented in R using, for example, the `coxph` function in the `survival` package [293, 295].

7.2   Exploratory Data Analysis of the Survival Data in NHANES

7.2.1   Data Structure

We now illustrate the survival data analysis concepts using examples from the NHANES data collected between 2011-2014. As mentioned in Section 1.2.1, the NHANES study was linked to death certificate records from the National Death Index from the National Center for Health Statistics (NCHS). At the time this book was written, the linked mortality files (LMF) were available for public use up to December 31, 2019. Therefore, December 31, 2019 is the administrative censoring induced by the mortality data release date. For the ith study participant, the observed outcomes are (Y_i, δ_i). For censored individuals (individuals who were still alive on December 31, 2019), $\delta_i = 0$ and Y_i is the time in years between the date when data were collected for study participant i (date when the participant entered the study) and December 31, 2019. For deceased individuals (individuals who died before December 31, 2019), $\delta_i = 1$ and Y_i is the time between the date when data were collected for study participant i and the time of death on their death certificate.

We start by illustrating the traditional structure of survival data and we then explain the specific differences associated with high-dimensional functional predictors, such as objective physical activity data obtained from accelerometers.

7.2.1.1   Traditional Survival Analysis

The data set below illustrates the structure of the NHANES data after merging NHANES with the National Death Index. Each study participant is uniquely identified by their sequence number (SEQN) and, for presentation purposes, we only display five covariates: age, body mass index (BMI), race, gender, and coronary heart disease (CHD). The complete NHANES data set contains more variables that are not shown. The survival data (Y_i, δ_i) are stored in the `time` and `event` columns, respectively. For example, the first study participant, SEQN 64759, died ($\delta_1 = 1$), indicated by `event = 1`, $Y_1 = T_1 = 3.33$ years after entering the study, indicated by `time = 3.33`. This is a Non-Hispanic white male who was 80 years old, had a BMI of 24.8, and did not have CHD when he entered the study. The fourth study participant, SEQN 75630, was alive ($\delta_4 = 0$) on December 31, 2019, indicated by `event = 0`, 6.33 years ($Y_4 = C_4 = 6.33$) after entering the study, indicated by `time =`

	SEQN	time	event	age	BMI	race	gender	CHD
1	64759	3.33	1	80	24.8	Non-Hispanic White	Male	No
2	70962	8.17	0	40	45.0	Non-Hispanic Black	Male	No
3	73561	0.75	1	73	19.7	Non-Hispanic White	Female	No
4	75630	6.33	0	59	28.3	Mexican American	Female	No
5	77244	6.58	0	42	25.9	Non-Hispanic Black	Male	No
6	82820	1.50	1	78	26.9	Non-Hispanic White	Male	No

FIGURE 7.1: The structure of the NHANES data for six study participants. Each participant is uniquely identified by their sequence number (SEQN). Each row displays the survival outcome information (time, event) and a subset of predictors: age, BMI, gender, and coronary heart disease indicator (CHD) for one study participant.

6.33. This is a Mexican American female who was 59 years old, had a BMI of 28.3 and did not have CHD when she entered the study.

Figure 7.2 provides a visualization of the survival data structure for the same six study participants. The left panel displays the sequence number (SEQN) of each study participant. The middle panel displays the age, sex, BMI, race, and CHD diagnosis information of the individual at the time of enrollment into the study. This is a different visualization of the same information shown in Figure 7.1. The right panel in Figure 7.2 displays the corresponding survival information. A red bar with an "x" sign at the end indicates that the study participant died before December 31, 2019. The length of the bar corresponds to the time between when the study participant enrolled in the study and when they died. For

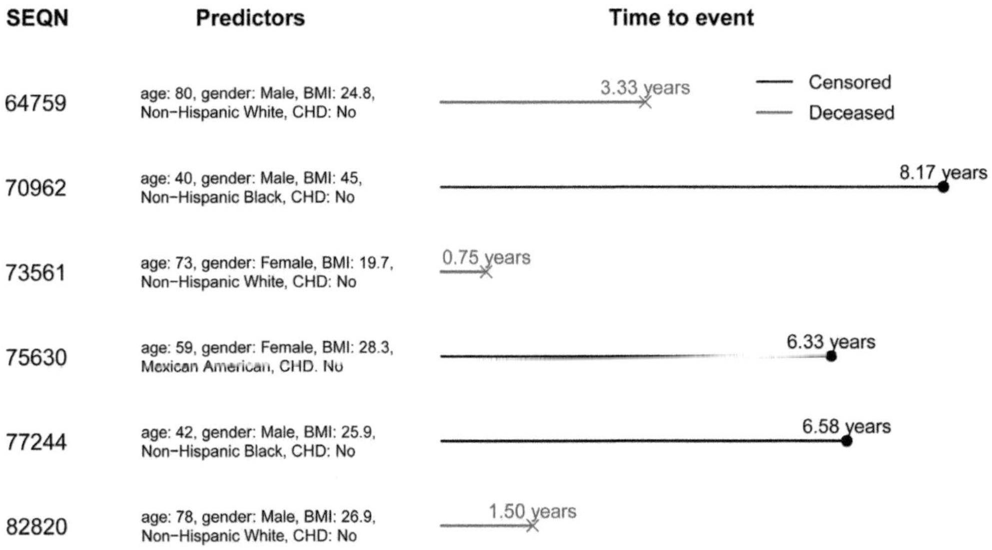

FIGURE 7.2: The left panel displays the sequence number (SEQN) of each study participant. The middle panel displays some baseline predictors: age, BMI, gender, and coronary heart disease indicator (CHD). The right panel displays whether the study participant died (red line) or was still alive (black line) as of December 31, 2019. If the person died, the survival time from entering the study is indicated. If the person was still alive, the time between entering the study and December 31, 2019 (censoring time) is indicated.

example, the first study participant was 80 years old at the time of enrollment and died 3.33 years later. A black bar with a "•" sign at the end indicates that the study participant was alive on December 31, 2019. For example, the fourth study participant was 59 years old at the time of enrollment and was alive 6.33 years later on December 31, 2019. Traditional survival analysis is concerned with analyzing such data structures where risk factors are scalar (e.g., age, BMI). Functional survival data analysis is also concerned with accounting for high-dimensional predictors, such as the minute-level accelerometry data. The next section describes the data structure for survival data with a combination of traditional and functional predictors.

7.2.1.2 Survival Analysis with Functional Predictors

We now describe the specialized structure of the survival data with functional predictors. Consider the NHANES 2011-2014 studies, where study participants were asked to wear a wrist-worn device (ActiGraph GT3X+) for up to 7 consecutive days. The accelerometry data were collected and processed into minute-level MIMS units. The NHANES accelerometry data were originally provided in the "long format" [168], which is memory intensive. We follow the processing pipeline described in [168] and convert the accelerometry data from long to wide format. More precisely, each day of each study participant is stored as a row; here "wide format" refers to the fact that the data contain many columns, each column corresponding to a minute of the day.

As introduced in Section 1.2.1, the NHANES data set contains 12,610 study participants, where the accelerometry data are stored as a matrix with 1,440 columns. Survival data were available for 8,713 of the 12,610 study participants. For the purpose of this chapter, the activity data is further reduced by taking the by-column (minute-specific) average of the data for each study participant. The resulting data consists of a vector of 1,440 minutes for each study participant and a $8,713 \times 1,440$ matrix for all study participants. This approach ignores the day-to-day, within-person variability of the data, but makes the data easier to work with. In Chapter 8, we discuss how to directly model the multi-day (multi-level) accelerometry data using multilevel functional data analysis methods.

Understanding the data structure is a major step towards successfully using functional survival models. Figure 7.3 provides an abbreviated version of the data structure for data frame `nhanes_df_surv`. The first component is identical to that described in Section 7.2.1.1. In our notation this is a data frame, so the familiar variables such as SEQN

```
'data.frame':   8713 obs. of  10 variables:
 $ SEQN  : num  62161 62164 62169 62174 62177 ...
 $ time  : num  7.67 7.42 8.67 1.75 7.08 0.58 8 7.58 7.5 8.25 ...
 $ event : int  0 0 0 1 0 1 0 0 0 0 ...
 $ age   : num  22 44 21 80 51 80 35 26 30 70 ...
 $ BMI   : num  23.3 23.2 20.1 33.9 20.1 28.5 27.9 22.1 22.4 NA ...
 $ race  : Factor w/ 6 levels "Mexican American",..: 3 3 5 3 5 3 3 4 5 4 ...
 $ gender: Factor w/ 2 levels "Male","Female": 1 2 1 1 1 1 1 1 2 1 ...
 $ CHD   : Factor w/ 4 levels "No","Yes","Refused",..: 1 1 1 1 1 1 1 1 1 1 ...
 $ TMIMS : num  13194 13411 9991 8579 11918 ...
 $ MIMS  : 'AsIs' num [1:8713, 1:1440] 1.11 1.92 5.85 5.42 6.14 ...
  ..- attr(*, "dimnames")=List of 2
  .. ..$ : chr [1:8713] "62161" "62164" "62169" "62174" ...
  .. ..$ : chr [1:1440] "MIN0001" "MIN0002" "MIN0003" "MIN0004" ...
```

FIGURE 7.3: An example of the survival data structure in R where accelerometry data is a matrix of observations stored as a single column of the data frame.

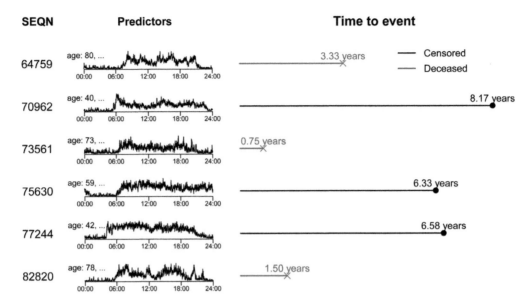

FIGURE 7.4: The left panel displays the sequence number (SEQN) of each study participant. The middle panel displays the average physical activity data from midnight to midnight for six study participants (age at time of study enrollment also shown). The right panel displays whether the study participant died (red line) or was still alive (black line) as of December 31, 2019. If the person died, the time from the physical activity study is indicated. If the person was still alive, the time study enrollment and December 31, 2019 (censoring time) is indicated.

and `event` are columns in the data frame and can be accessed as `nhanes_df_surv$SEQN` and `nhanes_df_surv$event`, respectively. The $8{,}713 \times 1{,}440$ dimensional accelerometry data matrix is stored as an entry in the data frame and can be accessed as `nhanes_df_surv$MIMS`. This can be done using the `I()` function, which inhibits the conversion of objects. Additional functional predictors, if available, could be stored as additional single columns of the data frame. This is the standard input format for the R `refund` package.

Figure 7.4 provides a visual representation of the data described in Figure 7.3. Also, the information displayed is similar to that shown in Figure 7.2. The difference is that physical activity information data was added to the middle panel. The left panel still contains the `SEQN` number, which provides the primary correspondence key for data pertaining to individuals. The right panel contains the survival data and is identical to the right panel in Figure 7.2.

Given this data structure, it is useful to consider the options that would make sense for fitting such data. Indeed, simply plugging in the high-dimensional predictors with complex correlations (e.g., physical activity data) in the right-hand side of the Cox proportional hazards model (7.2) is not feasible. One solution could be to extract low-dimensional summaries of the data (e.g., mean, standard deviation) and use them as predictors in the Cox model. Another solution could be to decompose the functional data using principal components (possibly smoothed) and plug in the principal component scores in the Cox regression [139, 155, 252]. As in standard functional regression, the shape and interpretation of the functional coefficient depends strongly on the number of principal components. Therefore, in addition to these approaches, we will consider models that directly estimate the functional coefficient using penalized regression. To build the intuition, we further explore the

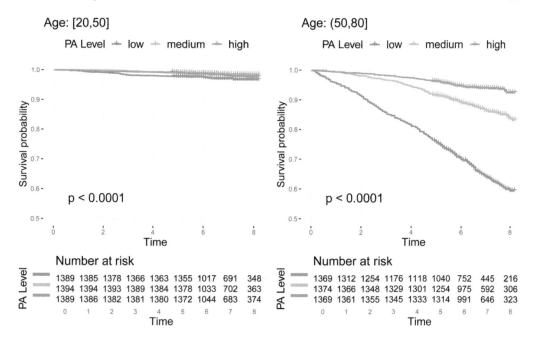

FIGURE 7.5: The Kaplan-Meier Curves of NHANES 2011-2014 population stratified by physical activity intensity and age. Left panel: age 20-50; right panel: age 50-80. Within each age group, study participants were combined in three subgroups based on the tertiles of the total MIMS (TMIMS). KM survival probability estimates for low TMIMS (red curve), medium TMIMS (green curve) and high TMIMS (red curve) are displayed. The tables below each panel display the number of study participants at risk as a function of time by PA group.

data and provide a progression of models that build upon the standard Cox proportional hazards model.

7.2.2 Kaplan-Meier Estimators

We further explore the NHANES 2011-2014 survival data by investigating the Kaplan-Meyer (KM) estimators [141] of the survival function for specific subgroups. Figure 7.5 displays results for different age groups at baseline: 20-50 years old (left panel), and 50-80 years old (right panel). Within each age group, the population was divided into three subgroups corresponding to three levels of physical activity intensity. More precisely, the total MIMS (TMIMS) was obtained for each individual as a proxy for the total volume of their activity. TMIMS was calculated for each day and then averaged across days with valid accelerometry data within individuals. The tertiles of TMIMS were obtained within each age group separately. These age-specific tertiles were used to obtain three sub-groups within each age group corresponding to low (red KM curve), moderate (green KM curve), and high (blue KM curve) physical activity intensity, respectively.

The information about the KM estimators in each panel is supplemented with the number of study participants who are still at risk after a given number of years. For example, for the 50- to 80-year-old age group there were 1,369 study participants in the low PA subgroup and 1,374 in the medium PA subgroup at the time of recruitment into the study. In the same age group there were 1,118 study participants who were still alive 4 years later in the low PA group. In contrast, there were 1,333 study participants who were still alive 4 years later in the high PA group.

For the study participants between the ages of 20 and 50 the probability of survival is very high at all times irrespective of the physical activity level group. Moreover, there are no visual differences between the KM curves. This is reassuring, as it corresponds to what is expected among young adults. For study participants between the ages of 50 and 80 the survival probabilities are much lower and there are substantial differences between the three PA subgroups. The survival probability in the high PA group is higher than in the moderate PA group and much higher than in the low PA group.

7.2.3 Results for the Standard Cox Models

The exploratory data visualization in Section 7.2.1 indicates that age and objective summaries of physical activity may be associated with survival time. We now investigate three Cox survival models with an increasing number of covariates. The first model (M1) has three predictors: age, as a continuous covariate, gender, as a binary covariate, and race as a categorical covariate with six categories: Mexican American, Non-Hispanic White, Non-Hispanic Black, Non-Hispanic Asian, Other Hispanic, and Other Race. The reference category for gender is Male, while for race it is Mexican American. The second model (M2) also includes body mass index (BMI) as a continuous covariate and coronary heart disease (CHD), as a categorical covariate with two categories No and Yes. The reference category for CHD is No CHD. The third model (M3) adds total MIMS (TMIMS), as a global measure of objective physical activity. All covariates are used as recorded in NHANES and the definition and categories, especially for gender and race, may be refined in future NHANES studies. For this analysis, we focus on study participants who were older than 50 and had no missing data for these selected covariates. The data set included 4,180 study participants with an average follow-up time of 6.47 years (minimum = 0 years, maximum = 9.25 years).

Table 7.1 provides the point estimator and associated p-values for all covariates in models M1-3. Age is statistically significant in all models and indicates that older individuals are at higher risk of dying. However, the effect is substantially attenuated when one accounts for

TABLE 7.1: Model 1 shows the coefficient and p-values from a Cox proportional hazard model assessing the association between age, gender, race and risk of death. Model 2 expands model 1 by adding BMI and CHD status to the model. Model 3 expands model 2 by adding total MIMS (TMIMS) to the model. The baseline level of race is Mexican American, and the baseline level of CHD is No (not diagnosed with CHD at the time of interview).

Models	M1		M2		M3	
C-index	0.754		0.760		0.783	
	β	p	β	p	β	p
age	**0.102**	**<0.001**	**0.098**	**<0.001**	**0.071**	**<0.001**
gender: Female	**−0.273**	**<0.001**	**−0.214**	**0.005**	−0.060	0.442
race: Non-Hispanic White	0.211	0.196	0.183	0.263	0.072	0.659
race: Non-Hispanic Black	0.180	0.294	0.210	0.222	0.084	0.626
race: Non-Hispanic Asian	−0.390	0.090	−0.412	0.075	−0.469	0.043
race: Other Hispanic	−0.170	0.432	−0.163	0.450	−0.149	0.489
race: Other Race	0.543	0.070	0.519	0.084	0.437	0.146
BMI			−0.006	0.384	**−0.025**	**<0.001**
CHD: Yes			**0.568**	**<0.001**	**0.441**	**<0.001**
TMIMS					**−0.0002**	**<0.001**

TMIMS (model M3). Indeed, the hazard is estimated to increase by $\exp(0.102) = 1.108$ (or 10.8%) for a 1 year increase in age in M1. In M3 the increase in hazard is $\exp(0.071) = 1.073$ (or 7.3%) for individuals with the same level of TMIMS (and other covariates). This result shows that there is a strong association among the variables age, TMIMS, and mortality. It also suggests physical activity, a modifiable risk factor, may be a target for interventions designed to reduce mortality.

The point estimator for "gender: Female" is negative in all models, which corresponds to a lower risk of mortality. The result is not surprising as it is well known that women live longer than men. Indeed, in the US the life expectancy at birth for females has been about 5 years longer than for males; see Figure 1 in [7] on Life expectancy at birth by sex in the US 2000–2020. In our data, this effect is not statistically significant in models M3 (p-value=0.442). The reduction in the point estimate for the log hazard is substantial when accounting for CHD. However, the reduction in the point estimator is extraordinary when accounting for TMIMS.

In these data, none of the race categories had a significantly different mortality hazard compared to the reference category Mexican American.

Results for BMI are qualitatively different in models M2 and M3. Indeed, in M2 there is no statistically significant association (at significance level $\alpha = 0.05$) between BMI and hazard of mortality. However, in model M3, which adjusts for TMIMS, there is an estimated protective effect of increased BMI. Such effects are known in the literature as the "obesity paradox" [215, 220, 225]. Our models were not designed to address this paradox, but it does suggest that (1) after accounting for age, gender, race, and CHD the obesity paradox is not statistically significant; and (2) after accounting for an objective summary of physical activity, the obesity paradox is statistically significant with a large change in the point estimator. This suggests a strong dependence between BMI, objective physical activity, and mortality risk.

Both models M2 and M3 indicate that having and reporting a history of CHD (category Yes) is a strong, statistically significant predictor of mortality. The hazard ratio in model M2 is estimated to be $\exp(0.568) = 1.76$ relative to study participants in the CHD No category. In M3 the increase in hazard is $\exp(0.441) = 1.55$ compared to individuals of the same age, gender, race and TMIMS levels. The reduction in the estimated hazard rate between model M2 and M3 is due to the introduction of the objective physical activity summary TMIMS.

TMIMS is highly significant (p-value< 0.001) after accounting for age, gender, race, BMI, and CHD. Including this variable substantially affects the estimated effect and interpretation of all other variables, except race. This indicates that objective measures of physical activity are important predictors of mortality and that they interact in complex ways with other traditional mortality risk factors. The effect on hazard ratio is not directly interpretable from the point estimator because MIMS are not expressed in standard units. However, the estimator indicates a covariate-adjusted hazard ratio of $\exp\{-0.0002 \times (Q_3 - Q_1)\} = 0.36$, when comparing a study participant at the third quartile with one at the first quartile of TMIMS. As a comparison, the covariate-adjusted hazard ratio is $\exp\{0.071 \times (Q_1 - Q_3)\} = 0.48$ when comparing a study participant at the first quartile with one at the third quartile of age.

In model M3 we combine the objective physical activity data with other predictors by taking the summation of activity intensity values across time of day. While the newly created TMIMS variable is interpretable under the Cox proportional hazard model, one can argue that we lose substantial information on within-day physical activity variability through such compression. One solution is to calculate the total MIMS in two-hour windows (0-2 AM, 2-4 AM, ...) and use them as 12 separate predictors in a Cox proportional hazard model. However, the interpretation of these predictors becomes challenging due to their intrinsic correlation between adjacent hours. For example, the TMIMS at 2-4 AM is very close to the TMIMS at 4-6 AM for most study participants. In addition, we still lose the variability within each two-hour window.

Another solution is to directly model the minute-level physical activity data, as they can be viewed as functional data observed on a regular grid (minute) of the domain (time of day from midnight to midnight). However, this leads to both methodological and computational challenges, since it is unreasonable and impractical to simply include minute-level intensity values as 1440 predictors in a Cox proportional hazard model. To solve these challenges, we introduce penalized functional Cox models in the next section, which model the functional predictors observed at baseline and survival outcome through penalized splines. We also describe the detailed implementation of these models using the `refund` and `mgcv` software.

7.3   Cox Regression with Baseline Functional Predictors

Motivated by the NHANES data, we consider the case when we have both scalar and functional predictors measured at baseline, along with a possibly censored time to event outcome. For example, Figure 7.4 displays data from the NHANES 2011-2014. The middle panels contain the average MIMS at the minute level for six study participants. The x-axis is time from midnight (00:00) to midnight (24:00). The middle panels also contain information about the age at the time when physical activity was measured. Suppose we wish to model the association between MIMS profiles (our functional predictor) and time to non-accidental, all-cause mortality.

7.3.1   Linear Functional Cox Model

To model such data, we extend the scalar-on-function regression model discussed in Chapter 4 to time-to-event outcomes by combining the classical Cox regression model with the linear functional model. Notation is the same with that introduced in Section 7.1 with the addition of the functional predictor $W_i : S \to \mathbb{R}$ for each study participant, $i = 1, \ldots, n$. In NHANES, $W_i(s)$ is the average over valid days of MIMS at minute s of the day for study participant i. While the methods are written for a function of time, the same approach can be applied to any other univariate or multivariate functional domain. With this notation, we introduce the following functional linear Cox model

$$\log \lambda_i \{t | \mathbf{Z}_i, W_i(\cdot)\} = \log\{\lambda_0(t)\} + \mathbf{Z}_i^t \gamma + \int_S W_i(s)\beta(s)ds \ . \tag{7.6}$$

In this model neither $W_i(s)$ nor $\beta(s)$ depend on t, the domain of the event time process. This can be a bit confusing, as $W_i(s)$ is a function of time, though s refers to time within the baseline visit day, whereas t refers to time from the baseline visit. Therefore, the proportional hazards assumption of the Cox model is reasonable and could be tested.

Interpreting the association between the functional predictor and the log hazard follows directly from the interpretation of the functional linear model. More precisely, one interpretation is that the term $\exp \int_S \beta(s)ds$ corresponds to the multiplicative increase in one's hazard of death if the entire covariate function, $W_i(s)$, was increased by one (with $\mathbf{Z}_i$ held constant). More generally, $\beta(s)$ is a weight function for $W_i(s)$ to obtain its overall contribution towards the hazard of mortality over the functional domain S. It may be helpful to center $W_i(s)$ such that $\beta(s)$ can be interpreted as a unit change in the log hazard relative to some reference population (e.g., the study population average).

Surprisingly, there are few published methods for estimating Model (7.5). In particular, [94, 155, 238] proposed different versions of the "linear functional Cox model", which included a linear functional term of the form $\int_S W_i(s)\beta(s)ds$ in the log-hazard expression to capture the effect of the functional covariate $\{W_i(s) : s \in S\}$. Here we focus on estimating the linear functional Cox model based on penalized regression splines, though we also touch on the functional principal components basis method proposed by [155].

7.3.1.1 Estimation

The first step of the estimation procedure is to expand the functional coefficient as $\beta(s) = \sum_{k=1}^K \beta_k B_k(s)$, where $B_1(s), \ldots, B_K(s)$ is any functional basis. We will use spline bases and penalize the roughness of the functional coefficient $\beta(\cdot)$ via a quadratic penalty on the β_k, $k = 1, \ldots, K$ parameters, though other approaches have been used in the literature.

With this basis expansion, model (7.5) becomes

$$\log \lambda_i\{t|\mathbf{Z}_i, W_i(\cdot)\} = \log\{\lambda_0(t)\} + \mathbf{Z}_i^t\boldsymbol{\gamma} + \int_S W_i(s)\sum_{k=1}^K \beta_k B_k(s)ds . \tag{7.7}$$

A popular alternative is to expand both the functional predictor, $W_i(s)$, and the functional coefficient, $\beta(s)$, in the space spanned by the first K functional principal components of the functional predictor. With this approach, model (7.7) becomes a standard Cox regression on the first K scores for each participant, estimated via standard Cox regression software (e.g., the `survival` package). This approach tends to perform well in terms of prediction, though it can provide functional coefficient estimates that can vary wildly with the choice of the number of principal components. We avoid this problem by expanding the functional coefficient in a rich enough basis and then penalizing the roughness of the coefficient directly. The philosophy is similar to the semiparametric smoothing [258], but applied to functional data. This is a subtle point that is likely to become increasingly accepted.

We now provide the details of our approach, which builds on the penalized Cox model introduced by [94]. Recall that we only observe $W_i(s)$ at a finite number of points $s_1, \ldots, s_p$. Therefore, the following approximation can be used for model (7.7)

$$\log \lambda_i\{t|\mathbf{Z}_i, W_i(\cdot)\} = \log\{\lambda_0(t)\} + \mathbf{Z}_i^t\boldsymbol{\gamma} + \int_S W_i(s)\sum_{k=1}^K \beta_k B_k(s)ds$$

$$\approx \log\{\lambda_0(t)\} + \mathbf{Z}_i^t\boldsymbol{\gamma} + \sum_{j=1}^p q_j W_i(s_j)\sum_{k=1}^K \beta_k B_k(s_j)$$

$$= \log\{\lambda_0(t)\} + \mathbf{Z}_i^t\boldsymbol{\gamma} + \sum_{k=1}^K \beta_k \left\{\sum_{j=1}^p q_j W_i(s_j)B_k(s_j)\right\} , \tag{7.8}$$

where q_j is the quadrature weight. Note that $\sum_{j=1}^p q_j W_i(s_j)B_k(s_j) = C_{ik}$ is a random variable that depends on the study participant, i, and on the basis function number, k. This becomes a standard Cox regression on the covariates $\mathbf{Z}_i$ and $\mathbf{C}_i = (C_{i1}, \ldots, C_{iK})^t$. Because K is typically large, maximizing the Cox partial likelihood results in substantial overfitting. The penalized spline approach addresses overfitting by adding a penalty on the curvature of $\beta(s)$. This penalty is added to the Cox partial log likelihood of (7.3), resulting in the penalized partial log likelihood

$$l(\boldsymbol{\gamma}, \boldsymbol{\beta}) = \sum_{i:\delta_i=1}[(\mathbf{Z}_i^t\boldsymbol{\gamma} + \mathbf{C}_i^t\boldsymbol{\beta}) - \log\{\sum_{j:Y_j \geq Y_i}\exp(\mathbf{Z}_j^t\boldsymbol{\gamma} + \mathbf{C}_j^t\boldsymbol{\beta})\}] - \lambda P(\boldsymbol{\beta}) . \tag{7.9}$$

For a given fixed basis there are multiple choices for the penalty term. The `mgcv` package offers a wide variety of bases and penalty terms, including quadratic penalties of the form $\lambda P(\boldsymbol{\beta}) = \lambda \boldsymbol{\beta} \mathbf{D} \boldsymbol{\beta}^t$ where $\mathbf{D}$ is a known matrix and λ is a scalar smoothing parameter. This form can easily be extended to multiple smoothing parameters. Penalties of this form are quite flexible and include difference penalties associated with penalized B-splines [71], second derivative penalties associated with cubic regression splines and thin plate regression splines [221, 258, 303, 313] and many more [319]. The balance between model fit (partial log likelihood) and the smoothness of $\beta(\cdot)$ (penalty term) is controlled by the tuning parameter λ, which can be fixed or estimated from the data. There are many methods for smoothing parameter selection in nonparametric regression, though fewer were developed and evaluated for the linear functional Cox model. For example, [94] proposed to use an AIC-based criterion for selecting λ. Here, we use the marginal likelihood approach described in [322].

From a user's perspective, the challenge is how to make the transition from writing complex models to actually setting-up the software syntax that optimizes the correct penalized log likelihood. This is a crucial step that needs special attention in line with our philosophy: "writing complex models is easy, but choosing the appropriate modeling components and implementing them in reproducible software is not."

As with scalar-on-function regression for exponential family outcomes, the `refund` package makes model estimation easy. Suppose we have a data frame of the form presented in Figure 7.3 and we wish to fit the linear functional Cox model with age, BMI, gender and CHD as scalar predictors, and participant average minute-level MIMS as a functional predictor. This model can be fit using the `refund::pfr()` function as follows:

```
#Fit a linear functional Cox regression
fit_lfcm_pfr <- pfr(time ~ age + BMI + gender + CHD +
              lf(MIMS, bs = "cc", k = 30),
              weights = event, data = nhanes_df_surv,
              family = cox.ph())
```

The `refund` package uses `mgcv` for the model estimation routine and survival outcomes are specified differently than in most other R survival regression functions. Specifically, `mgcv` currently only accepts right censored survival data indicated by the `event` variable specified in the `weights` argument. Otherwise, syntactically the model is written exactly as it would be for any other response distribution outcome. That is, we specify time to event (`time`) as the outcome, which is a linear function of scalar predictors `age`, `BMI`, `gender`, `CHD`, and a linear functional term of the functional data `MIMS` stored as a matrix within the data frame `nhanes_df_surv`. The functional parameter $\beta(\cdot)$ is modeled using cyclic splines (`bs = "cc"`) since time of day, the functional domain, is cyclical, using $K - 30$ basis functions (`k = 30`). The last component is to specify the statistical model using the `family = cox.ph()` argument.

As with Chapters 4-6, although the utility of the `refund` package makes model fitting straightforward, it is instructive to understand what `refund` is doing "under the hood." To see this, we fit the same model using `mgcv::gam()` directly. To do so, in addition to the data inputs required for estimation via `pfr`, we require (1) the matrix associated with the functional domain $\mathbf{S} = \mathbf{1}_n \otimes \mathbf{s}^t$ where $\mathbf{s}^t = [s_1, \ldots, s_p]$ is the row vector containing the domain of the observed functions (in the case of the NHANES data, $p = 1440$ and $\mathbf{s}^t = [1, \ldots, 1440]/60$) and $\otimes$ denotes the Kronecker product; and (2) a matrix containing the quadrature weights associated with each functional predictor multiplied elementwise by the functional predictor $\mathbf{W_L} = \mathbf{W} \odot \mathbf{L}$ where $\mathbf{W}$ is the matrix with each row containing one participants' MIMS profiles and $\mathbf{L} = \mathbf{1}_n \otimes \mathbf{q}^t$ with $\mathbf{q}^t = [q_1, \ldots, q_p]$ is the row vector containing the quadrature weights. The code below constructs these matrices and adds them to the data frame which we will pass to `mgcv`. Note that these required data inputs are precisely the same as were required for fitting scalar-on-function regression in Chapter 4 with one functional predictor.

```
nS <- ncol(nhanes_df_surv$MIMS) # dimension of the functional predictor
N <- nrow(nhanes_df_surv) # number of samples
svec <- seq(0, 1, len = nS) # observed points on the functional domain
#Quadrature weights for Riemann integration
lvec <- matrix(1 / nS, 1440, 1)
L <- kronecker(matrix(1, N, 1),t(lvec))
S <- kronecker(matrix(1, N, 1), t(svec))
nhanes_df_surv$S <- I(S)
#Pointwise product of w_i(s) and the quadrature weights, "L"
nhanes_df_surv$W_L <- I(nhanes_df_surv$MIMS * L)
```

In our example, the functions are observed on a regular, equally spaced grid, so constructing the matrices of interest is relatively straightforward. The choice of the functional domain is effectively arbitrary, and here we choose $S = [0, 1]$ to match the default behavior of `refund::pfr()`. Therefore, we start with an equally spaced grid on $[0, 1]$ stored in the `svec` variable. This vector corresponds to the grid $s_j, j = 1, \ldots, p$, where p corresponds to `nS` in the R code. Given our choice of S, the quadrature weights associated with Riemann integration are simply $1/1440$ (1440 observations per function) and are stored in the `lvec` variable. This vector corresponds to the quadrature weights $q_j, j = 1, \ldots, p$. These vectors are common to all study participants, and they are simply stored by row as a matrix in our data frame as the variables `S` and `L` variables, respectively. The last step is to create a matrix corresponding to the pointwise product of the quadrature weights and the functional predictor, stored as `W_L`. The matrix `W_L` is the $n \times p$ dimensional matrix with the (i, j)th entry equal to $q_j W_i(s_j)$. Only minor changes to the code above would be necessary for irregular data or other quadrature weights (e.g., trapezoidal, Simpson's rule). With these data objects constructed, we can fit the linear functional Cox model using `mgcv` directly.

```
#Fit the LFCM using mgcv
fit_lfcm_mgcv <- gam(time ~ age + BMI + gender + CHD +
                     s(S, bs = "cc", k = 30, by = W_L),
                     weights = event, data = nhanes_df_surv,
                     family = cox.ph())
```

This syntax is similar to that used by `refund::pfr()` with the exception of the code used to construct the linear functional term. We now connect the `mgcv` syntax to the penalized partial log likelihood for our Cox model. First, note that the `s()` function is the `mgcv` smooth term constructor. By specifying `s(S, bs = "cc", k = 30, ...)`, `mgcv` effectively adds the term $\sum_{k=1}^{K=30} \beta_k B_k(s_j)$ to the linear predictor where the basis, $B_k(\cdot)$, contains cyclic cubic splines (see `?smooth.terms`; other bases can be specified using the `bs` argument). The final component is the `by` argument within the call to `s()`, which instructs `mgcv` to multiply each term specified in the smooth constructor by the corresponding object, in this case `W_L`. Recall that `W_L` is an $n \times p$ dimensional matrix with the (i, j)th entry equal to $q_j W_i(s_j)$ in the notation of (7.8). Together, this syntax adds the term

$$\sum_{j=1}^{p} \{\sum_{k=1}^{K} \beta_k B_k(s_j)\}\{W_i(s_j)q_j\} = \sum_{k=1}^{K} \beta_k \sum_{j=1}^{p} \{q_j W_i(s_j) B_k(s_j)\}$$

to the linear predictor, which corresponds to the functional component in model (7.8). The final component is the penalization. This is added automatically to the log likelihood by

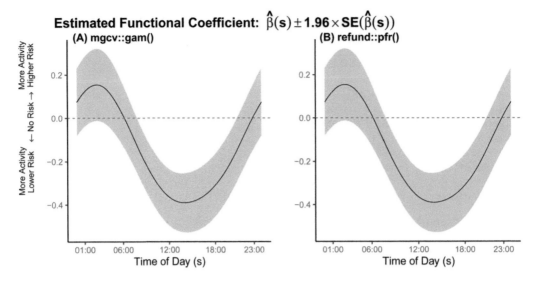

FIGURE 7.6: Estimated functional coefficient, $\widehat{\beta}(s)$, obtained by calling (A) `mgcv::gam()` directly; and from (B) `refund::pfr()`. Point estimates are shown as solid black lines and the shaded regions correspond to 95% pointwise unadjusted confidence intervals. A red dashed line corresponds to a null effect of $\beta(s) = 0$ for every s.

the `s()` function, though an unpenalized fit could be specified by supplying the argument `fx = FALSE` to the call to `s()`.

We plot the results of our two models and demonstrate their nearly identical estimates for $\widehat{\beta}(s)$ in Figure 7.6. Although the plots look very similar, they are not identical because `pfr` uses trapezoidal integration by default instead of midpoint Riemann integration. A version of the plots presented in Figure 7.6 can be created quickly using `plot.gam()`, a plot function specifically for the `gam` class objects. Notice that the plot function for `pfr` internally calls `plot.gam()` as well. In the following sections, we describe how to manually extract point estimates and their corresponding standard error estimates for the functional coefficient, as well as predicted survival probabilities. This will allow us to make personalized plots outside of the `plot.gam()` function, such as Figure 7.6, and allow the extraction of important components that can be used for other inferential purposes.

7.3.1.2   Inference on the Functional Coefficient

Inference on Cox regression models may be performed in a number of ways. Frequently these inferences are based on the limiting behavior of maximum likelihood coefficient estimators. Specifically, we use the assumption that $\widehat{\beta}$ have an approximately normal distribution with variance $\text{Var}(\widehat{\beta}) \approx \widehat{\mathcal{I}}_{\beta}^{-1}$, the inverse of the observed Fisher Information matrix (negative second derivative of the log-likelihood evaluated at $\widehat{\beta}$). For a fixed smoothing parameter, this result holds for scalar-on-function regression models. Alternatively, a non-parametric bootstrap approach may be used to estimate variability without conditioning on the estimated smoothing parameter(s) of the model. We discuss below how to apply both approaches to perform unadjusted and correlation and multiplicity adjusted (CMA) inference on the functional coefficient. All approaches presented use the `mgcv` fitted object. The reason is that, as of this writing, the `refund` package does not support the required functionality in a more straightforward way than `mgcv` to extract the relevant quantities.

Unadjusted Pointwise Inference

The asymptotic normality of spline coefficients, conditional on the smoothing parameter, allows for simple construction of point-wise confidence intervals for $s \in S$. Thus, for a fixed $s \in S$, we can obtain the estimated variance of $\widehat{\beta}(s)$ as

$$\mathrm{Var}\{\widehat{\beta}(s)\} = \mathrm{Var}\{\boldsymbol{B}(s)\widehat{\boldsymbol{\beta}}\}$$
$$= \mathbf{B}(s)\mathrm{Var}(\widehat{\beta})\mathbf{B}(s)^t ,$$

where $\mathbf{B}(s) = [B_1(s), \ldots, B_K(s)]$ is the row vector containing the basis functions used to model $\beta(s)$ evaluated at s. These confidence intervals retain the properties of smooth terms estimated via penalized regression splines in generalized additive models [216, 258]. That is, the coverage is nominal when averaged across the domain, S, but may be above or below the nominal level for any given s. As such, users should be cautious of over-interpreting inference confidence intervals constructed using this procedure.

The `pfr()` function has a `coef.pfr()` method, which allows for easy extraction of both coefficient estimates and the associated standard errors. Obtaining estimated standard errors from the `mgcv::gam()` fit requires more work, but can be done using the functionality of the `predict.gam` method. The first step with both approaches is to create a vector of points on the functional domain, s_{pred}, where we want to make predictions. From there, a simple call to `coef.pfr()` provides the `pfr()` results. For the `mgcv::gam()` fit, a deeper dive is necessary into the functionality of `predict.gam()`. Specifically, the `predict.gam()` function allows users to obtain contributions to the linear predictor associated with each row of a new data frame supplied to the function. First, recall the form of the linear predictor presented in (7.13), for a given s_j. If we create a data frame with $W_i(s_j) = 1$ and $q_j = 1$, then the contribution for the "new" data is simply $\widehat{\beta}(s_j) = \sum_{k=1}^{K} \widehat{\beta}_k B_k(s_j)$. Thus, we need only create a new data frame according to this procedure with each row corresponding to each element of s_{pred}. The code below shows how to obtain the fits for `pfr()` and `mgcv`.

```
#Establish the set of "s" values we wish to predict on
#Number of points on the functional domain (s) to predict
slen_pred <- 100
#Actual s values used for prediction
s_pred <- seq(0, 1, len = slen_pred)
#pfr: estimated coefficients and standard errors
coef_lfcm_pfr <- coef(fit_lfcm_pfr, , n = slen_pred, useVc = FALSE)
#mgcv: estimated coefficients and standard errors
df_lfcm_gam <- data.frame("S" = s_pred, "W_L" = 1,
          "Age" = 50, "BMI" = 22, gender = "Male", CHD = "No")
coef_lfcm_gam <- predict(fit_lfcm_mgcv, newdata = df_lfcm_gam, type = "terms",
          se.fit = TRUE)
#pull hat(beta)
beta_hat <- coef_lfcm_gam$fit[,"s(S):W_L"]
#pull se[hat(beta)]
se_beta_hat <- coef_lfcm_gam$se.fit[,"s(S):W_L"]
```

A few points are worth noting. First consider the call to `coef.pfr()`. The n argument specifies the number of equally spaced points on the functional domain to evaluate on. Setting `n = slen_pred` corresponds to evaluations on `sind_pred`. It is possible to specify these points directly, but we prefer this more direct approach. Specifying `useVc=FALSE` is done to be consistent with the default behavior of `predict.gam()` which, by default, returns variance estimates conditional on the estimated smoothing parameter. For some models fit by ML/REML, adjustments for uncertainty in the smoothing parameter can be done. Moving to the `mgcv::gam()` fit, consider the data frame `df_lfcm_gam`, which is supplied to the `predict.gam()` function. This data frame has two columns, with each row corresponding to a point $s \in s_{\text{pred}}$ (syntactically denoted as `sind_pred`.) The first column in `df_lfcm_gam` specifies the value of `smat`, which is the name of the matrix of the functional domain used in model, as each element of s_{pred}. The second column, `wlmat`, corresponds to the term $q_j W_i(s_j)$, which we set to 1 per the formula described in the previous paragraph.

Correlation and Multiplicity Adjusted (CMA) Inference

The pointwise inference presented for the functional linear Cox model suffers from the same problems which are present in the functional linear model for scalar-on-function regression described in Chapter 4. The procedure presented here very closely mirrors the approaches of Chapter 4. For completeness we provide an abbreviated discussion here, as well as code for obtaining confidence intervals and p-values, though we refer readers to 4.5.2 for an expanded discussion on key points. Briefly, unadjusted pointwise inference, when interpreted simultaneously across the functional domain, suffers from the statistical problems associated with multiple comparisons when tests are correlated (inflated type I error). Correlation and multiplicity adjusted inference provides a principled means for maintaining power while accounting for the correlated nature of the test statistics along the domain.

Suppose we wish to make inference on a finite set of points $s_{\text{pred}} \subset \mathcal{S}$. The hypothesis tests we consider here are either global tests with a null hypothesis of

$$H_0 : \beta(s) = 0, \ \forall \ s \in s_{\text{pred}}$$

or pointwise hypothesis tests

$$H_0 : \beta(s_0) = 0, \ s_0 \in \mathbb{R} \subset s_{\text{pred}}$$

while maintaining a family-wise error rate (FWER) of α for all $s \in s_{\text{pred}}$. Below we present four approaches to performing these hypothesis tests. The discussion here is somewhat abbreviated and we refer readers to 4.5.2 for further discussion.

Simulations from the Spline Parameter Distribution

Recall that, conditional on the estimated smoothing parameter, the joint distribution of the spline coefficients is approximately normal with mean $\widehat{\boldsymbol{\beta}}$ and variance $\text{Var}(\widehat{\boldsymbol{\beta}})$. Premultiplying by the collection of basis functions evaluated at s_{pred}, we obtain $\widehat{\beta}(s_{\text{pred}}) = \mathbf{B}(s_{\text{pred}})\widehat{\boldsymbol{\beta}}$. Therefore, simulating from the approximate distribution of $\widehat{\beta}(s_{\text{pred}})$ requires only simulating from a relatively low-dimensional multivariate normal random vector. This is done B times, calculating the simulated test statistic $d^b, b = 1, \ldots, B$, where $d^b = \max_{s \in s_{\text{pred}}} |\beta^b(s) - \widehat{\beta}(s)| / \text{SE}\{\widehat{\beta}(s)\}$. The $1 - \alpha$ quantile of the B simulated test statistics corresponds to the value such that

$$\text{Pr}_{H_0:\beta(s)=0}[\{|\widehat{\beta}(s)| / \text{SE}\{\widehat{\beta}(s)\} \geq q(\mathbf{C}_\beta, 1 - \alpha) : s \in s_{\text{pred}}\}] \approx \alpha \,,$$

where the approximation is a result of the variability inherent in simulation and the asymptotic normality of $\widehat{\boldsymbol{\beta}}$. Inverting the p-value, we obtain $1 - \alpha$ correlation and multiplicity adjusted confidence intervals of the form

$$\widehat{\beta}(s) \pm q(\mathbf{C}_\beta, 1 - \alpha)\text{SE}\{\widehat{\beta}(s)\}$$

for all $s \in s_{\text{pred}}$.

```
# set the seed
set.seed(99845)
#Get the design matrix associated with s_pred
lpmat <- predict(fit_lfcm_mgcv), newdata = df_lfcm_gam, type = "lpmatrix")
#Get the column indices associated with the functional term
inx_beta <- which(grepl("s\\(S\\):W_L\\.[0-9]+", dimnames(lpmat)[[2]]))
#Get the design matrix A associated with beta(s)
Bmat <- lpmat[,inx_beta]
#Get Var spline coefficients (beta_sp) associated with beta(u,s)
beta_sp <- coef(fit_lfcm_mgcv)[inx_beta]
Vbeta_sp <- vcov(fit_lfcm_mgcv)[inx_beta, inx_beta]
#Number of bootstrap samples (B)
nboot <- 1e7
#Set up container for bootstrap
beta_mat_boot <- matrix(NA, nboot, length(s_pred))
#Do the bootstrap
for(i in 1:nboot){
  beta_sp_i <- MASS::mvrnorm(n = 1, mu = beta_sp, Sigma = Vbeta_sp)
  beta_mat_boot[i,] <- Bmat %*% beta_sp_i
}

#Find the max statistic
dvec <- apply(beta_mat_boot, 1, function(x) max(abs(x - beta_hat) / se_beta_hat))
#Get 95% global confidence band
Z_global <- quantile(dvec, 0.95)
beta_hat_LB_global <- beta_hat - Z_global * se_beta_hat
beta_hat_UB_global <- beta_hat + Z_global * se_beta_hat
## get pointwise p-values
# quantiles of the max statistic to search over
qs_vec <- seq(0, 1, len = 1e6)
# get the corresponding quantiles of the max statistic
qs_vec_d <- quantile(dvec, qs_vec, na.rm = TRUE)
# Find whether the bands at each quantile of the max statistic contain 0
cover_1_0 <-
        vapply(qs_vec_d,
               function(x){
               (beta_hat - x * se_beta_hat) < 0 &
               (beta_hat + x * se_beta_hat > 0)
        },numeric(length(s_pred)))

# get the p-value
p_val_lg <-
        apply(cover_1_0,1,
              function(x){
              inx_sup_1_p = which(x==1)
              ifelse(length(inx_sup_1_p)== 0, 1 / nboot, 1 - min(qs_vec[inx_sup_1_p]))
        })
```

Figure 7.7 presents the results from the procedure above. The left panel of Figure 7.7 plots the pointwise CMA p-values (light gray line) and the pointwise unadjusted p-values (dark gray line). The right panel of Figure 7.7 plots the 95% CMA confidence bands (light gray shaded region) and the unadjusted 95% pointwise confidence intervals (dark gray shaded region). We see that the majority of the times of day where the pointwise intervals suggested statistical significance continue to be statistically significant at $\alpha = 0.05$.

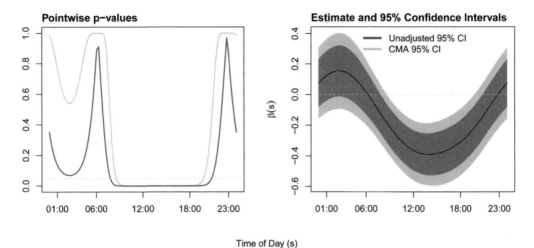

FIGURE 7.7: Pointwise CMA inference for SoFR based on simulations from the distribution of spline coefficients. BMI is the outcome and PA functions are the predictors. Left panel: estimated pointwise unadjusted (dark gray) and CMA (light gray) p-values denoted by $p_{\mathrm{pCMA}}(s)$. Right panel: the 95% pointwise unadjusted (dark gray) and CMA confidence intervals for $\beta(s)$.

Specifically, increased activity during the period between approximately 9 AM and 7 PM is significantly associated with a decreased risk of mortality.

Nonparametric Bootstrap of the Max Absolute Statistic

The procedure for performing the non-parametric bootstrap involves resampling study participants from the observed data with replacement, allowing variability in all estimated parameters to be assessed, including the smoothing parameters. Within each resample dataset the model is (1) re-estimated; (2) the estimated coefficient is evaluated on the grid of interest; and (3) distributional summaries are calculated across the re-sampled coefficients (in this case the maximum absolute statistic, d^b described in the parametric bootstrap). Key differences in this approach are that the nonparametric bootstrap does not assume normality of $\widehat{\beta}$, nor does it condition on the estimated smoothing parameter obtained from the initial model fit. However, as stated in Chapter 4, the non-parametric bootstrap implicitly requires an assumption of spherical symmetry of the distribution of $\beta(s_{\mathrm{pred}})$. As of this writing, it remains an open problem how to perform inference when this implicit assumption is violated.

The code below illustrates how to perform the non-parametric bootstrap with repeated model estimation done using `refund::pfr()` instead of `mgcv::gam()`. As with the SoFR implementation presented in Chapter 4, either function is equally easy to use for estimation given the wide format storage of the data, though `mgcv::gam` is more convenient for obtaining coefficient predictions on an arbitrary grid. We reiterate here that this is not an inherent limitation of `pfr`, but rather a design decision made when designing the `predict.pfr` method.

```
#Set up container for bootstrap
beta_mat_boot_np <- matrix(NA, nboot, length(s_pred))
#Number of participants to resample
N <- nrow(df_mgcv)
#Do the bootstrap
for(i in 1:nboot){
  inx_i <- sample(1:N, size = N, replace = TRUE)
  df_i <- df_analysis[inx_i,]
  fit_lfcm_gam_i <- gam(time ~ age + BMI + gender + CHD +
                              s(S, by = W_L, bs = "tp", k = 10),
                    method = "REML", data = df_i, weights=df_i$event,
                    family=cox.ph)
  beta_mat_boot_np[i,] <- predict(fit_lfcm_gam_i, newdata = df_lfcm_gam,
                              type = "terms")[,[,"s(S):W_L"]]
}

#Find the max statistic
dvec_np <- apply(beta_mat_boot_np, 1, function(x)max(abs(x-beta_hat)/se_beta_hat))
#Get 95% global confidence band
Z_global <- quantile(dvec, 0.95)
beta_hat_LB_global_np <- beta_hat - Z_global_np * se_beta_hat
beta_hat_UB_global_np <- beta_hat + Z_global_np * se_beta_hat
```

Figure 7.8 presents the results of the non-parametric bootstrap using the same format as Figure 7.7, with pointwise CMA and unadjusted p-values in the left panel (light/dark gray, respectively) and 95% CMA confidence bands and unadjusted pointwise confidence intervals in the right panel (light/dark gray, respectively). Interestingly, the nonparametric bootstrap produces narrower global confidence bands than the parametric bootstrap, with a global 95% multiplier of ≈ 2.76 versus ≈ 2.92 for the parametric bootstrap. It is unclear why the nonparametric bootstrap produces narrower confidence bands at this time, but practical interpretations are largely similar. The p-values associated with the global null hypothesis are estimated to be smaller than one divided by the number of bootstraps in both cases, which suggests a highly significant association between diurnal physical activity patterns and mortality.

Analytic Solution to the Parametric Bootstrap Approximation

In Section 2.4.4 we described an approach based on the PCA decomposition of the covariance operator, which may seem like a good idea. *However, we reiterate that the problems discussed in the context of SoFR presented in Chapter 4 persist in this context and we do not recommend this approach at this time.* Once again, we highlight the point that methods which superficially seem compatible may in practice yield very different results. We thus advise readers to consider implementing multiple methods which target the same null hypothesis (e.g., parametric and nonparametric bootstrap) to check for agreement between multiple approaches. The text below contains an abbreviated version of the text contained in 4.5.2 for completeness.

Suppose that the assumption of multivariate normality of the spline coefficients, $\widehat{\boldsymbol{\beta}}$, is reasonable. It follows that $\widehat{\beta}(\mathbf{s}_{\mathrm{pred}}) = \mathbf{B}(\mathbf{s}_{\mathrm{pred}})\widehat{\boldsymbol{\beta}}$ has a multivariate normal distribution with covariance matrix $\mathrm{Var}\{\widehat{\beta}(\mathbf{s}_{\mathrm{pred}})\} = \mathbf{B}(\mathbf{s}_{\mathrm{pred}})\mathrm{Var}(\widehat{\boldsymbol{\beta}})\mathbf{B}^t(\mathbf{s}_{\mathrm{pred}})$.

By the same argument discussed in Section 2.4.1

$$[\{\widehat{\beta}(s) - \beta(s)\}/\mathrm{SE}\{\widehat{\beta}(s)\} : s \in \mathbf{s}_{\mathrm{pred}}] \sim N(0, \mathbf{C}_\beta) ,$$

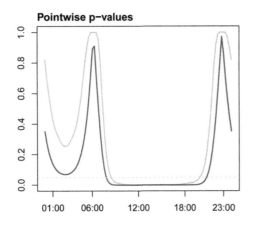
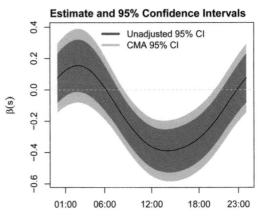

Time of Day (s)

FIGURE 7.8: Pointwise CMA inference for the linear functional Cox model based on the nonparametric bootstrap of the max statistic. Left panel: estimated pointwise unadjusted (dark gray) and CMA (light gray) p-values denoted by $p_{\mathrm{pCMA}}(s)$. Right panel: the 95% pointwise unadjusted (dark gray) and CMA confidence intervals for $\beta(s)$.

where $\mathbf{C}_\beta$ is the correlation matrix corresponding to the covariance matrix $\mathrm{Var}\{\widehat{\beta}(\mathbf{s}_{\mathrm{pred}})\} = \mathbf{B}(\mathbf{s}_{\mathrm{pred}})\mathrm{Var}(\widehat{\boldsymbol{\beta}})\mathbf{B}^t(\mathbf{s}_{\mathrm{pred}})$. As we discussed in Section 2.4.1, we need to find a value $q(\mathbf{C}_\beta, 1 - \alpha)$ such that

$$P\{q(\mathbf{C}_\beta, 1 - \alpha) \times \mathbf{e} \leq \mathbf{X} \leq q(\mathbf{C}_\beta, 1 - \alpha) \times \mathbf{e}\} = 1 - \alpha \ ,$$

where $\mathbf{X} \sim N(0, \mathbf{C}_\beta)$ and $e = (1, \dots, 1)^t$ is the $|\mathbf{s}_{\mathrm{pred}}| \times 1$ dimensional vector of ones. Once $q(\mathbf{C}_\beta, 1 - \alpha)$ is available, we can obtain a CMA $1 - \alpha$ level confidence interval for $\beta(s)$ as

$$\widehat{\beta}(s) \pm q(\mathbf{C}_\beta, 1 - \alpha)\mathrm{SE}\{\widehat{\beta}(s)\} : \forall\ s \in s \in \mathbf{s}_{\mathrm{pred}} \ .$$

However, because its rank of $\widehat{\beta}(\mathbf{s}_{\mathrm{pred}}) = \mathbf{B}(\mathbf{s}_{\mathrm{pred}})\widehat{\boldsymbol{\beta}}$ is at most K, it has a degenerate normal distribution where K is the number of basis used to estimate $\beta(s)$. Since we have evaluated $\widehat{\beta}(s)$ on a grid of $|\mathbf{s}_{\mathrm{pred}}| = 100$, the covariance and correlation matrices of $\widehat{\beta}(\mathbf{s}_{\mathrm{pred}})$ are 100 dimensional. A simple example for understanding this problem is to consider the case where $\widehat{\beta}(s) = c + s$ for $c \in \mathbb{R}$. Then $\mathrm{rank}(\mathbf{C}_\beta) \leq 2$ for any choice of $\mathbf{s}_{\mathrm{pred}}$.

Recall that if $\widehat{\beta}(\mathbf{s}_{\mathrm{pred}})$ has a degenerate multivariate normal distribution of rank $m \leq K$, then there exists some random vector of independent standard normal random variables $\mathbf{Q} \in \mathbb{R}^m$ such that

$$\widehat{\beta}^t(\mathbf{s}_{\mathrm{pred}}) = \mathbf{Q}^t\mathbf{D} \ ,$$

with $\mathbf{D}^t\mathbf{D} = \mathbf{B}(\mathbf{s}_{\mathrm{pred}})\mathrm{Var}(\widehat{\boldsymbol{\beta}})\mathbf{B}^t(\mathbf{s}_{\mathrm{pred}})$. If we find m and a matrix $\mathbf{D}$ with these properties, the problem is theoretically solved. Consider the eigendecomposition $\mathbf{B}(\mathbf{s}_{\mathrm{pred}})\mathrm{Var}(\widehat{\boldsymbol{\beta}})\mathbf{B}^t(\mathbf{s}_{\mathrm{pred}}) = \mathbf{U}\mathbf{\Lambda}\mathbf{U}^t$, where $\mathbf{\Lambda}$ is a diagonal matrix of eigenvalues and the matrix $\mathbf{U}\mathbf{U}^t = \mathbf{I}_{|\mathbf{u}_{\mathrm{pred}}| \times |\mathbf{s}_{\mathrm{pred}}|}$ is an orthonormal matrix with the kth column being the eigenvector corresponding to the kth eigenvalue. Note that all eigenvalues $\lambda_k = 0$ for $k > m$ and

$$\mathbf{B}(\mathbf{s}_{\mathrm{pred}})\mathrm{Var}(\widehat{\boldsymbol{\beta}})\mathbf{B}^t(\mathbf{s}_{\mathrm{pred}}) = \mathbf{U}_m\mathbf{\Lambda}_m\mathbf{U}_m^t \ ,$$

where $\mathbf{U}_m$ is the $|\mathbf{s}_{\text{pred}}| \times m$ dimensional matrix obtained by taking the first m columns of $\mathbf{U}$ and $\mathbf{\Lambda}_m$ is the $m \times m$ dimensional diagonal matrix with the first m eigenvalues on the main diagonal. If we define $\mathbf{D}^t = \mathbf{U}_m \mathbf{\Lambda}_m^{1/2}$ and

$$\{\widehat{\beta}^t(\mathbf{s}_{\text{pred}}) - \beta^t(\mathbf{s}_{\text{pred}})\}\mathbf{D}^{-1} = \mathbf{Q}^t \,,$$

where $\mathbf{D}^{-1} = \mathbf{U}_m \mathbf{\Lambda}_m^{-1/2}$ is the Moore Penrose right inverse of $\mathbf{D}$ ($\mathbf{D}$ is not square, but is full row rank). Under the null hypothesis, $\beta(\mathbf{s}_{\text{pred}}) = 0$, thus we can obtain a p-value of no effect by calculating

$$1 - \mathrm{P}(\max\{|\mathbf{Q}|\} \leq \max\{|\widehat{\beta}^t(\mathbf{s}_{\text{pred}})\mathbf{D}^{-1}|\}) \,,$$

where $\mathbf{Q}$ is a vector of independent standard normal random variables. This calculation is fairly straightforward, though we use `mvtnorm::pmvnorm` for convenience. The mathematics is clear and nice and it would be perfect if it worked. Below we show how to calculate these quantities.

```
beta <- coef(fit_lfcm_mgcv)[inx_beta]
#Get Var(hat(beta))
Vbeta <- vcov(fit_lfcm_mgcv)[inx_beta, inx_beta]
Vbeta <- Bmat %*% Vbeta %*% t(Bmat)
#Get eigenfunctions via svd
eVbeta <- svd(Vbeta, nu = ncol(Bmat), nv = ncol(Bmat))
#Get only positive eigenvalues
inx_evals_pos <- which(eVbeta$d >= 1e-6)
#Get "m" - dimension of lower dimensional RV
m <- length(inx_evals_pos)
#Get associated eigenvectors
U_m <- eVbeta$v[, inx_evals_pos]
#Get D transpose
D_t <- U_m %*% diag(sqrt(eVbeta$d[inx_evals_pos]))
#Get D inverse
D_inv <- MASS::ginv(t(D_t))
#Get Q
Q <- t(Bmat %*% beta) %*% D_inv
#Get max statistic
Zmax <- max(abs(Q))
#Get p-value
p_val <- 1 - pmvnorm(lower = rep(-Zmax, m), upper = rep(Zmax, m),
                     mean = rep(0, m), cor = diag(1, m))
```

The resulting p-value is numerically 0. Unlike in Chapter 4, the results here largely appear to agree with both the bootstrap approaches and the `mgcv` summary output, though given the very small values of the p-values in this context, it is hard to verify them via simulations in this scenario. This is an example of why we say that the p-values from these methods sometimes agree, but we urge caution in general.

Multivariate Normal Approximation

Another approach which has less theoretical justification, but seems to work in practice, is to ignore the fact that the distribution of $\widehat{\beta}(\mathbf{s}_{\text{pred}})$ in our example is degenerate multivariate normal. Rather, the solution is to slightly jitter the covariance matrix to ensure that the implied correlation function is positive definite. Then we may obtain inference using software for calculating quantiles of a multivariate normal random vector (e.g., `mvtnorm::pmvnorm`). The code to do so is provided below.

```
#Get hat(beta)/SE(hat(beta))
beta_hat_std <- beta_hat / se_beta_hat
#Get Var(hat(beta)/SE(hat(beta)))
Vbeta_hat <- Bmat %*% Vbeta %*% t(Bmat)
#Jitter to make positive definite
Vbeta_hat_PD <- Matrix::nearPD(Vbeta_hat)$mat
#Get correlation function
Cbeta_hat_PD <- cov2cor(matrix(Vbeta_hat_PD, 100, 100, byrow = FALSE))
#Get max statistic
Zmax <- max(abs(beta_hat_std))
#p-value
p_val <- 1 - pmvnorm(lower = rep(-Zmax, 100), upper = rep(Zmax, 100),
                     mean = rep(0, 100), cor = Cbeta_hat_PD)
```

The resulting p-value is 2.9×10^{-6}, which is very close to that obtained from the parametric bootstrap. In principle, this method should yield the same result as the exact solution, though we find with some consistency that it does not. In our experience this approach provides results which align very well with the parametric bootstrap, though the theoretical justification for jittering of the covariance function is, as of this writing, not justified. To construct a 95% CMA confidence interval, one may use the `mvtnorm::qmvnorm` function, illustrated below. Unsurprisingly, the resulting global multiplier is almost identical to that obtained from the parametric bootstrap. For that reason we do not plot the results or interpret further.

```
Z_global <- qmvnorm(0.95, mean = rep(0,100), cor = Cbeta_hat_PD, tail =
"both.tails")$quantile
```

7.3.1.3 Survival Curve Prediction

Next, we discuss how to obtain predicted survival curves from the functional linear Cox model (FLCM). The formula presented in Section 7.1 for calculating survival probabilities given baseline covariates applies directly to the FLCM. Specifically, we have $\widehat{S}_i(t|\mathbf{Z}_i, \mathbf{W}_i) = \exp\{-\exp(\mathbf{Z}_i^t \gamma + \int_S W_i(s)\beta(s)ds)\Lambda_0(t)\}$. Thus, given the estimated model parameters $\widehat{\gamma}$, $\widehat{\beta}$ and the Breslow estimate of $\widehat{\Lambda}_0(t)$, we can obtain predicted survival probabilities by plugging the estimated quantities into (7.12), obtaining

$$\widehat{S}_i(t|\mathbf{Z}_i, \mathbf{W}_i) = \exp\{-\exp(Z_i^t \widehat{\gamma} + \sum_{k=1}^{K} \widehat{\beta}_k \sum_{j=1}^{p} q_j W_i(s_j) B_k(s_j))\widehat{\Lambda}_0(t)\}$$

for $t \in \mathcal{T}$. Extracting predicted survival probabilities requires users to present new data in the same form as was supplied to the estimating function, both for `pfr()` and `mgcv::gam()`, with one row for each survival time of interest. In contrast to obtaining predictions for the functional coefficient, $\beta(\cdot)$, the procedures for obtaining predicted survival probabilities are similar for `pfr()` and `mgcv::gam()`. Therefore, we show how to obtain them using the `mgcv::gam()` function.

Suppose that we were interested in obtaining predicted survival curves for individuals with the unique identifiers `SEQN` of 80124 and 62742 from the FLCM on a fine grid over the first four years of follow-up. Study participant 62742 is a 70-year-old Other Hispanic male with a BMI of 28.8. Study participant 80124 is a 70-year-old White male with a BMI of

32.7 and with CHD. These two participants were chosen to match on age, gender, and to have similar BMI, differing primarily on their activity profile $W_i(s)$. As a result, differences in estimated survival probabilities will be primarily associated with the differences in their average activity levels. The code below shows how to obtain predicted survival probabilities for these two individuals.

```
#Set up grid for predicted probabilities
#Range of times to obtain predicted survival probabilities for
t_max <- 4
t_min <- 0
#Density of the grid for predicted probabilities
nt_pred <- 100
#Equally spaced grid of times to predict on
tind_pred <- seq(t_min, t_max, len = nt_pred)

#Get predicted survival probabilities for each participant
#Participants of interest
ids_plt <- c(80124, 62742)
#Row indices associated with participants
inx_ids <- which(nhanes_df_surv$SEQN %in% ids_plt)
#Set up a data frame to make predictions on
#Get data associated with each participant
#Repeat for each element of tind_pred
df_plt <- nhanes_df_surv[rep(inx_ids, each = nt_pred),]
#Add in prediction times
df_plt$time <- rep(tind_pred, length(ids_plt))
#Get predictions
surv_preds <- predict(fit_lfcm_mgcv, newdata = df_plt,
            type = "response", se.fit = TRUE)
#Add in auxilary data for plotting, create 95% CIs based on SEs
surv_preds <- surv_preds %>%
            data.frame() %>%
            mutate(SEQN = df_plt$SEQN,
            time = df_plt$time,
            LB = pmax(0, fit - qnorm(0.975) * se.fit),
            UB = pmin(1, fit + qnorm(0.975) * se.fit))
```

Going through the code above step-by-step, we first choose a grid on which to obtain predicted survival probabilities. Here, we specify an equally spaced grid (`tind_pred`) over 0 (`t_min`) to 4 (`t_max`) years. Next, we specify the participants we wish to make predictions for, SEQN 80124 and 62742. Next, we need to create a dataframe with one row containing each individual's baseline covariate values and each prediction time (`df_plt`). Then we obtain survival predictions using the `predict` function with `type = "response"` to get predictions on the response (survival probability) scale and `se.fit = TRUE` to obtain the corresponding standard errors. While standard errors on the covariate-dependent portion of the log hazard can be obtained from using the Fisher information matrix from the partial log likelihood, inference on survival probabilities requires one to incorporate uncertainty in the estimate for the cumulative baseline hazard function $\widehat{\Lambda}(t)$. There are many methods for obtaining standard errors on survival curves from a Cox model, though the method used by `mgcv` is described in [151]. Note that this process would involve additional steps in the presence of either time-varying covariates or time-varying effects.

Using the data frame `surv_preds` created above, we can create plots of estimated survival probabilities with unadjusted pointwise 95% confidence intervals for each of the study participants. Figure 7.9 displays the ten-minute binned MIMS diurnal profiles, $W_i(s)$, in the

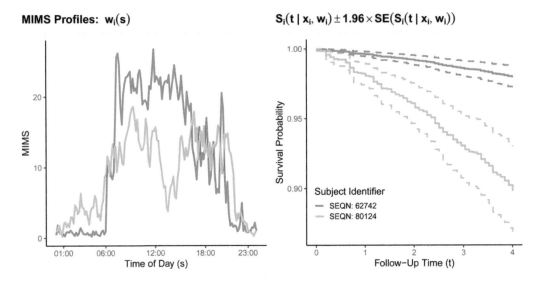

FIGURE 7.9: MIMS profiles (left panel) and estimated survival curves (right panel) obtained from the fitted FLCM for two participants, SEQN 62742 (red lines) and SEQN 80124 (blue lines). MIMS profiles were binned into 10-minute intervals prior to plotting for readability. Estimated survival probabilities are presented as solid lines with 95% confidence intervals as dashed lines.

left panel and the corresponding estimated survival curves as a function of time in the right panel for each of SEQN 62742 (red curves) and 80124 (blue curves). We see that SEQN 80124 was overall notably less active during the day, and more active at night. Recall that the shape of $\widehat{\beta}(s)$ presented in Figure 7.6 implies that higher activity, particularly during mid-day, is associated with lower risk. This effect on estimated survival probabilities is reflected in the right panel of Figure 7.9, with lower estimated survival probability of SEQN 80124. Given that these individuals were matched on the other covariates included in our Cox model, these differences are almost entirely attributable to differences in physical activity patterns.

7.3.2 Smooth Effects of Traditional and Functional Predictors

One of the advantages of using the mgcv package for estimating our functional regression models is the ability to fit models with non-linear associations between non-functional predictors. For example, we may believe that there is a non-linear association between age and risk of mortality. In addition, there is evidence to suggest a non-linear relationship between BMI and risk of mortality. Specifically, studies have found that having either an extremely low BMI or high BMI is associated with increased risk of all-cause and cause-specific mortality [17]. We can easily extend the linear functional Cox model to include such non-linear effects. Below, we describe how such a model can be estimated using mgcv::gam.

Suppose that an individual's covariate vector of non-functional predictors is $\mathbf{Z}_i = [\mathbf{Z}_{i,1}^t, \mathbf{Z}_{i,2}^t]^t$, where $\mathbf{Z}_{i,1}$ is a vector of length Q_1 containing the covariates for which we assume a linear effect, and $\mathbf{Z}_{i,2}$ is a vector of length Q_2 containing the covariates for which we assume a non-linear (additive) effect. Here $Q = Q_1 + Q_2$ is the total number of scalar covariates. Model (7.6) then becomes

$$\log \lambda_i\{t|\mathbf{Z}_i, W_i(\cdot)\} = \log\{\lambda_0(t)\} + \sum_{q=1}^{Q_1} \gamma_q Z_{iq} + \sum_{q=Q_1+1}^{Q} f_q(Z_{iq}) + \int_S W_i(s)\beta(s)ds . \quad (7.10)$$

Estimating model (7.10) using the framework of penalized splines to estimate non-linear effects proceeds in a similar fashion as was described for the linear functional Cox model in Section 7.3.1. Specifically, we apply a set of bases to both the functional coefficient, $\beta(s)$, and the non-linear effects of non-functional covariates, $f_q(Z_{iq})$, and penalize the curvature of the estimated effects through an additive penalty on the log partial likelihood. Model (7.10) becomes

$$\log \lambda_i\{t|\mathbf{Z}_i, W_i(\cdot)\} = \log\{\lambda_0(t)\} + \sum_{q=1}^{Q_1} \gamma_q Z_{iq} + \sum_{q=Q_1+1}^{Q} \sum_{k=1}^{K_q} \alpha_{kq} B_{k,q}(Z_{iq})$$

$$+ \int_S W_i(s) \sum_{k=1}^{K_\beta} \beta_k B_k(s)ds ,$$

where K_q is the number of basis functions used to model $f_q(\cdot)$ and α_{kq} and $B_{k,q}$ are the corresponding spline coefficients and basis functions, respectively. Approximating the integral term numerically, combining terms, and using vector notation, the model can be re-written as

$$\log \lambda_i\{t|\mathbf{Z}_i, W_i(\cdot)\} = \log\{\lambda_0(t)\} + \mathbf{Z}_{i,1}^t \boldsymbol{\gamma} + \sum_{q=Q_1+1}^{Q} \mathbf{B}_{i,q}^t \boldsymbol{\alpha}_q + \mathbf{C}_i^t \boldsymbol{\beta} ,$$

where $\mathbf{Z}_{i,1} = (Z_{i1}, \ldots, Z_{iQ_1})^t$, $\boldsymbol{\gamma} = (\gamma_1, \ldots, \gamma_{Q_1})^t$ correspond to the linear covariate effect $\sum_{q=1}^{Q_1} \gamma_q Z_{iq}$, $\mathbf{B}_{i,q} = \{B_{1,q}(Z_{iq}), \ldots, B_{K_q,q}(Z_{iq})\}^t$ and $\boldsymbol{\alpha}_q = (\alpha_{1q}, \ldots, \alpha_{K_q q})^t$ correspond to the $Q_2 = Q - Q_1$ nonparametric effects of scalar covariates $\sum_{q=Q_1+1}^{Q} f_q(Z_{iq})$, while $\mathbf{C}_i$ and $\boldsymbol{\beta}$ are defined as in Section 7.3.1.1 and correspond to the functional predictor $\int_S W_i(s)\beta(s)ds$. The corresponding penalized log partial likelihood to be maximized is then

$$l(\boldsymbol{\beta}, \boldsymbol{\xi}) = \sum_{i:\delta_i=1} [\eta_i - \log\{\sum_{j:Y_j \geq Y_i} \exp(\eta_i)\}] - \sum_{q=Q_1+1}^{Q} \lambda_q P(\boldsymbol{\alpha}_q) - \lambda_\beta P(\boldsymbol{\beta}) , \quad (7.11)$$

where $\eta_i = \mathbf{Z}_{i,1}^t \boldsymbol{\gamma} + \sum_{q=Q_1+1}^{Q} \mathbf{B}_{i,q}^t \boldsymbol{\alpha}_q + \mathbf{C}_i^t \boldsymbol{\beta}$, $P(\boldsymbol{\alpha}_q)$ is a known penalty structure matrix and λ_q is an unknown smoothing parameter for the nonparametric function $f_q(\cdot)$, while $P(\boldsymbol{\beta})$ is a known penalty structure matrix and λ_β is an unknown smoothing parameter for the functional parameter $\beta(\cdot)$. All smoothing parameters are estimated using a marginal likelihood approach.

In our data example we set $Q = 4$ and allow two predictors ($Q_1 = 2$) to have a linear effect on the log hazard (gender and CHD) and two predictors ($Q_2 = 2$) to have a smooth nonparametric effect on the log hazard (age and BMI). The code to fit the LFCM with additive smooth terms for age and BMI via `refund::pfr` is presented below. Given the correlation between age, movement, and weight, we are interested in how these estimated associations change when MIMS profiles are excluded from the model. Thus, we also estimate an additive Cox model which excludes the physical activity (no functional predictor) for comparison.

```
#Fit the additive Cox model with smooth effects for age and BMI
fit_acm <- gam(time ~
               s(age, bs = "cr", k = 30) +
               s(BMI, bs = "cr", k = 30) +
               gender + CHD,
               weights = event, data = nhanes_df_surv,
               family = cox.ph())
#Fit the additive Cox above with a linear functional effect of MIMS
fit_acm_lfcm <- pfr(time ~
               s(age, bs = "cr", k = 30) +
               s(BMI, bs = "cr", k = 30) +
               gender + CHD +,
               lf(MIMS, bs = "cc", k = 30),
               weights = event, data = nhanes_df_surv,
               family = cox.ph())
```

The resulting coefficient estimates from these models can be plotted quickly using the `plot.gam` function. Specifically, one could plot the coefficients using the following code.

```
#Set up a 2x3 matrix for plotting the coefficients
par(mfrow = c(2,3))
#Plot the age and BMI effects from the additive Cox model
plot(fit_acm)
#Create a blank plot to fill in the third column
plot.new()
#Plot the coefficients from additive functional Cox model
plot(fit_acm_lfcm)
```

With some additional effort for formatting, these plots become Figure 7.10. Figure 7.10(A) presents the estimated coefficients $f(\text{age})$, $f(\text{BMI})$ from the additive Cox model in the left and middle panels, respectively. Figure 7.10(B) presents the same coefficients estimated in the additive functional Cox model along with the estimated effect of physical activity, $\widehat{\beta}(s)$, in the right column. Estimated age and BMI effects are non-linear in both models, with the estimated BMI effect more strongly non-linear. In the additive Cox model, the lowest risk BMI is estimated to be around 30, which corresponds to the border between "overweight" and "obese" BMI. This effect, referred to as the "obesity paradox" has been observed in other studies, with potential confounding mechanisms proposed [163, 215, 225]. A full discussion of this topic is beyond the scope of our data application. Adjusting for participants' MIMS profiles (linear functional additive Cox model), the estimated effects of both age and BMI are attenuated (pulled toward 0). Interestingly, adjusting for individuals' activity patterns, the lowest risk BMI is shifted to close to around 40-45, though confidence intervals for the BMI effect in both models are quite wide. This may be due to the relatively short follow-up length for mortality data currently available in NHANES 2011-2014. A final observation is that, (non-linearly) adjusting for age and BMI, the estimated association of activity is attenuated slightly as compared to the results presented in Figure 7.6.

7.3.3   Additive Functional Cox Model

Although the linear functional Cox model provides an efficient and reasonable estimation framework to analyze the association between baseline functional predictors and survival

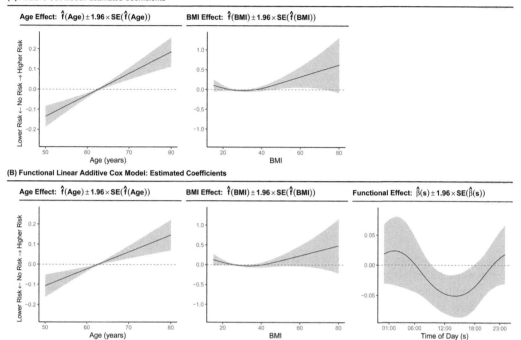

FIGURE 7.10: (A) Estimated associations between age (left panel) and BMI (right panel) and the log hazard of all-cause mortality in an additive Cox model. (B) Estimated associations between age (left panel) and BMI (right panel) and the log hazard of all-cause mortality in an additive Cox model. Point estimates (solid black lines) and 95% point-wise confidence intervals (gray shaded regions) are presented. Both models adjust for gender and congestive heart failure as linear effects.

outcome, one limitation is that it only allows a linear association between the functional predictor at each location of the domain and the log hazard. That is, for each $s \in S$, the association between $W_i(s)$ and log hazard is modeled through a fixed unknown parameter $\beta(s)$, which is the same across different values of $W_i(s)$. In this section we show how to extend the linear functional Cox model to the case when one is interested in allowing the effect to vary with the magnitude of $W_i(s)$.

The additive functional Cox model [56] replaces the linear functional $\int_S W_i(s)\beta(s)ds$ model with the functional term $\int_S F\{s, W_i(s)\}ds$, where $F(\cdot, \cdot)$ is an unknown bivariate smooth function to be estimated. The model becomes

$$\log \lambda_i\{t|\mathbf{Z}_i, W_i(\cdot)\} = \log\{\lambda_0(t)\} + \mathbf{Z}_i^t \gamma + \int_S F\{s, W_i(s)\}ds . \qquad (7.12)$$

To ensure identifiability, we impose the constraint $E[F\{s, W(s)\}] = 0$ for each $s \in S$; see [56] for the detailed discussions on identifiability. The functional term of the additive functional Cox model can be estimated using a tensor product of penalized splines, as introduced in [198]. Specifically, denote by $B_l^s(\cdot), B_k^w(\cdot), l = 1, \ldots, K_s, k = 1, \ldots, K_x$ two univariate splines basis. With this notation, the functional coefficient can be expanded as $F(\cdot, \cdot) = \sum_{l=1}^{K_s} \sum_{k=1}^{K_w} \beta_{lk} B_l^s(\cdot) B_k^w(\cdot)$, where $\{\beta_{lk} : l = 1, \ldots, K_s; k = 1, \ldots, K_w\}$ are the spline coefficients to be estimated. The model becomes

$$\log \lambda_i \{t | \mathbf{Z}_i, W_i(\cdot)\} = \log\{\lambda_0(t)\} + \mathbf{Z}_i^t \boldsymbol{\gamma} + \int_S \sum_{l=1}^{K_s} \sum_{k=1}^{K_w} \beta_{lk} B_l^s(s) B_k^w \{W_i(s)\} ds$$

$$\approx \log\{\lambda_0(t)\} + \mathbf{Z}_i^t \boldsymbol{\gamma} + \sum_{j=1}^{p} q_j \sum_{l=1}^{K_s} \sum_{k=1}^{K_w} \beta_{lk} B_l^s(s_j) B_k^w \{W_i(s_j)\}$$

$$= \log\{\lambda_0(t)\} + \mathbf{Z}_i^t \boldsymbol{\alpha} + \sum_{l=1}^{K_s} \sum_{k=1}^{K_w} \beta_{lk} \left[\sum_{j=1}^{p} q_j B_l^s(s_j) B_k^w \{W_i(s_j)\} \right] , \quad (7.13)$$

where q_j, $j = 1, \ldots, p$ are the quadrature weights. The expression

$$\sum_{j=1}^{p} q_j B_l^s(s_j) B_k^W \{W_i(s_j)\} = c_{ijk}$$

is a constant and the model reduces to a standard Cox regression model and the estimation framework follows from the linear functional Cox model introduced in Section 7.3.1. Note that, again, the sum in equation (7.13) is over all parameters β_{lk} with the weights c_{ijk}. Therefore, the exact same ideas used multiple times in this book can be applied here: use the `by` option in the `mgcv::gam` to implement this regression.

The R syntax to fit an additive functional Cox model using the `mgcv` package is shown below. Notice that here we use `MIMS_q` to denote quantiles of MIMS at each minute. For the NHANES data set, the model was fit using quantiles of MIMS instead of their absolute values. This step is necessary to ensure that data reasonably fills the bivariate domain spanned by the s and w directions; for more details on estimability and identifiability of these models see [56]. To specify the tensor product of two univariate penalized splines, we use the `ti()` function from the `mgcv` package. The first two arguments of `ti()` function, `MIMS_q` and `smat`, instruct the function to specify a tensor product of two univariate penalized splines on the domain spanned by $W_i(s)$ and s. The `by` argument instructs the package to multiply each specified term by object `lmat`, which corresponds to the quadrature elements, q_j, in the model. The `bs` and `k` arguments specify the type and number of each spline basis, respectively. To ensure identifiability, we need to specify `mc = TRUE` for the $W_i(s)$ direction, which imposes the marginal constraint in this direction and matches well with the identifiability constraint of this model. The rest of the syntax is the same as that used for fitting a linear functional Cox model.

```
#Fit the AFCM using mgcv
fit_afcm_mgcv <- gam(time ~ age + BMI + gender + CHD +
                ti(MIMS_q, smat, by = lmat, bs = c("cr", "cc"),
                k = c(Kx, Ks), mc = c(TRUE, FALSE)), weights = event,
                data = nhanes_df_surv, family = cox.ph())
```

In this application, we focus on study participants who were greater than or equal to 50 years old and had no missing data for age and BMI. The data set contains 4,207 study participants. For this specific example where the functional predictor is the minute-level physical activity intensity value, [56] contains some plots for detailed comparisons on the estimates before and after quantile transformation. Figure 7.11 displays the estimated surface $\widehat{F}\{s, W_i(s)\}$ of the additive functional Cox model using NHANES 2011–2014 data. The figure was created using the `vis.gam()` function and the code to reproduce Figure 7.11 is shown below.

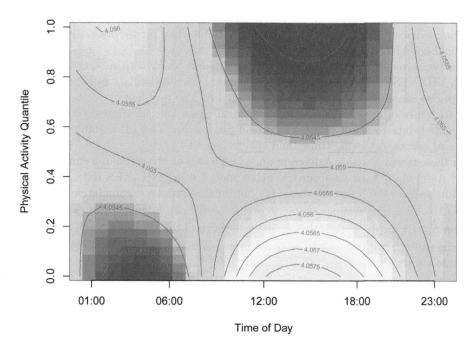

FIGURE 7.11: Estimated surface $\widehat{F}\{s, W_i(s)\}$ from the NHANES 2011-2014 data obtained by calling `mgcv::gam()` using the syntax above. The value $\widehat{F}\{s, W_i(s)\}$ decreases from green (highest) to blue (lowest hazard of all-cause mortality).

```
#Visualize the surface estimated using AFCM
vis.gam(fit_afcm_mgcv, view = c("smat", "MIMS_q"), plot.type = "contour",
        color = "topo", xaxt = "n",
        xlab = "Time of Day",
        ylab = "Physical Activity Quantile")
axis(side = 1, at = c(1, 6, 12, 18, 23) / 24,
     labels = c("01:00", "06:00", "12:00", "18:00", "23:00"))
```

The estimated surface indicates that lower physical activity quantile during the day and higher physical activity quantile at night are associated with a higher hazard of all-cause mortality. Specifically, being below the 35th percentile of physical activity intensity in the population during the daytime (8 AM to 10 PM) is associated with a higher hazard of mortality. This result is highly interpretable, is consistent with the lifestyle of the majority of the population, and suggests the benefit of having sleep without major interruptions at night and being active during the daytime. While the populations and the wearable device protocols differ between our book and [56], it is interesting to note that the analysis results from AFCM are quite similar.

7.4   Simulating Survival Data with Functional Predictors

Simulating survival data is important for evaluating and comparing the performance of various survival models. Simulating from parametric models is often used, but it remains difficult to match various parameters and distributions to realistic scenarios. In addition, it is not immediately clear how to conduct simulations when predictors are high-dimensional functional covariates.

Here we provide simple approaches that simulate data with a structure that closely mimics NHANES data. Indeed, we have been unable to use simulations based on parametric assumptions on survival times as discussed in [14]. The problem is due to (1) highly non-linear structure of associations between survival times and covariates; (2) high sensitivity of simulated data to small changes in model parameters; and (3) inability to preserve the balance of events, censoring times, and effect sizes.

For these reasons, we use a different approach based on the estimated cumulative baseline hazard $\widehat{\Lambda}_0(t) = \int_0^t \widehat{\lambda}_0(u)du$ from the fitted model using NHANES data. The simulation procedure follows the following steps: (1) obtain the estimated cumulative baseline hazard $\widehat{\Lambda}_0(t)$; (2) obtain the estimated linear predictor $\widehat{\eta}_i$; (3) calculate the estimated survival function $\widehat{S}_i(t) = \exp\{-e^{\widehat{\eta}_i}\widehat{\Lambda}_0(t)\}$; (4) simulate survival time $\widetilde{T}_i$ from $\widehat{S}_i(t)$; and (5) simulate censoring time $\widetilde{C}_i$ from the empirical distribution of censoring times in NHANES and obtain simulated survival data $(\widetilde{Y}_i, \widetilde{\delta}_i)$ from $\widetilde{T}_i$ and $\widetilde{C}_i$.

Below we provide the software steps associated with the simulations framework. Here we assume that the model was fit on the real data and that results are stored in the object `fit`. We also assume that new functional covariates are simulated and stored in the matrix `X_new`. The time to event information from the real data is stored in the vector `time` and the event indicator is stored in the vector `event`.

```
#Simulate survival data that mimic real data
## fit: the fitted model using gam() function of mgcv package
## X_new: simulated functional predictors
## time: time to event info from real data
## event: binary indicator of event from real data
```

We use the function `scam` [236] to conduct smoothing with monotonicity constraints. Scam stands for "Shape constrained additive models," not a choice of name that we endorse, but an R function that we gladly use. This allows us to obtain a non-decreasing cumulative baseline hazard. Below, the vector `t0` contains the event times and `H0_hat` contains the Breslow estimator of the cumulative hazard function. The next line uses the `scam` function to obtain an increasing fit to the cumulative hazard function stored in `H0_fit`. Here the option `bs = "mpi"` indicates that the function uses monotone increasing P-splines, while the `-1` indicates that the smooth is without an intercept. The remainder of the code chunk indicates how to predict the smooth estimator of the cumulative hazard function at an equally spaced grid of points between $[0, 10]$. Results are stored in the vector `H0_prd`.

```
#1. Estimate the cumulative baseline hazard
library(scam)
## derive raw estimates
t0 <- rev(fit$family$data$tr)
H0_hat <- rev(fit$family$data$h)
## smooth while imposing non-decreasing shape constraints
H0_fit <- scam(H0_hat ~ s(t0, bs = "mpi") - 1)
## set the time grid to evaluate cumulative baseline hazard
nt_pred <- 1000 # number of potential survival times
tgrid_sim <- seq(0, 10, len = nt_pred)
## derive final estimates on the grid
H0_prd <- pmax(0, predict(H0_fit, newdata = data.frame(t0 = tgrid_sim)))
```

The second step consists of using the functional predictors stored in the matrix X_new and the fitted survival model stored in `fit` to estimate the study participant-specific linear predictors. The elements in X_new could be the same as the ones used in fitting the model, or could be replaced with new trajectories simulated from a functional data generating mechanism. Here `nt` is the number of columns of X_new, which is the dimension of the functional predictors (study participants are stored by rows). The vector `tind` contains an equally spaced grid of points with a dimension equal to the dimension of the functional space. It is used to construct the Riemann sums that approximate the functional regression. The data frame `data_sim` contains the data structure where predictions are conducted. The `act_mat` element is the matrix of functional covariates, `lmat` is a matrix of dimension equal to the dimension of X_new with every entry equal to 1 / nt. The `tmat` is another matrix of dimension equal to the dimension of X_new, with the same rows corresponding to the vector `tind`. In this case we considered the case when we only have functional predictors, though additional terms could be added to account for standard covariates. The resulting linear predictors are then stored in the vector `eta_i`.

```
#2. Estimate the linear predictor
nt <- ncol(X_new)
tind <- seq(0, 1, len = nt)
data_sim <- data.frame(act_mat = I(X_new),
          lmat = I(matrix(1 / nt, ncol = nt, nrow = nrow(X_new))),
          tmat = I(matrix(tind, ncol = nt, nrow = nrow(X_new),
          byrow = TRUE)))
eta_i <- predict(fit, newdata = data_sim, type = "terms")
```

The third step uses the survival function $S_i\{t|\mathbf{Z}_i, W_i(\cdot)\} = \exp\{-\exp(\eta_i)\Lambda_0(t)\}$, where η_i is the linear predictor; see derivation of equation (7.4). In the code, the matrix `Si` contains the survival function $S_i\{t|\mathbf{Z}_i, W_i(\cdot)\}$ at every time point in the vector `tgrid.sim`. Each row in the matrix `Si` corresponds to a simulated study participant and every column corresponds to a particular time from baseline. The vector `eta_i` corresponds to the simulated linear predictor for all study participants and `H0_prd` is the vector containing the estimated cumulative hazard function.

```
#3. Estimate the survival function
Si <- exp(-(exp(eta_i) %*% H0_prd))
```

Step 4 consists of simulating survival times for each study participant based on the calculated survival function. This approach uses a simulation trick to obtain samples from a random variable with a given survival function $S(\cdot)$ with corresponding cdf $F(\cdot)$. Note that if $U \sim U[0,1]$, then $X = F^{-1}(U)$ is a random variable with survival function $S(\cdot)$. Indeed, for any $x \in \mathbb{R}$, $P\{F^{-1}(U) \leq x\} = P\{U \leq F(x)\} = F(x)$. The last equation can be rewritten as $P\{1 - F(x) \leq 1 - U\} = P\{S(x) \leq 1 - U\} = F(x)$. As $1 - U$ follows a uniform distribution on $[0,1]$ (just like U) this suggests the following algorithm for simulating random variables with survival function $S(\cdot)$: (1) simulate a random variable from the uniform distribution; and (2) identify the first x such that $S(x) < U$. Below we take advantage of this approach and we start by simulating a vector of uniform random variables `U` of length equal to the number of study participants. This is the $1 - U$ variable, but there is no need to do that as the effect is the same. The simulated survival times are stored in the vector `Ti` of the same length as the number of study participants. The `for` loop is a description of how to simulate survival times using the Monte Carlo simulation trick described above.

```
#4. Simulate survival times
U <- runif(nrow(X_new))
Ti <- rep(NA, nrow(X_new))
for(i in 1:nrow(X_new)){
    if(all(Si[i,] > U[i])){Ti[i] <- max(tgrid_sim) + 1}
    else{Ti[i] <- tgrid_sim[min(which(Si[i,] < U[i]))]}
}
```

Step 5 is relatively straightforward. The first component is to simulate censoring times for each study participant. The censoring times for each person who did not die can be obtained in the vector `time[event==0]`. Indeed, `event==0` is the indicator for all individuals who did not experience the event and the vector `time` contains the censoring times for these individuals. Thus, the vector `Ci` is obtained by sampling with replacement from the vector `time[event==0]` a number of censoring times equal to the number of study participants. The method avoids fitting a parametric model to the observed censoring time, which may be difficult. The observed survival data is the entry-wise minimum between the simulated censoring and survival time and is stored in the vector `Yi`. The indicator δ_i of death is stored in the vector `di` obtained by comparing entry-wise whether the survival time `Ti` is smaller than the censoring time `Ci`.

```
#5. Simulate censoring times and observed survival data
Ci <- sample(time[event == 0], size = nrow(X_new), replace = TRUE)
Yi <- pmin(Ci, Ti) # observed time to event
di <- as.numeric(Ti <= Ci) # binary event indicator
```

8

Multilevel Functional Data Analysis

Multilevel functional data are becoming increasingly common in many studies. Such data have all the characteristics of the traditional multilevel data, except that the individual measurement is not a scalar, but a function. For example, the NHANES 2011-2014 study collected minute-level accelerometer data for up to 7 consecutive days from each participant. This is a multilevel functional data set because for each participant (level-1) and each day of the week (level-2) a function is measured (MIMS values summarized at the minute level). Another example is the study of sleep electroencephalograms (EEGs) described in [62], where normalized sleep EEG δ-power curves were obtained at two visits for each participant enrolled in the Sleep Heart Health Study (SHHS) [239]. A far from exhaustive list of applications includes colon carcinogenesis studies [207, 279], brain tractography and morphology [98, 109], functional brain imaging through EEG [271], longitudinal mortality data from period life tables [41], pitch linguistic analysis [8], longitudinal EEG during a learning experiment [19], animal studies [278], end-stage renal disease hospitalizations in the US [178], functional plant phenotype [333], objective physical activity [175, 273], and continuous glucose monitoring [92, 270]. Each application raises unique methodological and computational challenges that contribute to a rapidly developing area of research. This leads to a substantial and diverse body of literature that cannot be completely addressed here. Instead, we focus on a specific group of methods and attempt to make connections with this vast and rapidly evolving literature. What we present is neither as exhaustive nor as inclusive as we would like it to be, but it is thematically and philosophically self-contained.

Compared to single-level functional data, multilevel functional data contain additional structure induced by known sampling mechanisms. Here we define multilevel functional data analysis (MFDA) as the analysis of functional processes with at least two levels of functional variability. This includes nested, crossed, and longitudinal functional data structures. Such data structures are different from traditional longitudinal data (where the individual observation is a scalar), which are analyzed using single-level functional approaches [283, 334]. They are also different from functional ANOVA [305], which allows for different functional means of groups, but have only one source of functional variability around these means.

Multilevel functional data raise many new methodological problems that we organize in the following three categories:

1. *Multilevel decomposition of variability*, which involves the estimation of volume and patterns of variability at different levels. While functional principal component analysis (FPCA) provides a solid framework of variability decomposition for single-level functional data, the extension to multilevel functional data is nontrivial. We introduce the multilevel functional principal component analysis (MFPCA) approach [58, 62] to address this problem. Structured functional principal component analysis (SFPCA) [273], an extension of MFPCA that provides solutions for nested and crossed functional data, is also discussed.

2. *Multilevel function-on-scalar regression*, which involves regressing multilevel functional responses on a set of scalar predictors. This is a natural extension of MF-PCA, where the outcome is a set of random curves observed within a measurement

unit (e.g., study participant) and the predictors are a set of scalar variables. This problem is increasingly common in different applications. For example, suppose that we are interested in modeling the effects of age, gender, and day of the week on physical activity intensity at different times of the day in the NHANES population. The outcome may be the collection of objective physical activity functions measured during multiple days (each day is a function, each day is a repetition of the measurement process). Functional mixed models have been introduced to address this type of structure for single-level [114] and multilevel functional data [27, 207]. The general idea is to start with the traditional mixed effects model structure and expand it to functional measurements. While a substantial literature exists for the estimation and inference in functional mixed models, we focus on two methods for which ready-to-use R software is available: Functional Additive Mixed Models (FAMM) [263] and Fast Univariate Inference (FUI) [57].

3. *Multilevel scalar-on-function regression*, which involves regressing a scalar response on a set of multilevel functional predictors. We extend the PFR approach used to fit a functional linear model in Chapter 4 by adding random terms and introduce the longitudinal penalized functional regression (LPFR) [103] method.

8.1 Data Structure in NHANES

We now describe the physical activity data in NHANES, which has a multilevel functional structure. A wrist-worn device was deployed in NHANES 2011-2014, and each participant was asked to wear it for seven consecutive days. The accelerometry data were collected from the device and released in minute-level MIMS, a measure of physical activity intensity. During data cleaning, some days were excluded for not meeting a set of quality standards; see Section 1.2.1 for details on exclusion criteria. Figure 8.1 displays the physical activity data structure in NHANES. Each column represents the physical activity profile of one study participant on different days of a week. The unique identifier (SEQN) of each participant is shown in the top row, ranging from 62161 to 83731. The number of days available varies by participant. For example, the physical activity data on Saturday for participant SEQN 62161 was excluded because it was considered to be of low quality. Therefore, study participant SEQN 62161 has only six days of data.

This is a multilevel functional and, specifically, two-level functional data set because: (1) the data were collected from multiple study participants (level-1) at multiple days of the week (level-2); and (2) for each study participant at each day of the week, the physical activity data were measured in MIMS (same unit) at each minute of a day (ordered) resulting in 1,440 observations arranged by time of day (functional). The dimension of the functional observation is 1,440, which can be viewed as high-dimensional. This dimension could easily increase if physical activity is summarized in smaller units of time (e.g., seconds instead of minutes).

Many questions arise from this multilevel functional data structure including: (1) What are the volume and patterns of participant- and day-level physical activity variability? and (2) How does the day of the week affect physical activity? Each of these questions corresponds to one of the research directions introduced at the beginning of this chapter. In the following sections, we discuss multilevel functional data methods designed to answer these questions.

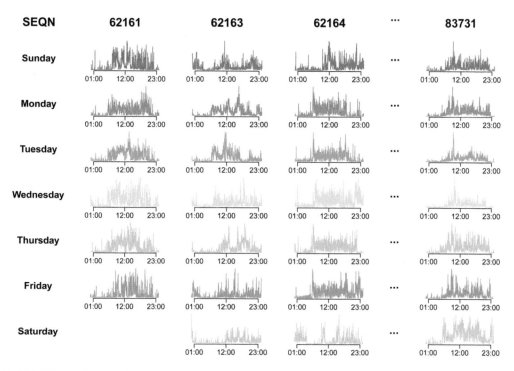

FIGURE 8.1: Physical activity data structure in NHANES 2011-2014. Each study participant is shown in one column and each row corresponds to a day of the week from Sunday to Saturday. The x-axis in each panel is time in one-minute increments from midnight to midnight.

8.2 Multilevel Functional Principal Component Analysis

8.2.1 Two-Level Functional Principal Component Analysis

Two-level functional data, having two levels of functional variability, is one of the simplest generalizations of single-level functional data. A canonical example of two-level functional data is when repeated functional measurements are obtained from the same study participant. For example, physical activity data collected every minute of the day from multiple days per NHANES participant.

We now focus on the first question arising from the NHANES data: What are the volume and patterns of participant- and day-level physical activity variability? To address this problem, we provide the technical details and practical implementations of multilevel functional PCA (MFPCA) [58, 62]. This is a generalization of functional principal component analysis (FPCA) to two-level functional data that accounts for the intrinsic correlation of functional observations within study participant. For presentation purposes, we use "subject" to refer to level-1 and "visit" to refer to level-2 observations, but the method is applicable to any two-level functional data structure.

8.2.1.1   Two-Level FPCA Model

For subject $i = 1, \ldots, n$ at visit $m \in \{1, \ldots, M_i\}$, denote by $W_{im}(s)$ the observed data at location $s \in \{s_1, \ldots, s_p\} \in S$, where S is an interval in $\mathbb{R}$. In the NHANES data set, $n = 12{,}610$, M_i varies between 3 and 7, $p = 1{,}440$, and $W_{im}(s)$ is the observed physical activity intensity for the mth day of the week of the ith study participant at minute s of the day. The total number of days is 79,910, which is less than $12{,}610 \times 7$ since some days were excluded due to the low quality of accelerometry data. The MFPCA model assumes a two-level structure with measurement error:

$$W_{im}(s) = X_{im}(s) + \epsilon_{im}(s) = \mu(s) + \eta_m(s) + U_i(s) + V_{im}(s) + \epsilon_{im}(s), \qquad (8.1)$$

where $\mu(s)$ is the population mean function, $\eta_m(s)$ is the mth visit-specific shift from $\mu(s)$, $U_i(s)$ is the ith subject-specific deviation, $V_{im}(s)$ is the mth visit-specific residual deviation from $U_i(s)$, and $\epsilon_{im}(s)$ is a white noise process with constant variance σ_ϵ^2 across S. In the NHANES study, $\mu(s)$ and $\eta_m(s)$ are treated as fixed functions, which is a reasonable assumption since NHANES contains over 12,000 study participants with multiple days of data. The random functions $U_i(s)$ capture the between-subject variation and are modeled as mean 0 Gaussian Processes with covariance function $K_U(s,t) = \text{cov}\{U_i(s), U_i(t)\}$. The random functions $V_{im}(s)$ capture the within-subject variation and are treated as mean 0 Gaussian Processes with covariance function $K_V(s,t) = \text{cov}\{V_{im}(s), V_{im}(t)\}$. We further assume that $U_i(s)$ and $V_{im}(s)$ are mutually uncorrelated. Model (8.1) is the "hierarchical functional model" introduced in [208], which is a particular case of the "functional mixed models" introduced in Section 8.3.

Define $K_X(s,t) = \text{cov}\{X_{im}(s), X_{im}(t)\}$ the total covariance function of the smoothed functional data, $X_{im}(s)$, which satisfies $K_X(s,t) = K_U(s,t) + K_V(s,t)$. Since the between-subject covariance function $K_U(s,t)$ is continuous symmetric non-negative definite, Mercer's theorem [199] ensures the eigendecomposition $K_U(s,t) = \sum_{k \geq 1} \lambda_k^{(1)} \phi_k^{(1)}(s) \phi_k^{(1)}(t)$, where $\lambda_1^{(1)} \geq \lambda_2^{(1)} \geq \cdots \geq 0$ are non-negative eigenvalues with associated orthonormal eigenfunctions $\phi_k^{(1)}(s)$. That is, $\int_S \phi_{k_1}^{(1)}(s) \phi_{k_2}^{(1)}(s) ds = \mathbf{1}_{\{k_1 = k_2\}}$ where $\mathbf{1}_{\{\cdot\}}$ is the indicator function. It follows from the Kosambi-Karhunen–Loève (KKL) theorem that $U_i(s) = \sum_{k \geq 1} \xi_{ik} \phi_k^{(1)}(s)$, where ξ_{ik} is the score of the ith subject on the kth principal component with mean 0 and variance $\lambda_k^{(1)}$. Similarly, the within-subject covariance function has eigendecomposition $K_V(s,t) = \sum_{k \geq 1} \lambda_k^{(2)} \psi_k^{(2)}(s) \psi_k^{(2)}(t)$, where $\lambda_1^{(2)} \geq \lambda_2^{(2)} \geq \cdots \geq 0$ are non-negative eigenvalues with associated orthonormal eigenfunctions $\psi_k^{(2)}(s)$, and $V_{im}(s) = \sum_{k \geq 1} \zeta_{imk} \psi_k^{(2)}(s)$, where ζ_{imk} are scores with mean 0 and variance $\lambda_k^{(2)}$ and are mutually uncorrelated. Model (8.1) now becomes

$$W_{im}(s) = \mu(s) + \eta_m(s) + \sum_{k \geq 1} \xi_{ik} \phi_k^{(1)}(s) + \sum_{k \geq 1} \zeta_{imk} \psi_k^{(2)}(s) + \epsilon_{im}(s), \qquad (8.2)$$

where $\mu(s), \eta_m(s), \phi_k^{(1)}(s), \psi_k^{(2)}(s)$ are fixed functional effects and ξ_{ik}, ζ_{imk} are uncorrelated random variables with mean 0. The main advantage of this decomposition is that the principal component decomposition at both levels substantially reduces the dimensionality of the problem.

8.2.1.2   Estimation of the Two-Level FPCA Model

To estimate the unknown fixed parameters and predict the random parameters in (8.2), MFPCA uses the following steps:

1. *Estimate the mean $\mu(s)$ and visit-specific functions $\eta_m(s)$ using univariate smoothers under the working independence assumption.* The choice of the smoother is flexible and has small effects on the estimator when the sample size is large, such as in the NHANES example. Popular smoothers include penalized spline smoothing [258] and local polynomial smoothing [76]. The penalized spline smoothing was adopted in the R implementation of MFPCA. Denote $\widetilde{W}_{im}(s) = W_{im}(s) - \hat{\mu}(s) - \hat{\eta}_m(s)$, where $\hat{\mu}(s)$ and $\hat{\eta}_m(s)$ are the estimators of $\mu(s)$ and $\eta_m(s)$, respectively.

2. *Construct method of moments (MoM) estimators of the total covariance function $K_X(s,t)$ and the between-subject covariance function $K_U(s,t)$.* If $\widetilde{W}_{im}(s)$ is the centered data obtained in Step 1, the estimator for $K_X(s,t)$ is $\widehat{G}_X(s,t) = \sum_{i=1}^{I}\sum_{j=1}^{J_i} \widetilde{W}_{im}(s)\widetilde{W}_{im}(t)/\sum_{i=1}^{I} M_i$, and the estimator for $K_U(s,t)$ is $\widehat{G}_U(s,t) = 2\sum_{i=1}^{n}\sum_{m_1<m_2} \widetilde{W}_{im_1}(s)\widetilde{W}_{im_2}(t)/\sum_{i=1}^{I} M_i(M_i-1)$. Note that $\widehat{G}_X(s,t)$ is not an unbiased estimator of $K_X(s,t)$ on the main diagonal ($s=t$) due to the measurement error, $\epsilon_{im}(s)$. Indeed, $\text{cov}\{W_{im}(s), W_{im}(t)\} = \text{cov}\{X_{im}(s), X_{im}(t)\} + \sigma_\epsilon^2 \mathbf{1}_{\{s=t\}}$. For $K_U(s,t)$ this is not a problem, since $\text{cov}\{W_{im_1}(s), W_{im_2}(t)\} = \text{cov}\{X_{im_1}(s), X_{im_2}(t)\}$.

3. *Obtain the smooth estimators $\widehat{K}_X(s,t)$ and $\widehat{K}_U(s,t)$ of $K_X(s,t)$ and $K_U(s,t)$, respectively.* This step is achieved by applying a bivariate smoother to the off-diagonal elements of $\widehat{G}_X(s,t)$ and the entire matrix of $\widehat{G}_U(s,t)$ obtained from Step 2, respectively. The idea of dropping diagonal elements of $\widehat{G}_X(s,t)$ was introduced by [283] to avoid the extra measurement error variance along the diagonal. The choice of bivariate smoother in this step is flexible including thin plate penalized splines [313] and fast bivariate penalized splines [330]. Finally, the within-subject covariance estimator of $K_V(s,t)$ is obtained from $\widehat{K}_V(s,t) = \widehat{K}_X(s,t) - \widehat{K}_U(s,t)$. To ensure that covariance estimators are positive semi-definite, their negative eigenvalues are truncated to 0; see, for example, [336].

4. *Conduct eigenanalysis on $\widehat{K}_U(s,t)$ to obtain $\widehat{\lambda}_k^{(1)}$ and $\widehat{\phi}_k^{(1)}(s)$, and on $\widehat{K}_V(s,t)$ to obtain $\widehat{\lambda}_k^{(2)}$ and $\widehat{\psi}_k^{(2)}(s)$.* In practice, this is achieved by performing eigendecompositions on the estimated covariance matrices. Computational details can be found in Section 2.1. An important decision is how to choose the number of eigenfunctions at each level. A simple approach is to use the percent of explained variance. Specifically, denote by P_1 the proportion of variability explained. The number of level-1 components is chosen as $N_1 = \min\{l : \rho_l^{(1)} \geq P_1\}$, where $\rho_l^{(1)} = \sum_{k=1}^{l} \lambda_k^{(1)}/\sum_{k\geq 1} \lambda_k^{(1)}$ is the proportion of the estimated variance using the first k components. The number of components at level-2, N_2, is chosen similarly.

5. *Estimate the measurement error variance σ_ϵ^2.* Given the smoothed estimates $\widehat{K}_X(s,s)$ obtained in Step 3 and the estimator $\widehat{G}_X(s,s)$ obtained in Step 2, the error variance is estimated as $\widehat{\sigma}_\epsilon^2 = \int_S \{\widehat{G}_X(s,s) - \widehat{K}_X(s,s)\} ds$.

6. *Predict the principal component scores ξ_{ik} and ζ_{imk} using Markov Chain Monte Carlo (MCMC) or best linear unbiased prediction (BLUP).* The score prediction is a more challenging problem in MFPCA than in the single-level FPCA. For single-level FPCA, the scores can be obtained by direct numerical integration, which is easy to implement. However, this approach cannot be directly extended to multilevel functional data because the functional bases at the two levels $\{\phi_k^{(1)}(s)\}$ and $\{\psi_k^{(2)}(s)\}$ are not mutually orthogonal. MFPCA addresses this problem using

mixed effects model inference. Indeed, assume that we have obtained the estimates of $\mu(s), \eta_m(s), N_1, N_2, \lambda_k^{(1)}, \lambda_k^{(2)}, \phi_k^{(1)}(s), \psi_k^{(2)}(s), \sigma_\epsilon^2$ following Steps 1-5 above. The MFPCA model becomes

$$W_{im}(s) = \mu(s) + \eta_m(s) + \sum_{k=1}^{N_1} \xi_{ik}\phi_k^{(1)}(s) + \sum_{k=1}^{N_2} \zeta_{imk}\psi_k^{(2)}(s) + \epsilon_{im}(s) \,, \qquad (8.3)$$

where $\xi_{ik} \sim N(0, \lambda_k^{(1)})$ and $\zeta_{imk} \sim N(0, \lambda_k^{(2)})$ can be viewed as mutually independent random effects and $\epsilon_{im}(s) \sim N(0, \sigma_\epsilon^2)$ are mutually independent residuals. If the number of principal components at level 1 and 2 is relatively small, the mixed effects model (8.3) remains relatively simple. The assumption of independence of random effects further reduces the methodological and computational complexity of the model. The Gaussian distribution assumption of scores (random effects) and errors is convenient, though other distributions could be assumed. As model (8.3) is a linear mixed model where ξ_{ik} and ζ_{imk} are random effects, standard mixed model inferential approaches can be used, including Bayesian MCMC [62] and BLUP [63].

The MFPCA is one of the simplest multilevel functional models. It also provides an explicit decomposition of variability using latent processes based on the same philosophy used in traditional mixed effects modeling. Understanding its structure and designing reasonable inferential approaches can help in more complex situations. An important characteristic of MFPCA is that each inferential step is practical and does not require highly specialized software. For example, we can use any type of univariate smoother to estimate the population mean function in Step 1 or bivariate smoother to estimate the covariance functions in Step 3. We have not identified major differences in performance between reasonable smoothers. However, here we focus on non-parametric smoothers obtained from a rich spline basis plus appropriate quadratic penalties.

8.2.1.3   Implementation in R

We now show how to implement MFPCA for the NHANES data set in R. As introduced in Section 8.1, physical activity data was collected for each participant in every minute of each eligible day. Figure 8.2 displays the data storage format in R. The data frame `nhanes_ml_df` consists of 79,910 rows and 6 columns. Each row represents the data for one day of the week for one study participant. The total number of days is 79,910 in this analysis. The `SEQN` column stores the unique study participant identifier in a vector. Notice that `SEQN` is repeated for each row (day) that corresponds to the same study participant. The `dayofwear` column stores the day-of-wear information in a vector, where, for example, a value of 2 corresponds to the second day of wearing the device. The `dayofweek` column stores the day of the week information in a vector, where "1" represents "Sunday," "2" represents "Monday," ..., and "7" represents "Saturday." For each study participant, the starting day of the week of wearing the device is not necessarily "1" (Sunday). The `MIMS` column stores the physical activity data in a matrix with 1,440 columns, each column corresponding to a minute of the day starting from midnight. This is achieved using the `I()` function. The column names are `"MIN0001"`, `"MIN0002"`, ..., `"MIN1440"`, representing the time of day from midnight to midnight. For example, for study participant `SEQN 62161` on Sunday, the `MIMS` value at 12:00-12:01 AM is stored in the first column (column name `"MIN0001"`) of the first row of `MIMS` matrix. For each study participant, the age and gender information are stored in `age` and `gender` columns, respectively. We will use this format to store and model multilevel functional data in R throughout this chapter.

```
'data.frame':   79910 obs. of  6 variables:
 $ SEQN     : num  62161 62161 62161 62161 62161 ...
 $ dayofwear: num  1 2 3 4 5 6 1 2 3 4 ...
 $ dayofweek: chr  "1" "2" "3" "4" ...
 $ MIMS     : 'AsIs' num [1:79910, 1:1440] 0.046 0.013 0.01 6.51 0.019 ...
  ..- attr(*, "dimnames")=List of 2
  .. ..$ : chr [1:79910] "62161" "62161" "62161" "62161" ...
  .. ..$ : chr [1:1440] "MIN0001" "MIN0002" "MIN0003" "MIN0004" ...
 $ age      : num  22 22 22 22 22 22 14 14 14 14 ...
 $ gender   : Factor w/ 2 levels "Male","Female": 1 1 1 1 1 1 1 1 1 1 ...
```

FIGURE 8.2: An example of the multilevel data structure in R, where accelerometry data is a matrix of observations stored as a single column of the data frame.

The code below shows how to implement MFPCA on this NHANES data set to decompose the within-subject and between-subject variability of physical activity using the `mfpca.face()` [58] function from the **refund** package. This function complements and substantially improves the previous function `mfpca.sc()` [62], especially from a computational perspective. The new function allows the use of much higher dimensional functions by incorporating fast covariance estimation (FACE, [331]) into the estimation; see [58] for technical details. For example, fitting MFPCA using `mfpca.face()` on the NHANES data takes less than a minute on a standard laptop, a remarkable reduction from the function `mfpca.sc()` [62], which took several days (computation was actually stopped after 24 hours).

```
#Fit MFPCA using mfpca.face
fit_mfpca <- mfpca.face(Y = nhanes_ml_df$MIMS,
                        id = nhanes_ml_df$SEQN,
                        visit = nhanes_ml_df$dayofweek)
```

The `Y` argument specifies a matrix of functions on a regular grid, corresponding to the multilevel functional dataset. The `id` argument identifies clusters in the data (in our case study participants). The `visit` argument specifies the visit information (in our case day of the week) for each study participant. For more details, please check the tutorial of `mfpca.face()` function.

8.2.1.4 NHANES Application Results

The estimated population mean function $\mu(s)$ and mean function for each day of the week $\mu(s) + \eta_m(s)$ from NHANES data are shown in Figure 8.3. The population mean function is shown as a black solid line. The weekend means (Saturday and Sunday) are shown as dashed lines, while the weekday curves are shown as dotted lines. In general, we observe lower physical activity intensity at night and higher physical activity during the day, which is consistent with the expected average patterns of physical activity in the population. The plot shows that, on average, people tend to be less active on Saturday and Sunday mornings (lower than average MIMS values between 5 AM and 10 AM) and more active on Friday and Saturday evenings (higher than average MIMS values between 9 PM and 2 AM). This weekday/weekend effect is consistent with what is expected about average patterns of physical activity in the US population.

For a percent variance explained (PVE) of 0.99 at each level, we identified 23 level-1 principal components and 31 level-2 principal components. The total explained between-subject variance is $\sum_{k_1=1}^{\infty} \lambda_{k_1}^{(1)} = 26.65$. The total explained within-subject variance is $\sum_{k_2=1}^{\infty} \lambda_{k_2}^{(2)} = 44.14$, which is nearly twice that of the between-subject variance. The

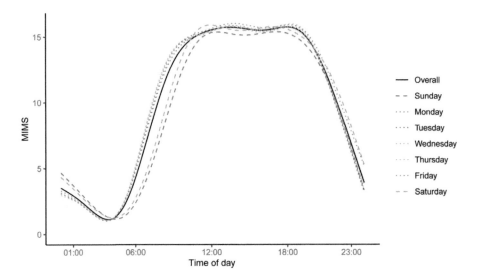

FIGURE 8.3: Estimated population mean function $\mu(s)$ and day-of-the-week-specific mean function $\mu(s) + \eta_m(s)$ in the NHANES 2011-2014 dataset using MFPCA. The population mean function is shown as a black solid line. The weekend-specific (Saturday, Sunday) curves are shown as dashed lines. The weekday-specific curves are shown as dotted lines.

estimated proportion of variability explained by level-1 variation is 0.38, defined as $\sum_{k_1=1}^{\infty} \lambda_{k_1}^{(1)} / (\sum_{k_1=1}^{\infty} \lambda_{k_1}^{(1)} + \sum_{k_2=1}^{\infty} \lambda_{k_2}^{(2)})$ in [62]. The estimated variance of the error term $\epsilon_{im}(s)$ is $\sigma_\epsilon^2 = 94.69$. The estimated top-three principal components of level 1 and level 2 in the NHANES dataset using MFPCA are shown in Figure 8.4 and Figure 8.6, respectively.

We first focus on level 1, which corresponds to the average individual-specific patterns of physical activity. The first level-1 principal component explains 55.45% of the total variability at level 1. It is positive between 6 AM and 11PM and negative at all other time of a day.

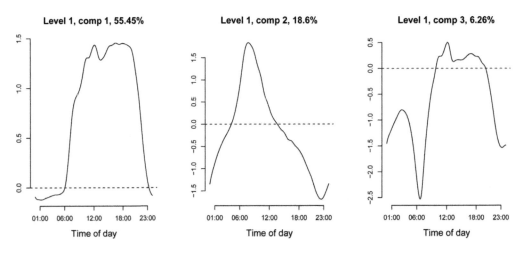

FIGURE 8.4: The first three estimated level-1 principal components in the NHANES 2011-2014 dataset using fast MFPCA. The proportion of variance explained out of the level-1 variance by each component is shown in the title of each panel.

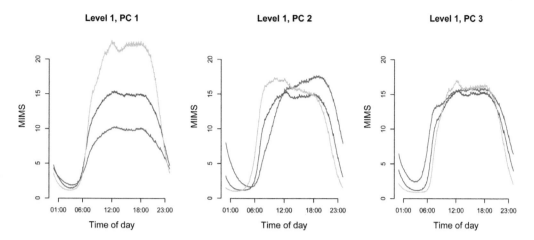

FIGURE 8.5: The average physical activity trajectories by tertile of the scores of the first three estimated level-1 principal components. For each panel, the red curve represents the average across individuals whose scores are in the first tertile, blue curve represents the average across individuals whose scores are in the second tertile, orange curve represents the average across individuals whose scores are in the third tertile.

The interpretation is that people with positive scores on this component are more active during the day and less active during the night. The second level-1 principal component explains 18.60% of the variability at this level and is negative only between 5 AM and 12 PM. People with positive scores on this component are less active in the morning and more active at other times of a day. The third level-1 principal component explains 6.26% of the variability at this level and is negative before 10 AM and after 8 PM and positive between 10 AM and 8 PM. People with positive scores on this component have lower activity during the night and early morning period, which could correspond to disrupted sleep, substantial changes from regular sleep hours, or night-shift work. The first three level-1 principal components explain roughly 80% of the variability at this level.

To further illustrate these estimated principal components in our data, Figure 8.5 shows the average physical activity curves by tertile of the scores of each principal component (the first PC to the third PC from left to right). For example, in the left panel, we partition the scores on the first PC into tertiles. The average minute-level physical activity intensity across individuals whose scores are in the first, second, and third tertile are shown in red, blue, and orange, respectively. From the left panel, we observe higher physical activity intensity during the day and lower at night for people whose scores are in the third tertile. These results are consistent with the left panel in Figure 8.4.

Similarly, the middle panel shows that individuals with scores in the highest tertile for the second level-1 PC (orange curve) tend to be more active in the morning with a slow decrease during the day and a very fast decrease in the evening. Study participants who are in the lowest tertile for the second level-1 PC (red curve) are, on average, less active in the morning but more active in the second part of the day compared with individuals with scores in the highest tertile for the second level-1 PC. Moreover, they tend to be more active during the night, which may indicate substantial change in their circadian patterns of activity.

The right panel in Figure 8.5 compares the average daily activity profiles by the tertiles of scores on the third level-1 principal component. The average profile of study participants

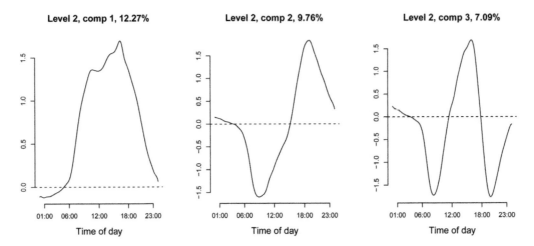

FIGURE 8.6: The first three estimated level-2 principal components in the NHANES 2011-2014 dataset using fast MFPCA. The proportion of variance explained out of the level-2 variance by each component is shown in the title of each panel.

in the highest tertile (orange) tends to be lower during the night, with a more rapid increase in physical activity intensity after 6 AM, generally higher PA between 10 AM and 6 PM, and a more pronounced decline after 8 PM.

Variability within level-2 principal components is much more evenly spread out and does not have the same quick drop in variance explained exhibited by level-1 principal components. This could be due to the fact that day-to-day differences may, at least in part, be due to de-synchronized physical activities. For example, a person may brush their teeth at 7 AM one morning and at 7:30 AM another. As we have discussed in Section 3.4 de-synchronization of functional data could lead to reductions in interpretability of principal components and slow decrease in the variance explained. In some sense, this is reasonable and indicates that the average subject-specific daily PA trajectories can be classified in a relatively small number of subgroups. In contrast, that may not be possible for the day-to-day variations.

Indeed, the first three level-2 principal components explain only 12.27%, 9.76%, and 7.09% of the variability at level 2, respectively. It is still instructive to plot these components and interpret them. The interpretation of level-2 principal components is different from that of level-1 principal components. Indeed, first level is focused on the variability patterns of activity between individuals, while the second level quantifies the variability from day to day within an individual. Notice, for example, that the first level-2 principal component is positive between 6 AM and 12 AM. This indicates that on days when a person has higher scores on this component, the individual is more active during that day and less active during that night compared to their average.

8.2.2 Structured Functional PCA

In Section 8.2.1, we described the MFPCA framework for two-level functional data. In many applications, though, the experimental design is more complicated. Structured functional principal component analysis (SFPCA, [273]) extends the MFPCA framework to structured functional data. Here "structured functional data" refers to the data collected from various experimental designs where the unit is a function; see [273] for a list of structured functional models where SFPCA is applicable. In this section, we focus on two more complex sampling scenarios: two-way crossed and three-way nested designs.

8.2.2.1   Two-Way Crossed Design

Let us consider again the MFPCA model described in (8.1)

$$W_{im}(s) = \mu(s) + \eta_m(s) + U_i(s) + V_{im}(s) + \epsilon_{im}(s). \tag{8.4}$$

In Section 8.2.1, we assumed that $\eta_m(s)$ is a fixed effect function and was estimated by smoothing the difference between the visit-specific average function and the population average function. This choice makes sense when the number of functional observations per study participant, M_i, is small and measurements are obtained after varying only one experimental condition (e.g., measuring physical activity data for multiple days within one individual). When multiple experimental conditions are changed there may be interest in analyzing the amount and patterns of variability between and within each experimental condition. For example, in clinical trials individuals can be repeatedly "crossed-over" from one treatment to another. In task fMRI studies [181], brain activity is monitored for study participants using repetitions of different tasks and rest. In a phonetic study described in [8, 273], the fundamental frequency (F0) of syllables from 19 nouns (experimental condition 1) were recorded from 8 native speakers (experimental condition 2) of the Luobozhai Qiang dialect in China.

To account for such data structures, model (8.4) can be modified to become a two-way crossed design if we assume that $\eta_m(s)$ and $U_i(s)$ are random. For notation consistency, we replace $\eta_m(s)$ with $Z_m(s)$ to denote a random effect function. The model can be rewritten as

$$W_{im}(s) = \mu(s) + U_i(s) + Z_m(s) + V_{im}(s) + \epsilon_{im}(s), \tag{8.5}$$

where $U_i(s)$ and $Z_m(s)$ are two uncorrelated processes with interaction $V_{im}(s)$. Model (8.5) is the functional equivalent of the two-way crossed-design random effects ANOVA model [152]. We use the name "crossed design" because the model allows crossing of the levels. Unsurprisingly, such modeling extensions create additional technical challenges. We now provide the technical details associated with fitting the crossed-designed functional model.

Since the fixed effect $\mu(s)$ can be estimated by smoothing the population mean function, without loss of generality we assume that the data are demeaned and focus on the random effects. For simplicity, we consider a noise-free model $W_{im}(s) = U_i(s) + Z_m(s) + V_{im}(s)$. The solution for accounting for the noise component, $\epsilon_{im}(s)$, is similar to what was done in the MFPCA approach and is described in detail in [273]. Denote $K_U(s,t) = E\{U_i(s)U_i(t)\}$, $K_Z(s,t) = E\{Z_m(s)Z_m(t)\}$, $K_V(s,t) = E\{V_{im}(s)V_{im}(t)\}$ as the covariance operators of mutually correlated mean-zero random processes $U_i(s), Z_m(s), V_{im}(s)$, respectively. Denote by $\phi_k^U(s), \phi_k^Z(s), \phi_k^V(s)$ the eigenfunctions of the corresponding covariance operators $\mathbf{K}_U, \mathbf{K}_Z, \mathbf{K}_V$. Using the Kosambi-Karhunen–Loève (KKL) expansion, the model becomes

$$W_{im}(s) = \sum_{k \geq 1} \xi_{ik}^U \phi_k^U(s) + \sum_{k \geq 1} \xi_{mk}^Z \phi_k^Z(s) + \sum_{k \geq 1} \xi_{imk}^V \phi_k^V(s), \tag{8.6}$$

where $\xi_{ik}^U, \xi_{jk}^Z, \xi_{ijk}^V$ are mutually independent random variables with mean 0 and variances $\lambda_k^Z, \lambda_k^V, \lambda_k^W$, respectively

The main differences between a two-way crossed design SFPCA and MFPCA are (1) the function $\eta_m(s)$ ($Z_m(s)$ after change of notation) is assumed to be a fixed effect in MFPCA and a random effect in SFPCA; and (2) MFPCA has two covariance operators, whereas two-way crossed design SFPCA has three. One major contribution of SFPCA was to show that all MoM estimators of these covariance operators have a "sandwich" form. To illustrate this idea, we next introduce some technical, but necessary, matrix notation.

Denote by $\mathbf{W} = (\mathbf{W}_{11}, \ldots, \mathbf{W}_{1M_1}, \ldots, \mathbf{W}_{n1}, \ldots, \mathbf{W}_{nM_n})$ the $p \times N$ matrix obtained by column binding $N = \sum_{i=1}^{n} M_i$ $p \times 1$ dimensional vectors $\mathbf{W}_{im} = \{W_{im}(s_1), \ldots, W_{im}(s_p)\}^t$. For the noise-free model, we have

$$
\begin{aligned}
&E[\{W_{im}(s) - W_{kl}(s)\}\{W_{im}(t) - W_{kl}(t)\}] \\
&= \begin{cases}
2\{K_V(s,t) + K_Z(s,t)\} & i = k, m \neq l\,; \\
2\{K_V(s,t) + K_U(s,t)\} & i \neq k, m = l\,; \\
2\{K_V(s,t) + K_U(s,t) + K_Z(s,t)\} & i \neq k, m \neq l\,.
\end{cases}
\end{aligned}
\tag{8.7}
$$

Let $\mathbf{H}_Z = 2(\mathbf{K}_V + \mathbf{K}_Z)$, $\mathbf{H}_U = 2(\mathbf{K}_V + \mathbf{K}_U)$, $\mathbf{H}_{UZ} = 2(\mathbf{K}_V + \mathbf{K}_U + \mathbf{K}_Z)$. For notation simplicity we assume $M_i = M$, though this assumption is not necessary. To account for missing data, define $n_{im} = 1$ if $W_{im}(s)$ is observed and 0 otherwise, and $n_{i0} = \sum_m n_{im}$, $n_{0m} = \sum_i n_{im}$, $N = \sum_{i,m} n_{im}$, $k_1 = \sum_i n_{i0}^2$, $k_2 = \sum_m n_{0m}^2$. We then define $\mathbf{D}_{N \times N} = \mathrm{diag}\{\mathbf{M}_1, \ldots, \mathbf{M}_n\}$ where $\mathbf{M}_i = n_{i0}\mathbf{I}_{n_{i0}}$, $\mathbf{E}_{n \times N} = \mathrm{diag}\{\mathbf{1}_{n_{10}}^t, \ldots, \mathbf{1}_{n_{n0}}^t\}$ where $\mathbf{1}_{n_{i0}} = (1, \ldots, 1)_{n_{i0}}^t$, $\mathbf{P}_{N \times N} = \mathrm{diag}\{\mathbf{P}_1, \ldots, \mathbf{P}_n\}$ where $(\mathbf{P}_i)_{n_{i0} \times n_{i0}} = \mathrm{diag}\{n_{01}, \ldots, n_{0n_{i0}}\}$, and $\mathbf{F}_{M \times N} = (\mathbf{f}_1, \ldots, \mathbf{f}_M)^t$ where $\mathbf{f}_j$ is a vector of value 1 on observations with $Z_m(s)$ and 0 otherwise. With this rather involved notation, the MoM estimators become

$$
\widehat{\mathbf{H}}_Z = \frac{1}{k_1 - N} \sum_i \sum_{m \neq l} (\mathbf{W}_{im} - \mathbf{W}_{il})(\mathbf{W}_{im} - \mathbf{W}_{il})^t = \frac{2}{k_1 - N} \mathbf{W}(\mathbf{D} - \mathbf{E}^t\mathbf{E})\mathbf{W}^t,
$$

$$
\widehat{\mathbf{H}}_U = \frac{1}{k_2 - N} \sum_{i \neq k} \sum_m (\mathbf{W}_{im} - \mathbf{W}_{km})(\mathbf{W}_{im} - \mathbf{W}_{km})^t = \frac{2}{k_2 - N} \mathbf{W}(\mathbf{P} - \mathbf{F}^t\mathbf{F})\mathbf{W}^t,
$$

$$
\widehat{\mathbf{H}}_{UZ} = \frac{1}{N^2 - k_1 - k_2 + N} \sum_{i \neq k} \sum_{m \neq l} (\mathbf{W}_{im} - \mathbf{W}_{kl})(\mathbf{W}_{im} - \mathbf{W}_{kl})^t
$$

$$
= \frac{2}{N^2 - k_1 - k_2 + N} \mathbf{W}(N\mathbf{I} - \mathbf{1}\mathbf{1}^t - \mathbf{D} + \mathbf{E}^t\mathbf{E} - \mathbf{P} + \mathbf{F}^t\mathbf{F})\mathbf{W}^t.
$$

$$\tag{8.8}$$

Therefore, the MoM estimators of covariance operators are simply

$$
\begin{aligned}
\widehat{\mathbf{K}}_Z &= (\widehat{\mathbf{H}}_{UZ} - \widehat{\mathbf{H}}_U)/2 := \mathbf{W}\mathbf{G}_Z\mathbf{W}^t\,; \\
\widehat{\mathbf{K}}_U &= (\widehat{\mathbf{H}}_{UZ} - \widehat{\mathbf{H}}_Z)/2 := \mathbf{W}\mathbf{G}_U\mathbf{W}^t\,; \\
\widehat{\mathbf{K}}_V &= (\widehat{\mathbf{H}}_Z + \widehat{\mathbf{H}}_U - \widehat{\mathbf{H}}_{ZV})/2 := \mathbf{W}\mathbf{G}_V\mathbf{W}^t\,,
\end{aligned}
\tag{8.9}
$$

which all have the "sandwich" form, $\mathbf{W}\mathbf{G}\mathbf{W}^t$. The matrix $\mathbf{G}$ for each process can be easily obtained from (8.8).

Here we provided the formula for the two-way crossed design. For more general multi-way crossed designs, the SFPCA paper [273] showed that the MoM estimators of covariance operators can always be written in "sandwich" form. After obtaining the covariance estimates, the eigenanalysis and score calculation of SFPCA follow the same principles as MFPCA.

8.2.2.2 Three-Way Nested Design

Another extension of the two-way nested functional design used in MFPCA occurs when there is a more complex hierarchy of the sampling mechanism. For example, [175, 273] describe three-way nested functional data in studies of continuously recorded objective physical activity (e.g., days are nested within weeks, which are nested within study participants). In a carcinogenesis study [9, 208] biomarkers were measured within cells that were nested within colonic crypts of rats that were further nested within diet groups. In a US study of hospitalization rates [178], dialysis hospitalizations were nested within dialysis facilities, which were further nested within geographic regions.

To model such data structures we introduce the three-way nested design. The de-meaned noise-free model is

$$W_{imk}(s) = U_i(s) + Z_{im}(s) + V_{imk}(s) , \qquad (8.10)$$

where $i = 1, \ldots, n$, $m = 1, \ldots, M_i$, $k = 1, \ldots, K_{im}$, and $U_i(s)$, $Z_{im}(s)$, and $V_{imk}(s)$ are three latent uncorrelated processes. Here we focus on the covariance estimation, as all other model fitting steps are similar to MFPCA. Note that

$$E[\{W_{imk}(s) - W_{luv}(s)\}\{W_{imk}(t) - W_{luv}(t)\}] \qquad (8.11)$$

$$= \begin{cases} 2K_V(s,t) & i = l, m = u, k \neq v; \\ 2\{K_V(s,t) + K_Z(s,t)\} & i = l, m \neq u; \\ 2\{K_V(s,t) + K_Z(s,t) + K_U(s,t)\} & i \neq l. \end{cases} \qquad (8.12)$$

Let $\mathbf{H}_V = 2\mathbf{K}_V$, $\mathbf{H}_Z = 2(\mathbf{K}_V + \mathbf{K}_Z)$, $\mathbf{H}_U = 2(\mathbf{K}_V + \mathbf{K}_Z + \mathbf{K}_U)$.

Denote by $\mathbf{W} = (\mathbf{W}_{111}, \ldots, \mathbf{W}_{11K_{11}}, \ldots, \mathbf{W}_{nM_n1}, \ldots, \mathbf{W}_{nM_nK_{nM_n}})$ the $p \times N$ matrix obtained by column binding $N = \sum_{i=1}^{n} \sum_{m=1}^{M_i} K_{im}$ $p \times 1$ dimensional vectors $\mathbf{W}_{ijk} = \{W_{ijk}(s_1), \ldots, W_{ijk}(s_p)\}^t$, $N_{i.} = \sum_{m=1}^{M_i} K_{im}$, $k_1 = \sum_{i,m} K_{im}^2$, $k_2 = \sum_i N_{i.}^2$. We then define $\mathbf{D}_1 = \text{diag}\{\mathbf{K}_{11}, \ldots, \mathbf{K}_{nM_n}\}$ where $\mathbf{K}_{im} = K_{im}\mathbf{I}_{K_{im}}$, $\mathbf{D}_2 = \text{diag}\{\mathbf{N}_1, \ldots, \mathbf{N}_n\}$ where $\mathbf{N}_i = N_{i.}\mathbf{I}_{N_{i.}}$, $\mathbf{E}_1 = \text{diag}\{\mathbf{1}_{K_{11}}^t, \ldots, \mathbf{1}_{K_{nM_n}}^t\}$, $\mathbf{E}_2 = \text{diag}\{\mathbf{1}_{N_{1.}}^t, \ldots, \mathbf{1}_{N_{n.}}^t\}$. With this notation, the MoM estimators can be written as

$$\widehat{\mathbf{H}}_V = \frac{1}{k_1 - N} \sum_{i,m} \sum_{k \neq v} (\mathbf{W}_{imk} - \mathbf{W}_{imv})(\mathbf{W}_{imk} - \mathbf{W}_{imv})^t$$

$$= \frac{2}{k_1 - N} \mathbf{W}(\mathbf{D}_1 - \mathbf{E}_1^t\mathbf{E}_1)\mathbf{W}^t,$$

$$\widehat{\mathbf{H}}_Z = \frac{1}{k_2 - k_1} \sum_i \sum_{m \neq u} \sum_{k,v} (\mathbf{W}_{imk} - \mathbf{W}_{iuv})(\mathbf{W}_{imk} - \mathbf{W}_{iuv})^t$$

$$\qquad\qquad (8.13)$$

$$= \frac{2}{k_2 - k_1} \mathbf{W}(\mathbf{D}_2 - \mathbf{E}_2^t\mathbf{E}_2 - \mathbf{D}_1 + \mathbf{E}_1^t\mathbf{E}_1)\mathbf{W}^t,$$

$$\widehat{\mathbf{H}}_U = \frac{1}{N^2 - k_2} \sum_{i \neq l} \sum_{m,u,k,v} (\mathbf{W}_{imk} - \mathbf{W}_{luv})(\mathbf{W}_{imk} - \mathbf{W}_{luv})^t$$

$$= \frac{2}{N^2 - k_2} \mathbf{W}(N\mathbf{I}_N - \mathbf{1}_N\mathbf{1}_N^t - \mathbf{D}_2 + \mathbf{E}_2^t\mathbf{E}_2)\mathbf{W}^t.$$

Hence, we obtain $\widehat{\mathbf{K}}_V = \widehat{\mathbf{H}}_V/2$, $\widehat{\mathbf{K}}_Z = (\widehat{\mathbf{H}}_Z - \widehat{\mathbf{H}}_V)/2$, and $\widehat{\mathbf{K}}_U = (\widehat{\mathbf{H}}_U - \widehat{\mathbf{H}}_Z)/2$. SFPCA also provided the formula of covariance estimators for general multi-way nested designs; see Appendix of [273] for details. The R code to fit a three-way nested model using SFPCA is available in the supplementary material of [273]. This R code is not deployed as software and may require additional work for specific applications. However, the modeling infrastructure exists.

8.3   Multilevel Functional Mixed Models

We now focus on the second question arising from NHANES: How does the day-of-wear affect physical activity? This question can be viewed as estimation of a possibly time varying,

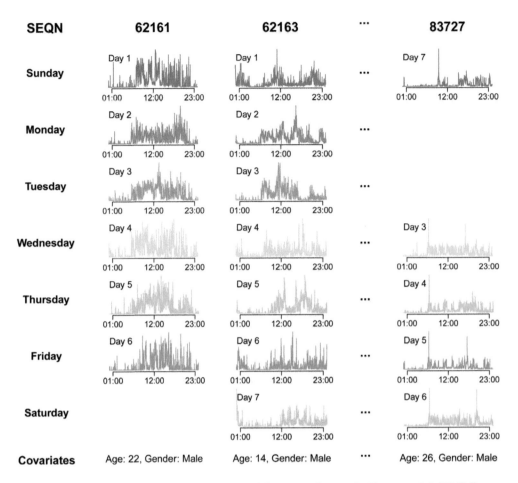

FIGURE 8.7: NHANES data with a multilevel functional mixed effects model (FMM) structure. Each column displays information for one participant, including physical activity data collected on each day of week. The day of wear is also displayed in each panel. For example, "Day 3" indicates the third day when the device was worn and could fall on any day of the week (e.g., Tuesday for study participant 62161 and Wednesday for study participant 83727). Age and gender of study participants are shown in the last row.

fixed effect that accounts for other variables (e.g., age, gender) and a complex structure of the residual functional variance (e.g., nesting of functional curves within study participants). Figure 8.7 provides an example, where the NHANES data has a multilevel functional structure and additional covariates. The last row displays two of the covariates, age and gender, for study participants. For example, participant SEQN 62161 was a 22-year-old male at the time of the study. In addition to the physical activity data displayed in Figure 8.1, NHANES also collected day-of-wear information for each individual. For example, participant SEQN 62163 started to wear the device on Sunday ("Day 1" displayed on Sunday's physical activity profile).

Functional mixed models (FMMs) are extensions of mixed effects models to functional data. They provide a useful framework that allows the explicit separation of different sources of observed variability: (1) fixed effects that may depend on the functional index (e.g., time); and (2) functional random effects that have a known structure (e.g., nested or crossed within the same sampling unit). The functional ANOVA model introduced in Section 8.2 is a particular case of FMM, where the visit indicator has a fixed time-varying effect and the functional residuals have a two-level nested structure.

Assume that the functional data are of the type $W_{im}(s)$ on an interval S, where $i = 1, \ldots, n$ is the index of the study participant and $m = 1, \ldots, M_i$ is the index of the visit. In this section we focus on two-level functional data, the simplest multilevel functional data. Let $\mathbf{X}_{im} = (X_{im1}, \ldots, X_{imQ})^t$ be the Q fixed effects variables and $\mathbf{Z}_{im} = (Z_{im1}, \ldots, Z_{imR})^t$ be the R random effects variables. The FMM can be written as

$$W_{im}(s) = \mathbf{X}_{im}^t \boldsymbol{\beta}(s) + \mathbf{Z}_{im}^t \mathbf{u}_i(s) + \epsilon_{im}(s), \tag{8.14}$$

where $\boldsymbol{\beta}(s) = \{\beta_1(s), \ldots, \beta_Q(s)\}^t$ are Q fixed effects functions and $\mathbf{u}_i(s) = \{u_{i1}(s), \ldots, u_{iR}(s)\}^t$ are the R random effects functions corresponding to the ith study participant.

The structure of the residuals $\epsilon_{im}(s)$ should be informed by and checked on the data. For example, if data can be assumed to be independent after accounting for the subject-specific mean, $\epsilon_{im}(s)$ can be assumed to be independent. However, in many studies that contain different levels of functional variability, this assumption is too stringent. Let us consider again the NHANES data and consider the de-meaned data $W_{im}(s) - \overline{W}_{i\cdot}(s)$, where $s = 1, \ldots, 1440$ and $\overline{W}_{i\cdot}(s) = \sum_{m=1}^{M_i} W_{im}(s)$ is the mean over M_i visits for study participant i. Figure 8.8 displays these de-meaned visit-specific functions for three NHANES study participants $i \in \{62161, 62163, 83727\}$. Here we abused the notation a bit, as i runs between 1 and the total number of study participants, whereas here we refer to the specific NHANES subject identifier. Alas, we hope that this is the biggest problem with our notation.

Note that even after subtraction, the functions still exhibit substantial structure and correlation patterns across the domain functional, especially during nighttime and daytime.

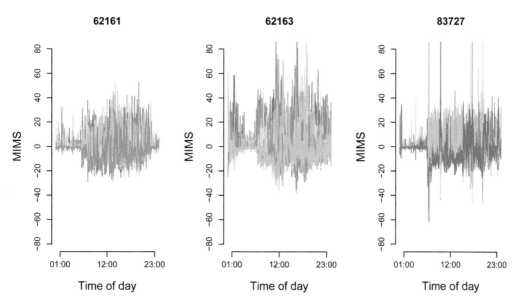

FIGURE 8.8: The visit-specific functions for three NHANES study participants after subtracting the subject-specific means. Each panel displays information for one participant. Within each panel, each line represents the function on one day of week.

Therefore, in the NHANES example, one cannot simply assume that after removing the study participant mean, the residuals are independent.

The outcome in (8.14) is a function, $W_{im}(s)$, and the component $\mathbf{X}_{im}^t \boldsymbol{\beta}(s)$ could be viewed as the "scalar regression" structure. Indeed, the components of $\mathbf{X}_{im}$ are scalar predictors (e.g., age and gender). For this reason, the model (8.14) can be viewed as an example of function-on-scalar regression [251]. An important difference here is that $W_{im}(s)$ is indexed both by i (e.g., study participant) and m (e.g., visit within study participant). Just as in standard mixed effects models, the component $\mathbf{Z}_{im}^t \mathbf{u}_i(s)$ captures the within-person variability and its structure follows the known sampling mechanisms. As discussed earlier, a major difference between model (8.14) and standard mixed effects models is that one cannot assume that $\epsilon_{im}(s)$ are independent across $s \in S$. Thus, the FMM model (8.14) is sometimes referred to in the literature as the "multilevel function-on-scalar regression model."

As discussed in Chapter 5, model (8.14) with independent $\epsilon_{im}(s)$ was known under different names and was first popularized by [242, 245] who introduced it as a functional linear model with a functional response and scalar covariates; see Chapter 13 in [245]. In Chapter 5 we used the FoSR nomenclature introduced by [251, 253], which refers directly to the type of outcome and predictor. It is difficult to pinpoint where these models originated, but they were first introduced as linear mixed effects (LME) models for longitudinal data [161], and then as functional models in recognition of new emerging datasets with denser and more complex sampling mechanisms [27, 79, 114, 243, 244]. The model with correlated $\epsilon_{im}(s)$ was likely introduced by [207], though other papers contained multilevel functional structures. Many methods have been developed to estimate fixed and random effects of FMM, such as [57, 106, 114, 207, 263]. Here we focus on two approaches with well-documented and easy-to-use software implementations, namely the Functional Additive Mixed Model (FAMM, [263]) and Fast Univariate Inference (FUI, [57]).

8.3.1 Functional Additive Mixed Models

Functional Additive Mixed Model (FAMM, [110, 263]) is an extensive framework for additive regression models of correlated functional responses. A list of functional models supported in FAMM was provided in Table 1 of [263].

For illustration purposes, consider the case when $Q = 2$, $\mathbf{X}_{im} = (1, X_{im})^t$, $R = 2$, and $\mathbf{Z}_{im} = (1, Z_{im})^t$. Models with additional fixed and random effects are estimated similarly. The model becomes

$$W_{im}(s) = \beta_0(s) + X_{im}\beta_1(s) + u_{i0}(s) + Z_{im}u_{i1}(s) + \epsilon_{im}(s). \quad (8.15)$$

To estimate (8.15), FAMM uses basis-expansion as a dimension reduction approach for all functional fixed (e.g., $\beta_0(s)$ and $\beta_1(s)$) and random (e.g., $u_{i0}(s)$ and $u_{i1}(s)$) effects parameters. In some applications, the residuals $\epsilon_{im}(s)$ are modeled as penalized splines. This is necessary when the visit-specific deviation from the study-specific mean has structure ($\epsilon_{im}(s)$ are not independent across s).

To start, each fixed effects function $\beta_q(s), q = 0, 1$ is approximated by the expansion of a set of chosen basis functions as $\beta_q(s) = \sum_{k=1}^{K_q^\beta} \beta_{qk} B_{qk}^\beta(s)$, where $B_{qk}^\beta(s)$, $k = 1, \ldots, K_q^\beta$ are spline bases (e.g., cubic spline, B-spline), and β_{qk} are the spline coefficients to be estimated.

The functional random effects are approximated as $u_{ir}(s) \approx \sum_{k=1}^{K_r^u} d_{irk} B_{rk}^u(s)$ for $r = 0, 1$, where $B_{rk}^u(s)$ are spline bases and d_{irk} are subject-specific spline coefficients. For penalized splines, K_q^β and K_q^u are usually selected to be large enough to capture the potential variability along the functional domain ([257]). Note that we have used the upper scripts β and u, respectively, for the number of knots and spline bases to distinguish between the expansion of the functional fixed and random effects, respectively. This is not strictly necessary and adds precision, but makes notation particularly messy.

With these choices the model becomes

$$
\begin{aligned}
W_{ij}(s) = \sum_{k=1}^{K_0^\beta} \beta_{0k} B_{0k}(s) + \sum_{k=1}^{K_1^\beta} \beta_{1k} X_{im} B_{1k}(s) + \\
\sum_{k=1}^{K_0^u} d_{i0k} B_{0k}^u(s) + \sum_{k=1}^{K_1^u} d_{i1k} Z_{ij} B_{1k}^u(s) + \epsilon_{ij}(s).
\end{aligned}
\tag{8.16}
$$

To avoid overfitting, a smoothing penalty is imposed on the spline coefficients for each function. The form of the penalty matrix varies by model assumptions and was discussed in [263]. The large number of basis functions and penalties make the estimation of (8.16) challenging. Just as with other approaches presented in this book, the approach to fitting is based on the equivalence between penalized splines and mixed models ([258]). Specifically, the spline coefficients $\beta_{0k}, \beta_{1k}, d_{i0k}, d_{i1k}$ in model (8.16) can be viewed as random effects in a mixed model; see Section 2.3.3. Therefore, model (8.16) can be fit using computational approaches for mixed effects models.

However, fitting (8.16) is not easy as the design matrix of (8.16) has a structure that cannot be simplified. Although FAMM is computationally infeasible for large data sets such as NHANES, in Section 8.3.3 we introduce the `pffr()` function in `refund` and provide the syntax for the NHANES application example. However, in many other applications with smaller data sets, FAMM is feasible, provides exceptional modeling flexibility, and provides a principled approach to inference in this context.

8.3.2 Fast Univariate Inference

Most methods proposed for multilevel FMM have used a type of basis expansion approach to account for the subject and residual functional variability. These methods account for the known and observed variability, but tend to be slow, especially when the dimension of the functional domain increases. When the target of inference are the fixed functional effects, [53] proposed estimation under the independence assumption for a two-level FMM. Effects were then smoothed and confidence bands were constructed using a bootstrap of study participants. The method was then extended to any multilevel FMM [223]. This idea was further extended to fast univariate inference (FUI, [57]), which uses point-wise mixed effects models to account for between-visit correlations *marginally*. A similar idea was proposed by [77] for single-level functional data, but the approach did not (1) account for between-visit correlations; (2) provide CMA confidence intervals; or (3) provide a simple summary for testing the global null hypothesis of no effect. FUI is designed to address these problems.

Specifically, FUI consists of three steps: (1) fitting massively univariate pointwise mixed-effects models at each location of the functional domain; (2) applying a smoother to the raw estimates along the functional domain; and (3) obtaining pointwise and CMA confidence intervals using analytical or bootstrap approaches. For consistency, we use the same formula (8.15) introduced in Section 8.3.1 to illustrate the model details.

Assume that the functional responses $W_{im}(s)$ are observed on a grid $\{s_1, \ldots, s_n\}$ on the functional domain S, where $i = 1, \ldots, n$ is the index of the subject, and $m = 1, \ldots, M_i$ is the index of the visit. For simplicity, we assume that data are Gaussian, though this assumption is not necessary as the inferential approach is applicable to both Gaussian and non-Gaussian data. In addition, we observe $\mathbf{X}_{im} = (1, x_{im})^t$ as the fixed and $\mathbf{Z}_{im} = (1, z_{im})^t$ as the random effects variables. The three-step inferential approach is as follows:

1. At each location $s_j \in S$, $j = 1, \ldots, p$, fit a pointwise linear mixed model

$$W_{im}(s_j) = \beta_0(s_j) + x_{im}\beta_1(s_j) + u_{i0}(s_j) + z_{im}u_{i1}(s_j) + \epsilon_{im}(s_j).$$

The major difference between this model and formula (8.15) is that here s_j is a specific location. Therefore, it is just a mixed model that can be fit using any software, such as the `lme4::lmer()` function [10]. For non-Gaussian data, the pointwise LMM estimate is replaced by the pointwise GLMM estimate. Denote the estimated fixed effects as $\widehat{\beta}_l(s_1), \ldots, \widehat{\beta}_l(s_p), l = 0, 1$ and random effects as $\widetilde{u}_{il}(s_1), \ldots, \widetilde{u}_{il}(s_p), l = 0, 1$.

2. Smoothing the estimated fixed effects and random effects along the functional domain. The choice of smoothers is flexible, including not smoothing and simply taking the average along the domain. The smoothed estimators are denoted as $\{\widehat{\beta}_l(s), s \in S\}$ and $\{\widehat{u}_{il}(s), s \in S\}$, respectively.

3. For Gaussian data, the pointwise and CMA confidence intervals for functional fixed effects can be obtained analytically. For both Gaussian and non-Gaussian data, inference could be conducted using the nonparametric bootstrap. Building prediction confidence intervals for functional random effects was still work in progress at the time of writing this book.

The key insight of FUI is to model between-subject and within-subject correlations separately. This marginal approach has several computational advantages. For example, fitting massively univariate LMMs in step 1 and bootstrapping inference in step 3 can be easily parallelized. Another advantage is that the method is "read-and-use," as it can be implemented by anyone who is familiar with mixed model software. In Section 8.3.3, we show how FUI can be implemented using the `fastFMM::fui()` function for the NHANES example.

8.3.3   NHANES Case Study

Consider the NHANES data introduced in Section 8.1. We are interested in the effect of age, gender, day of week, and day of wear on physical activity. For simplicity, here we focus on study participants who were between 18 and 30 years old at the time of wearing the device. There are 8,950 eligible days across 1,460 selected study participants in NHANES, so the data frame `nhanes_ml_df_fmm` consists of 8,950 rows and 6 columns. Each row represents the data collected on one eligible day for each selected participant. For some variables such as age and gender, their values did not change during the study. For example, study participant `SEQN 62161` was 22 years old at the time of the study and has six days of eligible physical activity data. Therefore, the value of the first six rows of the `age` column is 22.

In this study, we have four fixed effects variables, age, gender, day of wear, and day of week, so that $\mathbf{X}_{im} = (1, x_{im1}, x_{im2}, x_{im3}, x_{im4})^t$. For simplicity, we only include a random intercept in the model, so that $\mathbf{Z}_{im} = 1$. The code to implement FAMM on this NHANES data set using the `pffr()` function from `refund` package is shown below. The random intercept is specified as `s(SEQN, bs = "re")` following the `mgcv` syntax. To accelerate the computation of FAMM, we use the `"bam"` algorithm and specify `discrete = TRUE`. In this model, we use 30 penalized spline basis functions for the functional intercept and 10 penalized spline basis functions for the other functional effects. In R this was implemented as `bs.int = list(bs = "ps", k = 30, m = c(2, 1))` and `bs.yindex = list(bs = "ps", k = 10, m = c(2, 1))`, respectively.

```
#Fit FMM using FAMM
fit_pffr <- pffr(MIMS ~ age + gender + dayofwear + dayofweek +
                 s(SEQN, bs = "re"), data = nhanes_ml_df_fmm,
                 algorithm = "bam", discrete = TRUE,
                 bs.yindex = list(bs = "ps", k = 10, m = c(2, 1)),
                 bs.int = list(bs = "ps", k = 30, m = c(2, 1)))
```

This is a simplified implementation of FAMM, which assumes that $\epsilon_{im}(s)$ are uncorrelated across time s, which is a strong assumption. Indeed, in NHANES this essentially assumes that PA observations within a day are independent after removing the participant-specific mean. Unfortunately, even with the simplified (misspecified) model, FAMM could not run on the entire NHANES data set. One could likely use sub-sampling approaches and then combine the resulting fits by taking averages. The performance of such an approach is currently unknown.

To test the feasibility of this idea, FAMM was used on sub-samples of the NHANES data. This took around 2 minutes for 100 participants and 15 minutes for 200 participants using a standard laptop. However, for a sample of 1,000 participants, computation time increased substantially to over 6 hours (program was stopped after 6 hours due to memory problems). Thus, FAMM cannot be fit using existing software on the NHANES data set. More details on the computational limitations of current methods can be found in [57].

However, FUI worked well on the entire data set using a standard laptop, and took less than 2 minutes to obtain the point estimates. This is due to the fact that FUI fits 1440 univariate linear mixed models. This is computationally convenient and could be easily parallelized to further reduce computational time. For the jth minute of the day, the code to fit a linear mixed model using the `lme4::lmer()` function and extract fixed effects estimates is shown below.

```
#Fit a pointwise LMM in FUI
nhanes_ml_df_fmm$W <- unclass(df_fmm$MIMS[,1])
fit_lmm <- lmer(W ~ age + gender + dayofwear + dayofweek + (1 | SEQN),
                data = nhanes_ml_df_fmm)
betaTilde <- fixef(fit_lmm)
```

The point estimates obtained from these massively univariate mixed effects models can be smoothed using a wide variety of smoothing approaches. This is an advantage of the FUI approach, as it can be implemented by anyone who is familiar with mixed models.

Instead of manually implementing the FUI method step-by-step as shown above, the `fastFMM::fui()` function provides an integrated way of fitting FUI for the NHANES data set. The syntax is shown below.

```
#Fit FMM using the fui function
fit_fui <- fui(MIMS ~ age + gender + dayofwear + dayofweek + (1 | SEQN),
               data = nhanes_ml_df_fmm, family = "gaussian", var = TRUE)
```

Figure 8.9 displays the estimated functional fixed effects in NHANES together with the 95% pointwise unadjusted (darker gray shaded area) and correlation and multiplicity adjusted (CMA) (lighter shaded gray area) confidence intervals. Each panel denotes the effect of one continuous variable or one level of a categorical variable compared to baseline. For this younger population, as age increases there is a significant increase in physical activity in the morning. Compared to men, women are more active during the day, and less active at night. This is in direct contrast to the most cited paper in the field [296], but consistent with a growing body of literature. Compared to Sundays (the reference category), people have lower levels of activity on weekdays during the predawn hours (12 AM to

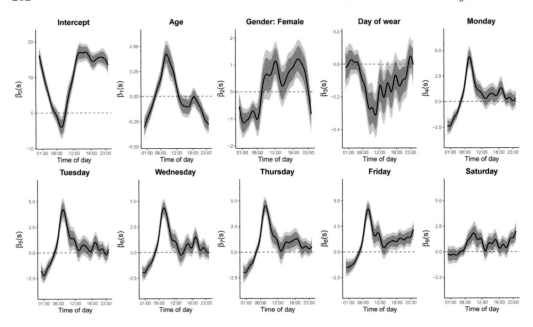

FIGURE 8.9: Estimated functional fixed effects in the NHANES case study using Fast Univariate Inference (FUI). Smoothed estimates are denoted using black solid lines. Pointwise unadjusted and correlation and multiplicity adjusted (CMA) 95% confidence intervals are shown as the dark and light gray shaded areas, respectively.

4 AM) and much higher levels of activity in the morning (6 AM to 11 AM). These results are highly interpretable and are consistent with previous findings using a different NHANES data set (NHANES 2003–2006) [57].

8.4   Multilevel Scalar-on-Function Regression

So far we have focused on functions as an outcome. However, consider the following question: What is the effect of physical activity on all-cause mortality? In this context, the outcome is scalar (dead/alive) and the predictor is a multilevel function (e.g., physical activity measured at every minute of the day for up to seven days). This is an example when the multilevel exposure function was observed once at "baseline." In some other applications, both functional data (e.g., physical activity) and outcomes (e.g., mental health status) may be observed at multiple visits. These two data structures lead to different approaches to inference. We introduce Generalized Multilevel Functional Regression (GMFR) to address the first data structure and Longitudinal Penalized Functional Regression (LPFR) to address the second data structure. Both approaches rely on the penalized functional regression (PFR) [102] introduced in Chapter 4.

8.4.1   Generalized Multilevel Functional Regression

Consider the case of multilevel functional predictors (e.g., physical activity) and one outcome (e.g., dead/alive after five years). For example, the mortality status of each NHANES study participant was collected up to December 31, 2019. In addition, demographic variables such as age, gender, BMI were collected at baseline and their values are invariant with the day of wearing the device.

Given that the physical activity data were collected across multiple days, one simple solution is to take the average at each minute of a day and model the compressed data as independent functional predictors using PFR. However, taking the average may reduce the predictive performance of functional covariates. The Generalized Multilevel Functional Regression (GMFR, [52]) provides a solution for such data structures.

Assume that for the ith subject the observed data are $\{Y_i, \mathbf{Z}_i, \{W_{im}(s), s \in S, m \in \{1, \ldots, M_i\}\}\}$, where Y_i is a continuous or discrete scalar outcome, $\mathbf{Z}_i$ is a vector of covariates, and $W_{im}(s)$ is the functional predictor at the mth visit. For simplicity we only introduce one functional predictor, though the estimation procedure can be easily generalized to multiple functional predictors. The GMFR model is

$$\begin{cases} W_{im}(s) = \mu(s) + \eta_m(s) + X_i(s) + U_{im}(s) + \epsilon_{im}(s), \\ Y_i \sim \mathrm{EF}(\mu_i, \phi), \\ g(\mu_i) = \mathbf{Z}_i^t \boldsymbol{\gamma} + \int_0^1 X_i(s)\beta(s)ds. \end{cases} \quad (8.17)$$

Here $\mathrm{EF}(\mu_i, \phi)$ denotes an "exponential family" distribution with mean μ_i and variance dispersion parameter ϕ. A closer look at model (8.17) reveals that it is a combination of multilevel variability decomposition and scalar-on-function regression. The first equation is similar to the two-level functional principal component analysis introduced in Section 8.2.1, where a subject-specific deviation is denoted as $X_i(s)$. The same term $X_i(s)$ also appears in the last equation, where we treat it as a functional predictor in a scalar-on-function regression model.

Fitting (8.17) is not straightforward. One solution is to first conduct MFPCA on the multilevel functional predictor, replace $X_i(s)$ with its estimate, and fit a standard scalar-on-function regression model. However, this two-stage procedure ignores the variability from MFPCA, which may induce bias when fitting the functional regression model. An alternative is to conduct a joint analysis using, for example, Bayesian posterior simulations. Detailed discussions on pros and cons of both methods can be found in [52].

8.4.2 Longitudinal Penalized Functional Regression

Consider the case of measuring functional data (e.g., physical activity) over multiple visits when outcomes are also measured. For such data structures, the longitudinal penalized functional regression (LPFR, [103]) can be used to extend PFR. Assume that for subject $i = 1, \ldots, n$ at visit $m \in \{1, \ldots, M_i\}$ we observe data $\{Y_{im}, \mathbf{X}_{im}, W_{imr}(s_r), s_r \in S_r\}$, where Y_{im} is a scalar outcome and $\mathbf{X}_{im} = (X_{im1}, \ldots, X_{imQ})^t$ are Q scalar predictors. In addition, we collect R functional predictors $W_{imr}(s), r = 1, \ldots, R$. The LPFR model is

$$\begin{cases} Y_{im} \sim \mathrm{EF}(\mu_{im}, \phi), \\ g(\mu_{im}) = \mathbf{X}_{im}^t \boldsymbol{\gamma} + \mathbf{Z}_{im}^t \mathbf{b}_i + \sum_{r=1}^R \int_{S_r} W_{imr}(s_r)\beta_r(s_r)ds_r, \end{cases} \quad (8.18)$$

where "$\mathrm{EF}(\mu_{ij}, \phi)$" denotes an exponential family distribution with mean μ_{ij} and dispersion ϕ. The parameters $\boldsymbol{\beta}$ are the fixed effects to be estimated, and $\mathbf{Z}_{im}^t \mathbf{b}_i$ is a standard random effects component where $\mathbf{b}_i \sim N(\mathbf{0}, \boldsymbol{\Sigma}_b)$, where $\mathbf{0}$ is a vector of zeros and $\boldsymbol{\Sigma}_b$ is the covariance of the random effects $\mathbf{b}_i$. The functional effects are quantified as $\beta_r(s)$ for the rth functional predictor.

In LPFR, the longitudinal correlation is modeled using random effect terms. Similar to PFR, functional terms are decomposed as a summation of weighted spline basis functions.

Therefore, formula (8.18) reduces to a generalized linear mixed model. The LPFR model is implemented in the `refund::lpfr()` function. An example based on a diffusion tensor imaging (DTI) in Multiple Sclerosis and healthy controls is provided in the `refund` package.

Another data structure was discussed in [177], who considered the case when the outcomes and functional predictors are observed longitudinally, but not at the same visits (asynchronously). Such data structures are increasingly prevalent in large observational studies.

9

Clustering of Functional Data

Clustering of functional data is an unsupervised exploratory analytic tool designed to identify subgroups of study participants that (1) share similar characteristics within subgroups; (2) exhibit substantial differences between subgroups; and (3) are interpretable. Like any unsupervised learning technique, the more supervised, the better.

Formally, if $\{W_i(s), s \in S\}$ are a sample of curves $i = 1, \ldots, n$, then clustering can be described as estimating a function $C : \{1, \ldots, n\} \to \{1, \ldots, K\}$, where both the $C(\cdot)$ function and the number of labels, K, are unknown. Typically, it is assumed that K is much smaller than n. In practice, the value of K is often fixed and the function $C(\cdot)$ is estimated conditional on K for a sequence of reasonable values. Once the function $C(\cdot)$ is estimated, it can be used as a categorical variable to simplify and illustrate latent functional data structures and conduct association analysis with other covariates. The main difference from multilevel functional modeling is that the number and subgroup membership is not known a priori and needs to be estimated.

9.1   Basic Concepts and Examples

Clustering is the problem of estimating the $C(\cdot)$ function. However, exactly how the function $C(\cdot)$ is estimated depends fundamentally on how similarity of curves and subgroups of curves is defined as well as the exact procedure of defining similarities between subgroups of curves. As functional data $\{W_i(s_1), \ldots, W_i(s_p)\}$ can be viewed as a set of n points in the $p = |S|$ dimensional space, clustering functional data can draw upon the vast number of clustering techniques used for multivariate data. There is a very rich literature on clustering multivariate data and here are some of the many monographs [4, 107, 118, 146, 197].

Clustering for functional data have some specific characteristics. For example, data are often observed with noise, which can affect the distances between functions. However, smoothing functional data is easier and better understood than smoothing multivariate data. This may lead to more powerful ways of identifying clusters, when clusters exist using a smoothing before clustering approach. Moreover, functional data has the same interpretation (units) along the functional domain, which allows for more consistent and easier to interpret clustering results. A good survey paper on clustering with functional data is [130] and an R resource page is maintained at https://bit.ly/3KmWJ9y. The site contains several functional clustering packages including `funFEM` [21], `funHDDC` [264], and `fdakma` [227]. This is an excellent resource that could be used in practice. However, in this chapter we take a different approach: (1) functional data are processed/smoothed using techniques described in this book; and (2) one of the many clustering approaches developed for multivariate data is used on the processed data. The advantage of this approach is that the multivariate clustering approaches provide a large and ever evolving set of clustering methods with high-quality associated software.

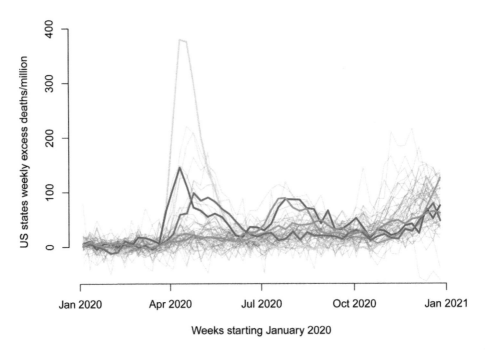

FIGURE 9.1: Each line represents the weekly excess mortality per one million residents for each state in the US and two territories (District of Columbia and Puerto Rico). Five states are emphasized: New Jersey (green), Louisiana (red), Maryland(blue), Texas (salmon), and California (plum).

Before going into technical details, consider the example of weekly excess mortality in 2020 for all states and two territories (District of Columbia and Puerto Rico) introduced in Chapter 1. Recall that for every week, the data represent the difference in total mortality between a specific week in 2020 and the corresponding week in 2019. Excess mortality is divided by the state population and multiplied by one million. Therefore, the resulting data are the weekly excess mortality per one million residents. Figure 9.1 displays these data, where each line corresponds to a state or territory. For presentation purposes five states are emphasized: New Jersey (green), Louisiana (red), Maryland(blue), Texas (salmon), and California (plum). The difference from Chapter 1 is that we are looking at the weekly excess mortality and not the cumulative excess mortality.

A quick inspection of Figure 9.1 indicates that (1) for long periods of time, many states seem to have similar excess mortality rate patterns (note the darker shades of gray that form due to trajectory overlaps in the interval $[-50, 50]$ excess deaths per one million residents); (2) some states, including New Jersey, Louisiana, and Maryland, have much higher peaks in the April–June, 2020 period with weekly excess mortality above 50 per one million residents; (3) some states, including Texas and Louisiana, have higher peaks in the July–August, 2020 periods with weekly excess mortality between 50 and 100 per one million residents; and (4) most states have a higher excess mortality in December 2020, with some states exceeding 100 excess deaths per week per one million residents. California (plum), which for most of the year tracked pretty closely with the median excess mortality, has a larger jump towards the end of the year.

Thus, visual inspection of such data suggests how the idea of "clustering" appears naturally in the context of functional data analysis. In some cases one can clearly observe different characteristics of a subgroup (e.g., states with a large increase in weekly excess

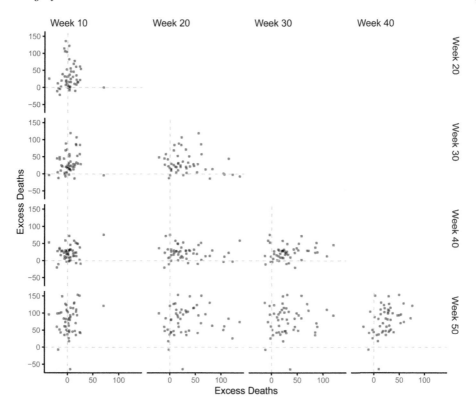

FIGURE 9.2: Each panel contains weekly excess mortality for each state in the US for one week (x-axis) versus another week (y-axis). Each dot is a state or territory. Weeks shown are 10 (ending March 7, 2020), 20 (ending May 16, 2020), 30 (ending July 25, 2020), 40 (ending October 3, 2020), and 50 (ending December 12, 2020).

mortality rate between April and June). In other cases, one may wonder whether additional subgroup structure may exist even when none is obvious on a line plot. Indeed, plots such as shown in Figure 9.1 can obscure temporal patterns simply due to over-plotting in areas of high density of observations.

To further explore the data structure, Figure 9.2 provides a different perspective on the same data shown in Figure 9.1. More precisely, each panel contains weekly excess mortality for each state in the US for one week (x-axis) versus another week (y-axis). Each dot is a state or territory. Weeks shown are 10 (ending March 7, 2020), 20 (ending May 16, 2020), 30 (ending July 25, 2020), 40 (ending October 3, 2020), and 50 (ending December 12, 2020). The panel columns correspond to data for weeks 10 (first column), 20 (second column), 30 (third column), and 40 (fourth column) on the x-axis. The panel rows correspond to data for weeks 20 (first row), 30 (second row), 40 (third row), and 50 (fourth row) are shown on the y-axis.

When week 10 is displayed on the x-axis (first panel column), almost all observations are on the left side of the plots. This indicates that for the week ending on March 7, 2020 there were few excess deaths in most states (mean 3.3, median 2.6) excess deaths per week per million residents. However, one point stands out (notice the one lone point to the right of the main point clouds). This point corresponds to North Dakota and indicates 70.6 excess deaths per one million residents. When looking more closely at North Dakota, four weeks before and after week 10, we see the following weekly excess mortality numbers 10.5, −17.0,

49.7, 10.5, 70.6, 39.2, 6.5, −30.1, −41.8. These are the data for consecutive weeks starting with the week ending on February 8, 2020 and ending with the week ending on April 4, 2020. Week 10 (ending on March 7, 2020) is an outlier, though the end of February through mid-March 2020, seems to correspond to an unusually high excess number of deaths in North Dakota.

The panel in the first row and column corresponds to week 10 versus week 20. One can visually identify several points close to the top of the point cloud. These are states or territories that experienced a much sharper increase in excess mortality rates at the beginning of the pandemic (week ending on April 16, 2020). In fact, for this week, there are 5 states or territories with an excess mortality rate larger than 100 per million residents: New Jersey (104.5), Connecticut (106.3), Delaware (114.5), Massachusetts (122.1) and District of Columbia (136.1).

The panel in the third row and column (week 30 on the x-axis and week 40 on the y-axis) suggests the presence of two or possibly three clusters. Indeed, there seems to be a larger group of states in the left lower corner. Another group of states have large excess mortality rates at week 30: South Carolina (84.7), Louisiana (87.0), Texas (88.8), Arizona (107.0), and Mississippi (119.3). The 5 states with the largest excess mortality rates at week 40 are: Wyoming (53.2), Missouri (58.4), District of Columbia (58.9), Arkansas (72.3), and North Dakota (75.8).

These simple exploratory techniques provide many insights, some of which were discussed and some that can be inferred from the plots. They also suggest that there may be additional structure in the data, especially in the form of subgroups. That is, some states may have trajectories with similar characteristics. So far, we have relied on intuition and visual inspection and have not addressed the problem of distance between trajectories or groups of trajectories.

9.2 Some Clustering Approaches

We consider several approaches to clustering that are widely used in practice: K-means, hierarchical clustering, and distributional clustering. This leaves a large number of approaches that will not be discussed including density models, such as DBSCAN [73] and OPTICS [5], graph-based models, such as HCS [120], or biclustering [117, 202, 235]. All these approaches can be extended to functional data and provide excellent opportunities for further exploration.

9.2.1 K-means

K-means [88, 119, 183, 189] is one of the oldest and most reliable techniques for data clustering and the function kmeans is part of the base stats library in R. In this section we will focus first on the actual implementation and then we will describe the basic ideas and implications for obtaining results.

9.2.1.1 Clustering States Using K-means

The weekly excess mortality data are stored in the matrix Wd, which has 52 rows, where each row corresponds to a state or territory, and 52 columns, where each column corresponds to a week from 2020. Conducting functional clustering using K-means is as simple as the following

```
#Cluster functional data using k-means
kmeans_CV19_3 <- kmeans(Wd, centers = 3)
#Extract the cluster indicators
cl_ind <- kmeans_CV19_3$cluster
#Extract the cluster means
cl_cen <- kmeans_CV19_3$centers
```

We have chosen a number of three clusters, though that can be changed. This is indicated by the variable `centers=3` in the call to the function `kmeans`. The `kmeans` function assumes that the data points (in our case, US states and territories) are stored by row.

To illustrate the results of clustering, Figure 9.3 displays the same data as Figure 9.1, though each state is shown in one of three colors, depending on its estimated cluster membership: purple (cluster 1), green (cluster 2), and orange (cluster 3). The centers of each cluster are shown in the same color, but using thicker lines.

This is a case when clustering seems to produce reasonable results. Indeed, cluster 1 (purple) tends to contain US states and territories that had (1) a very high rate of excess deaths in the April–June, 2020 period with maxima attained at the end of April or beginning of May; (2) a lower than average rate of excess deaths in the July–November 2020 period; and (3) a close to average rate of excess deaths in December 2020. Cluster 2 (green) tends to contain states and territories that had (1) a lower than average rate of excess deaths in April–June 2020; (2) a higher than average rate of excess deaths in July–October, though not as high as the rates in cluster 1 in the April–June 2020 period; and (3) a lower than average rate of excess deaths in December 2020. Cluster 3 (orange) tends to contain states and territories that had (1) a lower than average rate of excess deaths in April–June 2020;

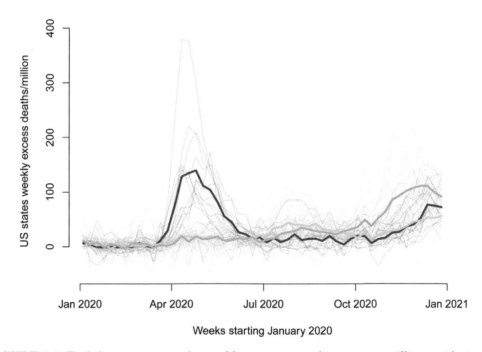

FIGURE 9.3: Each line represents the weekly excess mortality per one million residents for each state in the US and two territories (District of Columbia and Puerto Rico). Each state is clustered in one of three subgroups using K-means: purple (cluster 1), green (cluster 2), and orange (cluster 3). Thicker lines of the same color are the clusters means.

(2) an average rate of excess deaths in July–October; and (3) a higher than average rate of excess deaths in December 2020, with peaks that came close to those experienced by states in cluster 1 in the April–June 2020 period.

Note that even within clusters, there is substantial heterogeneity between states and territories. However, at least as a first visual inspection, each of the three groups seem to be more homogeneous than their combination.

It is of interest to identify which states belong to each cluster. One way is to enumerate them, but this is quite cumbersome even with just 52 states and territories. In cases when the number of observations is very large, alternative strategies are necessary. Here we choose to display the clusters on the US map.

Figure 9.4 displays the US map where each state and territory is colored according to their estimated cluster membership. Even though geographic information was not used in the clustering approach, the map indicates that clusters tend to have strong geographic co-localization. For example, cluster 1 (purple) includes states in the Northeastern United States, except Maine, Vermont, and New Hampshire. However, it also includes Michigan, Illinois, and Louisiana. A closer inspection of Figure 9.1 reveals that Louisiana had a very high excess mortality rate both in the April–June 2020 period (like the other states in cluster 1) and in the July–November 2020 period (like the other states in cluster 2). The reason for this cluster assignment is likely due to the fact that the first peak of the excess mortality rate in Louisiana is higher and closer to the mean of cluster 1. Cluster 2 (green) contains almost the entire Southeastern region of the US, Texas, Alaska, the Pacific West, Utah, Colorado, Arizona, as well as Maine, Vermont and New Hampshire. Cluster 3 (orange)

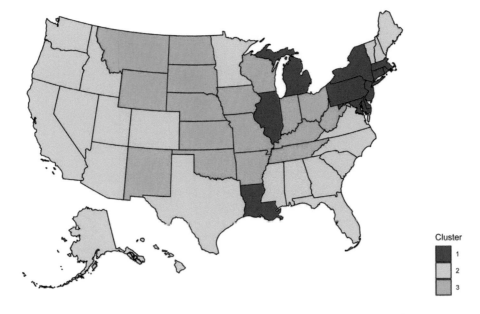

FIGURE 9.4: US map of three estimated clusters for weekly excess mortality rates in 2020. The method used was K-means on the data matrix where the observed data (without smoothing) were used. Clusters are the same as in Figure 9.3 and cluster colors were maintained: purple (cluster 1), green (cluster 2), and orange (cluster 3). Thicker lines indicate cluster centers.

contains a mix of states from the northern Intermountain Region and Midwest, except for Illinois and Michigan.

9.2.1.2 Background on K-means

The K-means algorithm starts with a sample of functions $W_i(s)$ stored in the $p \times 1$ dimensional vectors $\mathbf{W}_i$. The objective of the K-means algorithm is to solve the minimizing problem

$$\underset{C(\cdot)}{\arg\min} \sum_{k=1}^{K} \sum_{\{i:C(i)=k\}} ||\mathbf{W}_i - \mu_k||^2 \,, \tag{9.1}$$

where $C : \{1, \ldots, n\} \to \{1, \ldots, K\}$ is the cluster indicator functions and

$$\mu_k = \frac{\sum_{\{i:C(i)=k\}} \mathbf{w}_i}{|\{i : C(i) = k\}|}$$

is the mean of all vectors $\mathbf{w}_i$ in cluster k, that is with the property that $C(i) = k$, and $|\{i : C(i) = k\}|$ is the number of elements in cluster k.

For a given clustering of the data, $C(\cdot)$, the objective function in equation (9.1) is the sum over all clusters of the within-cluster sum of squares distances between functions and the corresponding cluster center (mean). A possible solution of the minimization problem could be to enumerate all partitions and calculate the sum of squares for each partition. However, this quickly runs into computational problems as the number of observations, n, and partitions, K, increases.

The general idea of the K-means algorithm is to: (1) start with a set of group centers; (2) identify all functions that are closest to these centers; (3) calculate the means of these groups; and (4) iterate.

There are many variations for each of these steps, which leads to a wide variety of K-means clustering approaches. For example, the default function kmeans in R uses a different heuristic for updating cluster membership. In particular, for a given clustering it searches functions that maximally reduce the within-cluster sum of squares by moving to a different cluster and updates the clusters accordingly. Another variation is on the metric used. For example changing the L_2 distance to L_1 distance in equation (9.1) leads to K-medians and K-medoids [132, 146]. It is worth noting that "medoids" refers to functions that exist in the original data set, whereas medians and means might not. This could be preferred for interpretation purposes because the cluster centers are actually functions. An alternative would be to use K-means or K-medians and use the closest functions to the estimated cluster centers as examples of "central behavior" within the cluster.

9.2.2 Hierarchical Clustering

Hierarchical clustering [74, 107, 118, 211, 310] is another popular unsupervised multivariate technique. As functional data can be viewed as multivariate data, hierarchical clustering can be applied directly to the observed functions. We show how to conduct hierarchical clustering first and then describe the basic analytic ideas.

9.2.2.1 Hierarchical Clustering of States

The "data points" in functional data analysis are the observed functions. Just as in the case of K-means, the data matrix Wd contains 52 rows, where each row corresponds to a state or territory, and 52 columns, where each column corresponds to a week from 2020. Conducting functional hierarchical clustering is as simple as the following

```
#Calculate the matrix of distances
dM <- dist(Wd) ^ 2
#Apply hierarchical clustering
hc <- hclust(dM, method = "ward.D2")
#Obtain k=5 clusters
cut_wardd2 <- cutree(hc, k = 5)
#Plot dendrogram
plot(hc, hang = -1)
```

The distance function `dist` is a function that transforms the data into a matrix of mutual distances between its rows. The default distance is `"euclidian"`, though many other distances can be used, including `"maximum"`, `"binary"` and `"minkowski"`. Here we use the square of the Euclidian distance. The matrix of mutual distances is stored in the matrix `dM`. The next step is to apply hierarchical clustering to the distance matrix `dM` using the R function `hclust`. The algorithm starts by aggregating the closest functions (points in $\mathbb{R}^p$) into clusters and then proceeds by aggregating clusters based on the mutual distance between clusters. Given two clusters of functions (points), there are many different ways of defining a distance between them. Here we use a method that minimizes the within-cluster variance, indicated as `method="ward.D2"`. We will discuss other methods for calculating distances between clusters in Section 9.2.2.2.

The next step is to cut the tree (dendrogram) and provide the estimated clustering for a given number of clusters; in our case, $K = 5$. This time we have chosen five clusters because when we considered only three clusters, there were two small clusters, one comprising New Jersey, one comprising North and South Dakota, and one comprising every other state. Therefore, to obtain a split more comparable to the one obtained using K-means we opted for $K = 5$.

The last line of code indicates how to obtain a plot of the hierarchical clustering shown in Figure 1.4. To obtain a prettier plot, we have used the package `dendextend` [90], which allowed for substantial customization. Please see the accompanying `github` repository for details on how to make publication-ready dendrogram plots. However, when conducting analysis, the `plot(hc)` option is fast and easy to use.

Another useful way to visualize the data is to combine the heatmap of the data with the estimated hierarchical clustering of the rows (in our case, states and territories). Figure 9.6 displays the entire data set stored in the matrix `Wd`, where states are displayed by row (note the labels on the right side of the plot). Colors correspond to the weekly excess mortality rates with stronger shades of yellow corresponding to higher rates. The hierarchical dendrogram from Figure 9.5 is shown on the left side of the plot. The colors of the estimated clusters is preserved and are shown from left to right in Figure 9.5 and from top to bottom in Figure 9.6. For rendering this plot we have used the functions `Heatmap` in the package `ComplexHeatmap` [111] as well as the package `circlize` [112]. A problem with rendering heatmaps is that outliers (such as the very high excess mortality rates in New Jersey at the beginning of the pandemic), can make heatmaps appear "washed out." To show details in the middle of the distribution, as well as the outliers, colors need to be mapped using color palettes with breaks that account for outliers.

In this type of hierarchical clustering, New Jersey stands out in one cluster and North and South Dakota in another. This is different from what we have observed using K-means clustering. We are already familiar with the trajectory of excess mortality in New Jersey, but North and South Dakota do stand out in Figure 9.6. Indeed, for most of the year the excess mortality rates in the two states were among the lowest in the country. However, the excess mortality rates in both states spiked dramatically between the middle of October to the beginning of December. The Sturgis Motorcycle Rally took place in South Dakota from

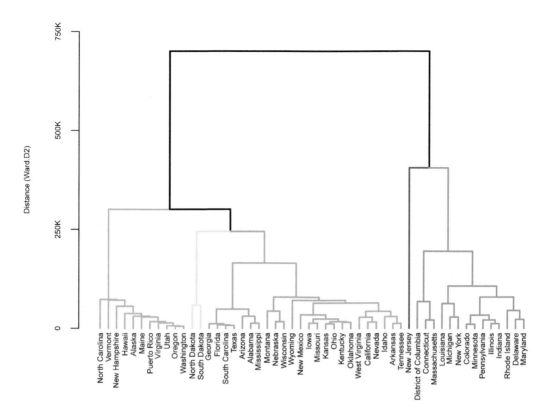

FIGURE 9.5: Hierarchical clustering of US states and territories of weekly excess mortality in 2020. Clusters are colored from left to right: `salmon` (cluster 1), `yellow` (cluster 2), `green` (cluster 3), `violet` (cluster 4), and `darkorange4` (cluster 5). Hierarchical clustering was used on the observed data matrix (no smoothing) with square Euclidian distance and Ward's between-cluster distance.

August 7-16 in 2020. This is consistent with the characterization of this event as a super spreader event as identified by the CDC response team [36]. Our results indicate that the local communities (North and South Dakota) might have been hit particularly hard.

Looking more closely at cluster 1 (`salmon`), it seems to contain states with lower overall weekly mortality rates during the entire year. It includes North Carolina, Vermont, New Hampshire, Hawaii, Alaska, Maine, Puerto Rico, Virginia, Utah, Oregon and Washington. Cluster 2 (`yellow`) contains two states North and South Dakota and was discussed earlier. Cluster 3 (`green`) is a very large cluster and contains both states that had a higher excess mortality rate during the summer (e.g., Florida, South Carolina, Texas) and states that experienced high rates of excess mortality towards the end of the year. Cluster 4 (`violet`) comprises only New Jersey. Cluster 5 (`dark orange`) contains states that had high excess mortality rates in the spring with a moderate increase in late summer and early winter.

Note that the `Heatmap` function contains the `cluster_rows` option, which re-arranges the rows to represent a particular cluster structure (e.g., estimated by the function `hclust`). The same thing can be done for clustering columns using the `cluster_columns` option. For the purpose of this plot we suppressed this option using `cluster_columns="FALSE"` to have the calendar time along the x-axis. However, it may be interesting to cluster the weeks to identify periods of time with similar behavior across time.

To better illustrate the geography of the clusters, Figure 9.7 displays the same clusters as Figures 9.5 and 9.6, but mapped onto the US map. The result is not identical to the

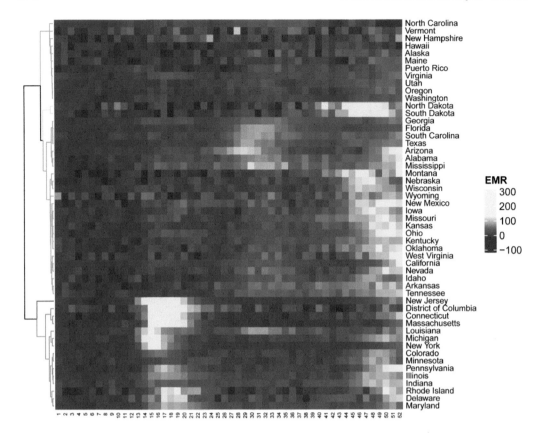

FIGURE 9.6: Heatmap of weekly excess mortality rates in US states and two territories. The dendrogram from Figure 9.5 is appended to the left of the figure with the same cluster colors as in Figure 9.5.

one obtained from K-means with three clusters, though some similarities are apparent. For example, cluster 5 in Figure 9.7 has substantial overlap with cluster 1 in Figure 9.4. This is the cluster with higher excess mortality rates in the spring of 2020. However, cluster 1 in Figure 9.7 contains a combination of states that were in clusters 1 and 2 in Figure 9.4. These are the states with low excess mortality rates for the entire year. Cluster 3 in Figure 9.7 is a very large cluster and combines many of the states in clusters 2 and 3 in Figure 9.4.

9.2.2.2 Background on Hierarchical Clustering

So far we have seen the results of hierarchical clustering and how visual representation of the data interfaces with critical analysis of results. We now provide the background necessary to understand how hierarchical clustering actually works. Essentially, there are two important ingredients for any hierarchical clustering algorithm: (1) a distance between individual functions; and (2) a distance between any two groups of functions. Once these two are available, the first clusters are defined as the individual functions. The algorithm proceeds by combining the two closest clusters, updating the mutual distances between clusters, and iterating. In our approach we used the square Euclidian distances between

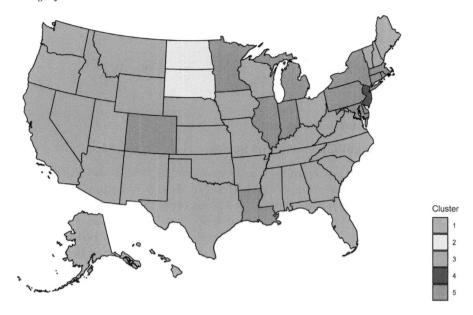

FIGURE 9.7: US map of five estimated clusters for weekly excess mortality rates in 2020. The method used was hierarchical clustering with square Euclidian distance between states and Ward's minimum variance method (`method="ward.D2"`) between groups. The observed data matrix (without smoothing) was used as input. Cluster colors are the same as in Figures 9.5 and 9.6.

functions:

$$d(i_1, i_2) = \sum_{j=1}^{p} \{W_{i_1}(s_j) - W_{i_2}(s_j)\}^2 = ||\mathbf{W}_{i_1} - \mathbf{W}_{i_2}||^2 \,.$$

However, many different types of distances can be used and many are already coded as part of the `dist` option for the `hclust` function in R. Of course, additional types of distances can be used. The `ward.D2` method for combining clusters uses the following distance between clusters C_{k_1} and C_{k_2}

$$d(k_1, k_2) = \sum_{i \in C_{k_1} \cup C_{k_1}} ||\mathbf{W}_i - \mu_{k_1, k_2}||^2 \,,$$

where μ_{k_1, k_2} is the mean of the $\mathbf{W}_i$ such that $i \in C_{k_1} \cup C_{k_1}$. This distance is the sum of squares of residuals after combining the two clusters. The idea is to first combine clusters that have small combined sum of squares (dispersion) and then continue aggregating clusters with the smallest combined sum of squares. Other well-known methods for combining clusters include the following

- Single linkage clustering where the distance between clusters is

$$d(k_1, k_2) = \min\{d(\mathbf{W}_{i_1}, \mathbf{W}_{i_2}) : i_1 \in C_{k_1}, i_2 \in C_{k_2}\}$$

- Complete linkage clustering where the distance between clusters is

$$d(k_1, k_2) = \max\{d(\mathbf{W}_{i_1}, \mathbf{W}_{i_2}) : i_1 \in C_{k_1}, i_2 \in C_{k_2}\}$$

- Unweighted pair group method with arithmetic mean (UPGMA) [277] where the distance between clusters is

$$d(k_1, k_2) = \frac{1}{|C_{k_1}||C_{k_2}|} \sum_{i_1 \in C_{k_1}} \sum_{i_2 \in C_{k_2}} d(\mathbf{W}_{i_1}, \mathbf{W}_{i_2})$$

9.2.3   Distributional Clustering

9.2.3.1   Distributional Clustering of States

Another popular clustering alternative is mixture modeling, which also contains many different sub-methods; see, for example [197]. Here we show how to implement a Gaussian mixture clustering approach implemented in the `mclust::Mclust` function in R. This is sometimes referred to as "latent mixture modeling," but in this book, almost everything is latent and we find this description quite imprecise. The good and bad news is that one does not need to understand much to conduct this analysis. This is good because sometimes everything we need is to quickly explore the data and see if something unusual comes up. It is also bad, because much of the complexity associated with various tuning parameters remains hidden.

Below we show how to conduct a Gaussian mixture clustering approach with the number of clusters estimated by the Bayesian Information Criterion (BIC) [266]. In our case the criterion estimated that there are 4 latent subgroups, which seems not unreasonable. Indeed, we have explored three groups for K-means and five groups for hierarchical clustering. So, it was quite a relief to see that BIC estimated four clusters. Again, either of these choices of number of clusters can probably be defended and there is nothing immutable about BIC estimators.

```
#Set the color palette
colset <- c("#E69A8DFF", "#F6D55C", "#2A9D8F", "#5F4B8BFF")
library(mclust)
#Center and scale the data
X <- as.data.frame(apply(Wd, 2, scale))
#Calculate BIC for distributional clustering
BIC <- mclustBIC(X)
#Fit GMM using EM algorithm with 4 clusters suggested by BIC
mod <- Mclust(X, x = BIC)
#Obtain clustering results
res <- mod$classification
```

To gain insight into the results, Figure 9.8 displays the geographical distribution of the four estimated clusters. It is somewhat reassuring that the clusters tend to be similar to those identified by K-means and hierarchical clustering. However, we recommend using multiple approaches and comparing their results. If results agree, investigate results using visualization and common sense. If results do not agree, investigate results using visualization and common sense.

We now provide the theoretical details for distributional clustering; there is nothing particularly "functional" about these methods, but we added them to provide a more complete view of methods.

9.2.3.2   Background on Distributional Clustering

Distributional clustering is also known as finite mixture clustering and latent class analysis. The assumptions are that there exist categorical random variables C_i, $i = 1, \ldots, n$ with K categories, which represent the cluster assignment. C_i are assumed to be mutually

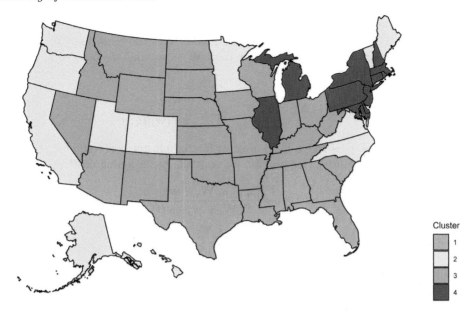

FIGURE 9.8: US map of four estimated clusters for weekly excess mortality rates in 2020. The method used was the Gaussian mixture model (implemented in the `Mclust()` function in `mclust` package) on the data matrix where the observed data (without smoothing) were used.

independent with $P(C_i = k|\pi_1,\ldots,\pi_K) = \pi_k$ and $\mathbf{W}_i|C_i = k$ follows a distribution with probability density function $\psi(\mathbf{W}_i|\boldsymbol{\theta}_k)$.

For simplicity we will assume that this distribution is the multivariate $N(\boldsymbol{\mu}_k,\sigma^2\mathbf{I}_p)$ distribution and will be denoted by $\phi(\mathbf{W}_i|\boldsymbol{\mu}_k,\sigma^2)$. Here $\mathbf{I}_p$ denotes the identity matrix of dimension $p \times p$, where p is the dimension of $\mathbf{W}_i$. Other distributions can be considered, as well, with minimal changes to the theoretical description of the approach.

With these assumptions it can be shown that the marginal pdf of $\mathbf{W}_i$ is

$$g_{\boldsymbol{\theta}}(\mathbf{w}_i) = \sum_{k=1}^{K} \pi_k\phi(\mathbf{W}_i|\boldsymbol{\mu}_k,\sigma^2) \,, \tag{9.2}$$

where $\boldsymbol{\theta} = (\pi_1,\ldots,\pi_K,\boldsymbol{\mu}_1,\ldots,\boldsymbol{\mu}_K,\sigma^2)$, $\pi_k \geq 0$ for $k = 1,\ldots,K$, and $\sum_{k=1}^{K} \pi_k = 1$. This distribution is a mixture of normal densities where the weights for each density, π_k, represent the proportion of study participants in each cluster.

Denote by $B = [0,1] \times \ldots \times [0,1]$ a neighborhood of $\mathbf{0}_p$ in $\mathbb{R}^p$. The proof for this result is

$$P(\mathbf{W}_i \in \mathbf{w}_i + \epsilon B) = \sum_{k=1}^{K} P(\mathbf{W}_i \in \mathbf{w}_i + \epsilon B, C_i = k)$$

$$= \sum_{k=1}^{K} P(\mathbf{W}_i \in \mathbf{w}_i + \epsilon B|C_i = k)P(C_i = k)$$

$$= \sum_{k=1}^{K} \pi_k P(\mathbf{W}_i \in \mathbf{w}_i + \epsilon B|C_i = k) \,.$$

Dividing both the left and right sides of the equation by ϵ and letting $\epsilon \to 0$ proves the result in equation (9.2). Using similar arguments we can show that the conditional distribution of the cluster assignment variables C_i given the data $\mathbf{W}_i = \mathbf{w}_i$ and the other parameters is a categorical distribution with probabilities

$$P(C_i = k|\mathbf{W}_i = \mathbf{w}_i, \boldsymbol{\theta}) = \frac{\pi_k \phi(\mathbf{w}_i|\boldsymbol{\mu}_k, \sigma^2)}{\sum_{l=1}^{K} \pi_l \phi(\mathbf{w}_i|\boldsymbol{\mu}_l, \sigma^2)} , \tag{9.3}$$

where $\boldsymbol{\mu} = (\boldsymbol{\mu}_1^t, \ldots, \boldsymbol{\mu}_K^t)^t$ is the collection of estimated cluster centers. The proof of this result follows from the identity (for brevity, we omit the conditioning on $\boldsymbol{\theta}$)

$$P(\mathbf{C}_i = k|\mathbf{W}_i \in \mathbf{w}_i + \epsilon B) = \frac{P(\mathbf{C}_i = k, \mathbf{W}_i \in \mathbf{w}_i + \epsilon B)}{P(\mathbf{W}_i \in \mathbf{w}_i + \epsilon B)}$$
$$= \frac{P(\mathbf{W}_i \in \mathbf{w}_i + \epsilon B)|\mathbf{C}_i = k)P(\mathbf{C}_i = k)}{\sum_{l=1}^{K} P(\mathbf{W}_i \in \mathbf{w}_i + \epsilon B)|\mathbf{C}_i = l)P(\mathbf{C}_i = l)} .$$

The result can be obtained by dividing both the left and right sides of the equation by ϵ and letting ϵ go to 0.

Note that this is a different way of allocating study participants to clusters. In K-means one typically assigns observations to the closest mean, $\boldsymbol{\mu}_k$. In mixture distribution clustering, observations are randomly assigned with the probabilities described in equation (9.3). As we will see, we can quantify the joint posterior distribution of these cluster indicators.

The standard approach for mixture distribution clustering is to use Expectation Maximization [61], where the cluster membership is treated as missing data. However, here we will take a Bayesian approach, which will allow a close investigation of full conditional distributions.

We start with setting the prior distribution for $\boldsymbol{\pi} = (\pi_1, \ldots, \pi_K)$ as a Dirichlet distribution denoted by $\text{Dir}(\alpha, \ldots, \alpha)$, where $\alpha > 0$. For the cluster centers we can assume that they are a priori independent with $\boldsymbol{\mu}_k \sim N(\mathbf{0}_p, \sigma_0^2 \mathbf{I}_p)$, where $\mathbf{0}_p$ is a p-dimensional vector of zeros, $\mathbf{I}_p$ is the p-dimensional identity matrix and σ_0 is large. We will discuss later restrictions on these priors, but keep this form for presentation purposes. Finally, the prior distribution for the $1/\sigma^2$ is $\Gamma(a, b)$ with small a and b. Here the parameterization of the $\Gamma(a, b)$ distribution is such that its mean is a/b and variance is a/b^2 (the `shape` and `rate` parameters in R).

With these assumptions, the full likelihood of the observed and missing data is

$$\{\prod_{i=1}^{n} [\mathbf{W}_i|C_i, \boldsymbol{\mu}, \sigma^2][C_i|\boldsymbol{\pi}]\}[\boldsymbol{\pi}][\boldsymbol{\mu}][\sigma^2] , \tag{9.4}$$

where we followed the Bayesian notation where the bracket sign $[\cdot]$ denotes the distribution and $[y|x]$ is the conditional distribution of variable y given variable x.

From this likelihood we can derive the full conditional distributions for each model parameter. As will be seen, all the full conditionals are closed form, making simulations from the posterior distribution relatively easy using Gibbs sampling [95]. Indeed, this avoids the Rosenbluth-Teller (also known as, with some substantial controversy, Metropolis-Hastings) algorithm [113, 200, 256].

A. *The full conditional of* $[C_i|others]$ *for* $i = 1, \ldots, n$ *is proportional to*

$$[\mathbf{W}_i|C_i, \boldsymbol{\mu}, \sigma^2][C_i|\boldsymbol{\pi}] \propto [C_i|\mathbf{w}_i, \boldsymbol{\mu}, \boldsymbol{\pi}, \sigma^2] ,$$

which is a categorical variable with K categories and probabilities for each category given by (9.3). Therefore, once the probabilities $\boldsymbol{\pi}$ are calculated, the cluster assignments C_i are simulated using the `sample` function in R.

B. *The full conditional of* $[\boldsymbol{\pi}|others]$ *is proportional to*

$$\prod_{i=1}^{n}[C_i|\boldsymbol{\pi}][\boldsymbol{\pi}] = \pi_1^{N_{C,1}}\ldots\pi_K^{N_{C,K}}\pi_1^{\alpha-1}\ldots\pi_K^{\alpha-1}$$

$$= \pi_1^{N_{C,1}+\alpha-1}\ldots\pi_K^{N_{C,K}+\alpha-1}$$

$$\propto \mathrm{Dir}(N_{C,1}+\alpha,\ldots,N_{C,K}+\alpha)\,,$$

where $N_{C,k} = |\{i : C_i = k\}|$ is the number of members in cluster k and $|A|$ denotes the number of elements in set A.

C. *The full conditional of* $[\boldsymbol{\mu}_k|others]$ *is proportional to*

$$\{\prod_{\{i:C_i=k\}}[\mathbf{W}_i|C_i,\boldsymbol{\mu},\sigma^2]\}[\boldsymbol{\mu}_k]\,.$$

Denote by $\mathbf{S}^{C,k} = \sum_{\{i:C_i=k\}}\mathbf{W}_i$ the sum of observations in cluster k. As all distributions in this product are Gaussian, the full conditional $[\boldsymbol{\mu}_k|others]$ is also Gaussian and given by

$$[\boldsymbol{\mu}_k|others] = N\left\{\frac{\mathbf{S}^{C,k}}{N^{C,k}+\sigma^2/\sigma_0^2}, \frac{1}{N^{C,k}/\sigma^2+1/\sigma_0^2}\mathbf{I}_p\right\}\,.$$

Note that when σ_0 is large relative to σ^2, the mean of the full conditional is approximately $\mathbf{S}^{C,k}/N^{C,k}$, the mean of the observations in cluster k. This is not dissimilar from the same step in the K-means algorithm. An important difference is that in this context the cluster means are treated as random variables and they have a joint posterior distribution given the data.

C. *The full conditional of* $[\boldsymbol{\sigma}^2|others]$ *is proportional to*

$$\{\prod_{i=1}^{n}[\mathbf{W}_i|C_i,\boldsymbol{\mu},\sigma^2]\}[\sigma^2]\,.$$

It can be shown that

$$\prod_{i=1}^{n}[\mathbf{W}_i|C_i,\boldsymbol{\mu},\sigma^2] \propto \left(\frac{1}{\sigma^2}\right)^{np/2}\exp\left\{-\frac{\sum_{i=1}^{n}||\mathbf{W}_i-\boldsymbol{\mu}_{C_i}||^2}{2\sigma^2}\}\right\}\,,$$

where $\boldsymbol{\mu}_{C_i} = S^{C,i}/|C_i|$. Moreover, we assumed that σ^2 has an inverse Gamma prior distribution with parameters a and b. Thus,

$$[\sigma^2] \propto \left(\frac{1}{\sigma^2}\right)^{-a-1}\exp(-b/\sigma^2)\,.$$

Multiplying these two distributions shows that

$$[\sigma^2|others] = \mathrm{Inverse\ Gamma}(a+np/2, b+\sum_{i=1}^{n}||\mathbf{W}_i-\boldsymbol{\mu}_{C_i}||^2/2)\,.$$

In some situations, the model is not identifiable or is very close to being not identifiable. Indeed, consider the case when data are simulated from a univariate normal distribution and we are fitting a mixture of two or more normal distributions. It is possible for some of the clusters to become empty. That is, $N^{C,k} = 0$. In this situation $S^{C,k}$ is undefined and the variance of $[\boldsymbol{\mu}_k | others]$ is σ_0^2, which is very large. This leads to substantial instabilities in the algorithm. Some solutions for stabilizing the Gibbs sampler include (1) impose restrictions on the prior of $\boldsymbol{\mu}_k$, $k = 1, \ldots, K$; (2) simulate from the Normal prior and take the closest observation $\mathbf{W}_i$ as the mean; and (3) analyze the simulated chains for high correlations between-cluster means.

9.3 Smoothing and Clustering

So far we have not used the functional characteristics of the functional data. As we have seen, in some situations it is reasonable to ignore the functional characteristics. However, when data are measured with large errors, distances between functions are amplified by the residual noise. This can lead to masking of the underlying clusters and meaningless results. Additionally, when data are sparse, the multivariate matrix of data does not even exist. Indeed, it is not obvious how to define the distance between two functions, one measured at three locations and another one observed at another ten locations. An elegant solution for both these problems is to (1) smooth the observed data via projections onto a basis; and (2) either cluster the smooth data or the projection coefficients. Here we will describe how to do that using functional PCA, though other approaches (e.g., using splines or Fourier basis) are also feasible.

9.3.1 FPCA Smoothing and Clustering

By this point we have everything we need to conduct functional smoothing using FPCA. Indeed, in Chapter 3 we introduced the methodology for FPCA smoothing. Here we use the powerful `fpca.face` function in the **refund** package to smooth the data. The R code is simple and self-explanatory.

```
#Load refund
library(refund)
#Set the time grid (weeks)
t <- 1:dim(Wd)[2]

#Apply functional PCA using FACE
results <- fpca.face(Y = Wd, Y.pred = Wd, contor = TRUE, argvals = t,
                     knots = 35, pve = 0.99, var = TRUE)
```

Recall that the data are stored in the 52×52 dimensional matrix `Wd`. The code starts by loading the **refund** package and setting the grid of observations where data are recorded. In this case, the grid is equally spaced because data are observed every week for 52 weeks. Applying FPCA is just one line of code (two lines in this text due to space restrictions). The first argument is the data `Y=Wd`, which is a matrix with study participants by row and functional observations by column. The second argument, `Y.pred=Wd`, indicates that we would like to obtain the smooth predictors of all input functions. The third argument `center=TRUE` indicates that PCA is conducted after centering the data. The argument `argvals=t` is self-explanatory, while `knots=35` is the maximum number of knots used in the univariate spline smoother. The argument `pve=0.99` indicates that the number of principal components used corresponds to a 0.99 proportion of variability explained (PVE) after smoothing the covariance operator. That is, percent variability explained after removing the noise variability. The argument `var=TRUE` returns the variability for each study participant, which can be used to obtain confidence intervals for the participant-specific function estimate.

```
#Obtain the eigenfunctions and eigenvalues
Phi <- results$efunctions
eigenvalues <- results$evalues

#Obtain the covariance and correlation matrices
cov_est <- Phi %*% diag(eigenvalues) %*% t(Phi)
cor_est <- cov2cor(cov_est)
```

The output of the FPCA smoothing is now stored in the list `results`. The code proceeds by extracting the smooth estimated eigenfunctions in the matrix `Phi`, which in our case is a 52×11 dimensional matrix with the eigenfunctions stored by columns. There are 11 eigenfunctions because they explain 99% of the variation after removing the noise variability. The estimated eigenvalues are stored in the vector `eigenvalues` in decreasing order and corresponding to the columns of `Phi`. The smooth covariance and correlation function estimators are stored in the `cov_mat` and `cor_mat` matrices, respectively.

Figure 9.9 displays the smooth covariance function estimator, where each row and column corresponds to a particular week in 2020. Note the strong high covariances among observations during the weeks 10-20 (March 7 to May 16) as well as among observations

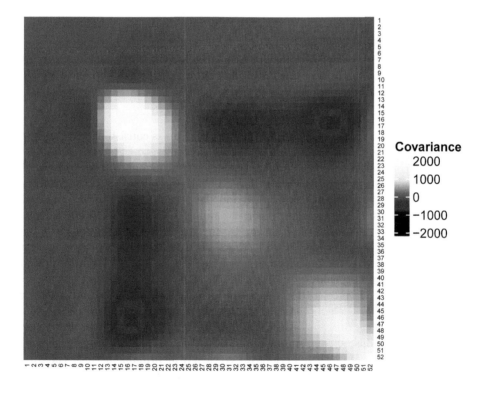

FIGURE 9.9: Smooth covariance matrix estimator for the 2020 weekly excess mortality rates in US states and territories. Each row and column corresponds to one of the 52 weeks of 2020 starting from January.

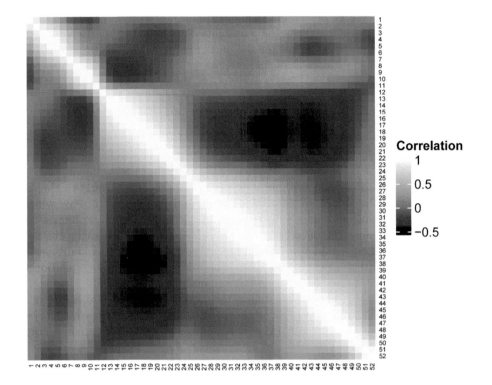

FIGURE 9.10: Smooth correlation matrix estimator for the 2020 weekly excess mortality rates in US states and territories. Each row and column corresponds to one of the 52 weeks of 2020 starting from January.

during the weeks 44-52 (October 31 to December 26). This is consistent with the spring and winter surges in excess mortality. The first surge affected more US states in the Northeast, whereas the second surge affected more states in the Intermountain and Midwest regions.

The covariance surface also indicates large negative covariances between observations during the weeks 10-20 (March 7 to May 16) and the period after; note the darker shades of blue to the right of the high covariances corresponding to the first surge in excess mortality rates. The covariance matrix is often driven by the size of the observations at a particular time. This is why the spring and winter surges are so clearly emphasized in Figure 9.9. However, this makes it hard to understand and visualize data correlation in periods when excess mortality is off its peaks.

To illustrate this, Figure 9.10 displays the correlation matrix stored in the cor_mat variable. Note that some characteristics of the covariance function are preserved, though the visuals are strikingly different. First, the main diagonal clearly indicates three periods of mutual high correlations (spring, summer, and fall/winter). Only two were clearly identified in the covariance plot. These three periods are so clearly identifiable because the correlations between the beginning and end of these periods quickly transition from strongly positive to close to zero or negative. This is illustrated by the almost rectangular shape of the high correlations in these three periods. From our previous analysis of the data this corresponds to the fact that states that have a large increase in excess mortality rates in one period tend to do better in the other periods. This is especially true for the spring surge, indicating that

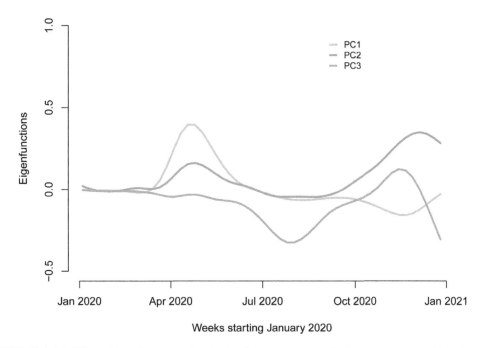

FIGURE 9.11: First three functional principal components of the excess mortality data in the US in 2020. The x-axis corresponds to time expressed in weeks starting with the first week in January.

states in the Northeast, which had the worst surge in excess mortality rates during spring, tended to do much better in the summer and fall.

Figure 9.11 displays the first three functional principal components (PCs), which explain 86% of the variability. The first PC (shown in green) has a substantial increase between March and May, is close to zero, and even becomes negative after September. This indicates that states that have large positive scores on this component had a large excess mortality rates in spring, though the excess mortality stayed at or below average for the remainder of 2020. New Jersey is the state with the largest score on this component. The second PC has a smaller bump in excess mortality in spring, a dip towards zero in the summer, and a larger bump in late fall and early winter. States with large positive scores on this component had a smaller excess mortality rate in the spring, stayed close to the average during the summer, and had a higher than average excess mortality rate in late fall and early winter. South Dakota is the state with the largest score on this component. The third PC decreases steadily from March to June, with a strong dip between the end of the June and September, a slow increase through November, and a fast decrease through December. States with large negative scores on this component had a steady, slow increase in excess mortality rate from March till June, a sudden increase in the summer, an average or below excess mortality rate in November, and a turn for the worse in the last few weeks of the year. Arizona is the state with the largest score on this component.

These analyses again emphasize the principal directions of variation in the functional space. They are quite interpretable and consistent with the other analyses of the data.

Each function can now be represented as the set of scores on the principal components. The idea is to use these scores for clustering instead of the original data. The R code below indicates how to obtain the principal scores, which are stored in PC_scores. This is a

52×11 dimensional matrix. K-means clustering using functional scores proceeds simply by replacing the data stored in `Wd` with `PC_scores[,1:3]`. Here we have used only the scores on the first 3 components, but this can be changed. In fact, the K-means clustering of the data and of the scores on the first three principal components is identical. This, however, is not true when using just the first principal component scores. Indeed, in this case, only four states are assigned to cluster 1 and 36 are assigned to cluster 2. This compares to twelve states in cluster 1 and 23 in cluster 2 when using the complete data. This is most likely due to over-smoothing of data using only one principal component.

```
#Obtain the scores and the functional predictors
PC_scores <- results$scores
#K-means clustering using the scores on the first three PCs
kmeans_CV19_3 <- kmeans(PC_scores[, 1:3], centers = 3)
cl_ind_sc <- kmeans_CV19_3$cluster
```

Figure 9.12 displays the principal component scores on the first three principal components. The top panel is the scatter plot for PC1 scores on the x-axis versus PC2 scores on the y-axis. The bottom panel is the scatter plot for PC1 scores on the x-axis versus PC3 scores on the y-axis. Colors correspond to the K-means cluster estimates, which are identical to those obtained from K-means of the original data. Colors are also identical to those used in Section 9.2.1.

Hierarchical and distributional clustering can be conducted similarly by replacing the `Wd` matrix with the `PC_scores[,1:3]` matrix. The general idea is that we can use any method for functional PCA and combine it with any method for multivariate clustering.

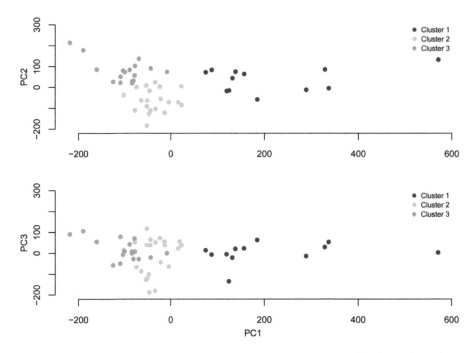

FIGURE 9.12: Scatter plots of principal component scores on the first three functional principal components of mortality data in the US in 2020. Principal component one scores are shown on the x-axis and the principal component scores on two and three are shown on the y-axis in the top and bottom panel, respectively.

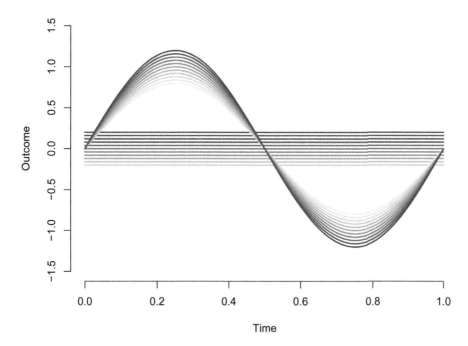

FIGURE 9.13: An example of synthetic functional data with two known groups shown in red and blue respectively. The first group consists of constant functions and the second group consists of sinusoidal functions with different amplitudes.

In this example, there is not much estimated residual variability. Indeed, the noise variance is estimated to be only $\sim 0.5\%$ of the overall variability. Thus, decomposing the observed covariance into principal components reduces the dimensionality, but does not substantially impact the smoothness of the data. So, when would functional smoothing actually improve results?

9.3.2 FPCA Smoothing and Clustering with Noisy Data

We consider a very simple example where functional data has two clusters. The first cluster is of functions $X_i : [0,1] \to \mathbb{R}$, $i = 1,\ldots,101$, where $s_j = (j-1)/101$, and $X_i(s_j) = -0.2 + 0.4*(j-1)/100$ for $j = 1,\ldots,101$. The second cluster is of functions $X_i : [0,1] \to \mathbb{R}$, $i = 102,\ldots,202$, where $X_i(s_j) = \{0.8 + 0.4*(j-1)/100\}\sin\{2\pi(j-1)/101\}$. Both clusters of functions are shown in Figure 9.13. The shades of red indicate functions in the first cluster, which are constant along the domain of the function. Darker shades of red correspond to higher values of the function. The shades of blue indicate functions in the second cluster. These functions are slightly amplified and attenuated sinus functions. Darker shades of blue indicate sinuses with larger amplitudes, while lighter shades of blue indicate sinuses with smaller amplitudes. The two clusters are visually distinctive and any clustering algorithm worth anything should be able to identify these two clusters.

For a sequence of standard deviations $\sigma = \{0,1,2,3,4,5,10\}$ for each simulation we generated data from the model

$$W_i(s_j) = X_i(s_j) + \sigma\epsilon_{ij},$$

where $\epsilon_{ij} \sim N(0,1)$ are mutually independent random errors. Therefore, for each simulation and value of σ the simulated data is a matrix $\mathbf{W}$ of dimension 202×101, where the first

101 rows correspond to the functions in cluster 1 and the remaining rows correspond to the functions in cluster 2. Larger values of σ correspond to more added noise and a value of zero corresponds to no noise. For each simulated data instance we applied two clustering approaches, both based on K-means with two clusters (the true number of clusters in the data). The first uses the raw, simulated data $\mathbf{W}$, while the second uses the scores on the functional principal components that explain 99% of the variability after smoothing the covariance function.

Figure 9.14 displays the misclassification rate for K-means clustering using the raw, unsmoothed data (green), and the scores obtained from functional PCA (red). The two lines are the averages over 500 simulations for each value of σ. The x-axis is the standard deviation of the added noise, σ. As expected, both approaches have zero misclassification rate when $\sigma = 0$ and very small and comparable misclassification rates when $\sigma = 1, 2$. However, for larger values of noise, clustering using FPCA has lower misclassification rates. For example, when $\sigma = 4$ the misclassification rate when using the raw data is 0.39 compared to 0.26 when using FPCA scores.

This simulation study indicates that when data have substantial noise it may be a good idea to conduct smoothing first using, for example, FPCA and then use clustering. From a practical perspective, we suggest applying both approaches, comparing results, and analyzing discrepancies between results. These discrepancies should be small when the noise is small, but could be substantial when noise is large.

There is still the small issue of how misclassification rate is actually defined. This sounds like a minor detail, but it is one that deserves some attention. Consider our case when we

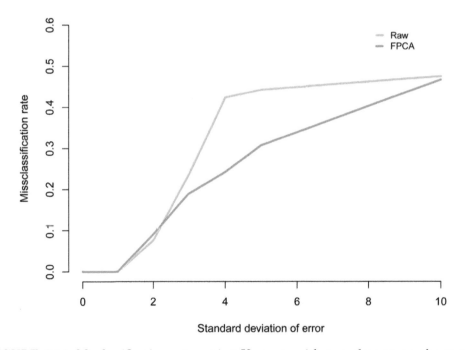

FIGURE 9.14: Misclassification rates using K-means with two clusters on the raw data (green) and FPCA scores with 99% of variance explained (red). True data are shown in Figure 9.13 and simulated data were obtained by adding independent Normal white noise with standard deviation σ at 101 equally spaced points around each curve. The x-axis shows σ and the y-axis shows the average misclassification rate over 500 simulations.

know the true labels, say $TL_1, \ldots, TL_n$, and we use a clustering algorithm that provides another set of labels, say $C_1, \ldots, C_n$. In our case there are two clusters and the first 101 true labels are "Red" and the last 101 labels are "Blue". The clustering algorithm provides labels $C_i \in \{1, 2\}$. There are two possibilities, one where "Red" corresponds to cluster "1" and one where it corresponds to cluster "2". The only way to know which makes more sense is to investigate if "Red"="1" or "Red"="2" leads to fewer misclassified individuals. We define the misclassification rate as

$$\min[A, B]/n \, ,$$

where $A = |\{i : TL_i = \text{"Red"}, C_i = \text{"1"}\}| + |\{i : TL_i = \text{"Blue"}, C_i = \text{"2"}\}|]$ and $B = |\{i : TL_i = \text{"Red"}, C_i = \text{"2"}\}| + |\{i : TL_i = \text{"Blue"}, C_i = \text{"1"}\}|]$, where $|S|$ denotes the number of elements in the set S. This definition can be extended to more than two clusters and the minimum would need to be taken over all permutations of estimated clusters.

9.3.3 FPCA Smoothing and Clustering with Sparse Data

Consider now the case when we have sparse data, where the number and location of observations may be different across individuals. For example, in the CD4 counts data there are 3 observations for the first study participant at times -9, -3, 3 months since seroconversion. For the second study participant there are 4 observations at times -3, 3, 9, and 15 months from seroconversion. Moreover, data are measured with substantial noise around the average CD4 dynamics for each individual. These data characteristics make it difficult to even define a distance between the study specific observations.

A reasonable solution is to use functional data analysis for sparse data, predict the smooth underlying curves, and cluster the curves using any multivariate clustering approach of the reconstructed trajectories. To predict each function we use the powerful `face.sparse` function in R. We also obtain the scores, which could be used for clustering, but we use the predicted functions for illustration.

In chapter 2 we have already discussed how the CD4 count data needs to be structured for the function `face.sparse`. Below we show how to extract what we need from the results of the function for clustering.

```
#Apply fpca.sparse to the CD4 count data
fit_face <- fpca.face(data, argvals.new = -20:40, newdata = data,
                      calculate.scores = TRUE, pve = 0.95)

#Obtain the scores
scores <- fit_face$scores$scores
#Store a replica of the data
data.h <- data
```

Note that the matrix of `scores` is 366×2 dimensional and could be used as an input for any clustering approach. However, here we want to illustrate how to predict and use the predicted functions for clustering. We provide code for how to extract the predicted function for one study participant. The approach is simply iterated over study participants to obtain all predictions.

The vector `temp_pred` contains the prediction of the CD4 counts at every month between 20 months before and 40 months after seroconversion (a total of 61 months). The same prediction can be done for every study participant. Figure 9.15 displays the predicted functions on this grid (-20 to 40) and the K-means clustering with three clusters of the predicted functions. Colors indicate the estimated clusters for each study participant. Thicker lines of a darker shade of the same color indicate the estimated cluster centers. Clusters seem to be characterized primarily by the mean log CD4 counts and how steep the decline is. Note that the cluster centers are closer at time -20 months compared to time 40 months.

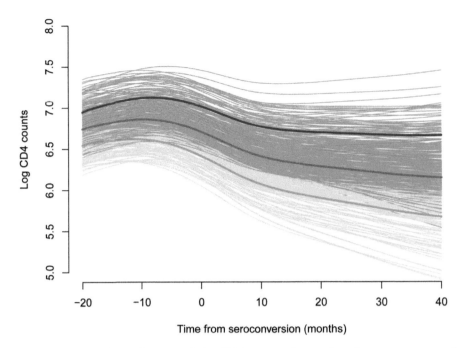

FIGURE 9.15: Smooth predictions of the CD4 counts data using the `fpca.sparse` function in R. Colors indicate the estimated clusters for each study participant. Thicker lines of a darker shade of the same color indicate the estimated cluster centers.

```
#Extract the id vector and the vector of unique ids
id <- data.h$subj
uid <- unique(id)
#Set the grid where to predict
seq <- -20:40
k <- length(seq)
#Extract one study participant
i <- 1
#Store a replica of the data
data.h <- data
#Select the i-th study participant
sel <- which(id == uid[i])
dati <- data.h[sel,]

#Set the framework for data prediction
dati_pred <- data.frame(y = rep(NA, nrow(dati) + k),
                argvals = c(rep(NA, nrow(dati)), seq)
                subj = rep(dati$subj[1], nrow(dati) + k))

#This is where the data is populated
dati_pred[1:nrow(dati),] <- dati
# Predict the data for study participant i
yhat2 <- predict(fit_face, dati_pred)

#Extract just the predictions on the grid
Ord <- nrow(dati) + 1:k
temp_pred <- yhat2$y.pred[Ord]
```

9.3.4   Clustering NHANES Data

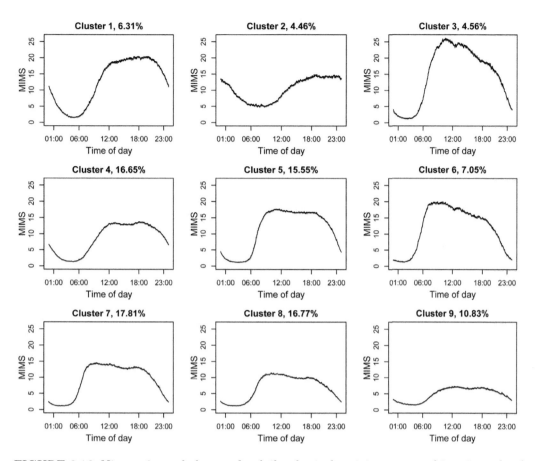

FIGURE 9.16: Nine estimated clusters for daily physical activity measured in minute-level MIMS in the NHANES 2011–2014 study. The method used was K-means on the physical activity matrix. The cluster was ordered based on the average age of study participants within the cluster. The proportion of participants in each cluster is shown on the title of each panel.

We also consider the case of NHANES data. Here we use the same NHANES dataset as that introduced in Chapter 7, which contains 8,713 study participants with time to all-cause-mortality information. To start, we take the study participant-specific average physical activity at every time point (minute) of the day over eligible days. These average trajectories are clustered using K-means with nine clusters. Clusters are ordered in terms of the average age of study participants within the cluster from youngest (Cluster 1) to oldest (Cluster 9). However, age was not used to perform K-means clustering. Labeling of clusters is not unique, which is why one needs to be precise when defining what is meant by "Cluster 4."

The cluster centers are displayed in Figure 9.16. Each cluster is accompanied by the percent study participants in that cluster. For example, 6.31% of the study participants are in Cluster 1 and 10.83% are in Cluster 9. While most clusters have a clear pattern of higher activity during the day, Cluster 2, which contains 4.46% of the population seems to correspond to individuals who, on average, have higher activity during the night. This may be primarily nigh-shift workers. There is a striking difference between the physical activity of individuals in Clusters 1 and 9. This is likely due to the fact that study participants in Cluster 9 are older, may suffer of more severe disease and/or may be at higher risk of death.

To further investigate the composition of each cluster, Table 9.1 displays some summary characteristics within each cluster. The first column provides the percent of individuals who died in each cluster by December 31, 2019. Notice that in Cluster 1 with an average age of 33.90 years, 2.2% of individuals died. This can be compared to 15.5% in Cluster 8 (average age 59.26 years) and 34.9% in Cluster 9 (average age 60.99 years). While the average age for Cluster 8 is about 1.7 years lower than in Cluster 9, this does not completely explain the large difference in mortality. Individuals in Cluster 9 have a smaller Poverty-Income Ratio (PIR) than those in Cluster 8. PIR is an index calculated by dividing family income by a poverty threshold specific to the family size. Smaller PIR corresponds to being poorer.

As discussed, the center of Cluster 2 has a very different shape from the other centers, which may indicate that Cluster 2 may include night-shift workers. This is further supported by the fact that the average age of individuals in this cluster is 35.36 years. This is much younger than the average age of the NHANES sample (48.76 years old). Study participants in Clusters 1, 3, and 5 have relatively lower BMI. In Figure 9.16 these three clusters correspond to the highest average physical activity intensity values during the day. In contrast, study participants in Cluster 9 have the highest proportion of all-cause mortality, highest average age and average BMI. Since the physical activity intensity average of this group is much lower across all times of the day, this suggests that it could be used to identify individuals who have elevated health and mortality risks.

Here we have used the means of physical activity profiles, though this is not necessary. Indeed, a multilevel functional approach can be used to decompose functional data at different levels (e.g., study participant average versus daily activity profile) and conduct clustering separately at each level; see, for example, [137, 269]. This has important implications as two study participants may belong to a particular cluster based on their average profile, while their day-to-day variation may place them in completely different clusters. Whether this is of any relevance to their health remains to be investigated. But, then again, something, somewhere, always does.

TABLE 9.1: Characteristics of study participants in each cluster as determined by K-means clustering on minute-level MIMS. Each row represents one cluster. Each column contains the summary of one variable, including the proportion of mortality as of December 31, 2019, average age, average BMI, and average poverty income ratio (PIR).

Cluster	Mortality (%)	Age (yrs)	BMI	PIR
1	2.2	33.90	27.34	1.85
2	4.6	35.36	29.03	1.77
3	2.0	40.98	27.50	1.89
4	5.3	41.64	28.73	2.28
5	2.7	43.99	27.80	2.44
6	3.9	49.06	28.08	2.55
7	6.5	52.74	29.34	2.94
8	15.5	59.26	30.28	2.67
9	34.9	60.99	30.57	2.17

Bibliography

[1] A. Aguilera, F. Ocaña, and M. Valderrama. Forecasting with unequally spaced data by a functional principal component approach. *Test*, 8(1):233–253, 1999.

[2] A. Ait-Saïdi, F. Ferraty, R. Kassa, and P. Vieu. Cross-validated estimations in the single-functional index model. *Statistics*, 42(6):475–494, 2008.

[3] H. Akaike. A new look at the statistical model identification. *IEEE Transactions on Automatic Control*, 19(6):716–723, 1974.

[4] M.R. Anderberg. *Cluster Analysis for Applications*. Academic Press: New York, 1973.

[5] M. Ankerst, M.M. Breunig, H.-P. Kriegel, and J. Sander. OPTICS: Ordering Points To Identify the Clustering Structure. *ACM SIGMOD international conference on Management of data. ACM Press*, pages 49–60, 1999.

[6] A. Antoniadis and T. Sapatinas. Estimation and inference in functional mixed-effects models. *Computational Statistics & Data Analysis*, 51(10):4793–4813, 2007.

[7] E. Arias, B. Tejada-Vera, and F. Ahmad. Provisional life expectancy estimates for January through June, 2020. *Vital Statistics Rapid Release: National Vital Statistics System*, 10:1–4, 2021.

[8] J.A.D. Aston, J.-M. Chiou, and J.P. Evans. Linguistic pitch analysis using functional principal component mixed effect models. *Journal of the Royal Statistical Society: Series C (Applied Statistics)*, 59(2):297–317, 2010.

[9] V. Baladandayuthapani, B.K. Mallick, M.Y. Hong, J.R. Lupton, N.D. Turner, and R.J. Carroll. Bayesian hierarchical spatially correlated functional data analysis with application to colon carcinogenesis. *Biometrics*, 64(1):64–73, 2008.

[10] D. Bates, M. Mächler, B. Bolker, and S. Walker. Fitting linear mixed-effects models using lme4. *Journal of Statistical Software*, 67(1):1–48, 2015.

[11] A. Bauer, F. Scheipl, H. Küchenhoff, and A.-A. Gabriel. An introduction to semiparametric function-on-scalar regression. *Statistical Modelling*, 18(3-4):346–364, 2018.

[12] A. Baíllo and A. Grané. Local linear regression for functional predictor and scalar response. *Journal of Multivariate Analysis*, 100(1):102–111, 2009.

[13] J.M. Beckers and R. Michel. EOF calculations and data filling from incomplete oceanographic datasets. *Journal of Atmospheric and Oceanic Technology*, 20, 12 2003.

[14] R. Bender, T. Augustin, and M. Blettner. Generating survival times to simulate cox proportional hazards models. *Statistics in Medicine*, 24(11):1713–1723, 2005.

[15] Y. Benjamini and Y. Hochberg. Controlling the false discovery rate: A practical and powerful approach to multiple testing. *Journal of the Royal Statistical Society: Series B (Methodological)*, 57(1):289–300, 1995.

291

[16] U. Beyaztas and H.L. Shang. On function-on-function regression: Partial least squares approach. *Environmental and Ecological Statistics*, 27:95–114, 2020.

[17] K. Bhaskaran, I. dos Santos-Silva, D.A. Leon, I.J. Douglas, and L. Smeeth. Association of BMI with overall and cause-specific mortality: A population-based cohort study of 3.6 million adults in the UK. *Lancet Diabetes & Endocrinology*, 6:944–953, 2018.

[18] P. Billingsley. *Convergence of Probability Measures, 2nd Edition*. Probability and Statistics. Wiley, 1999.

[19] J. Boland, D. Telesca, C. Sugar, S. Jeste, C. Goldbeck, and D. Senturk. A study of longitudinal trends in time-frequency transformations of EEG data during a learning experiment. *Computational Statistics & Data Analysis*, 167:107367, 2022.

[20] E. Borghi, M. de Onis, C. Garza, J. Van den Broeck, E. A. Frongillo, L. Grummer-Strawn, S. Van Buuren, H. Pan, L. Molinari, R. Martorell, A. W. Onyango, J. C. Martines, and for the WHO Multicentre Growth Reference Study Group. Construction of the World Health Organization child growth standards: Selection of methods for attained growth curves. *Statistics in Medicine*, 25(2):247–265, 2006.

[21] C. Bouveyron. *funFEM: Clustering in the Discriminative Functional Subspace*, 2021. R package version 1.2.

[22] N.E. Breslow. Discussion of the paper by D.R. Cox. *Journal of the Royal Statistical Society, Series B*, 34:216–217, 1972.

[23] S. Brockhaus and D. Rügamer. *FDboost: Boosting Functional Regression Models*, 2018.

[24] S. Brockhaus, D. Rügamer, and S. Greven. Boosting functional regression models with FDboost. *Journal of Statistical Software*, 94(10):1–50, 2020.

[25] S. Brockhaus, F. Scheipl, T. Hothorn, and S. Greven. The functional linear array model. *Statistical Modelling*, 15(3):279–300, 2015.

[26] B. Brumback, D. Ruppert, and M.P. Wand. Comment on variable selection and function estimation in additive nonparametric regression using data–based prior by Shively, Kohn, and Wood. *Journal of the American Statistical Association*, 94(447):794–797, 1999.

[27] B.A. Brumback and J.A. Rice. Smoothing spline models for the analysis of nested and crossed samples of curves. *Journal of the American Statistical Association*, 93(443):961–976, 1998.

[28] T.T. Cai and M. Yuan. Nonparametric covariance function estimation for functional and longitudinal data. *Technical Report*, 2010.

[29] E.J. Candes and Y. Plan. Matrix completion with noise. *Proceedings of the IEEE*, 98(6):925–936, 2010.

[30] E.J. Candes and T. Tao. The power of convex relaxation: Near-optimal matrix completion. *IEEE Transactions on Information Theory*, 56(5):2053–2080, 2010.

[31] H. Cardot, C. Crambes, and P. Sarda. Quantile regression when the covariates are functions. *Nonparametric Statistics*, 17(7):841–856, 2005.

[32] H. Cardot, A. Dessertaine, C. Goga, E. Josserand, and P. Lardin. Comparison of different sample designs and construction of confidence bands to estimate the mean of functional data: An illustration on electricity consumption. *Survey Methodology*, 39(2):283–301, 2013.

[33] H. Cardot, C. Goga, and P. Lardin. Uniform convergence and asymptotic confidence bands for model-assisted estimators of the mean of sampled functional data. *Electronic Journal of Statistics*, 7:562–596, 2013.

[34] H. Cardot, C. Goga, and P. Lardin. Variance estimation and asymptotic confidence bands for the mean estimator of sampled functional data with high entropy unequal probability sampling designs. *Scandinavian Journal of Statistics*, 41(2):516–534, 2014.

[35] B. Carpenter, A. Gelman, M.D. Hoffman, D. Lee, B. Goodrich, M. Betancourt, M. Brubaker, J. Guo, P. Li, and A. Riddell. Stan: A probabilistic programming language. *Journal of Statistical Software*, 76(1), 2017.

[36] R.J. Carter, D.A. Rose, R.T. Sabo, J. Clayton, J. Steinberg, M. Anderson, and CDC COVID-19 Response Team. Widespread severe acute respiratory syndrome coronavirus 2 transmission among attendees at a large motorcycle rally and their contacts, 30 US jurisdictions, August–September, 2020. *Clinical Infectious Diseases*, 73(Supplement 1):S106–S109, 04 2021.

[37] G. Casella and E.I. George. Explaining the Gibbs sampler. *The American Statistician*, 46(3):167–174, 1992.

[38] W. Checkley, L.D. Epstein, R.H. Gilman, R.E. Black, L. Cabrera, and C.R. Sterling. Effects of Cryptosporidium parvum infection in Peruvian children: Growth faltering and subsequent catch-up growth. *American Journal of Epidemiology*, 148:497–506, 1998.

[39] W. Checkley, L.D. Epstein, R.H. Gilman, L. Cabrera, and R.E. Black. Effects of acute diarrhea on linear growth in Peruvian children. *American Journal of Epidemiology*, 157:166–175, 2003.

[40] K. Chen and H.-G. Müller. Conditional quantile analysis when covariates are functions, with application to growth data. *Journal of the Royal Statistical Society Series B: Statistical Methodology*, 74(1):67–89, 2012.

[41] K. Chen and H.-G. Müller. Modeling repeated functional observations. *Journal of the American Statistical Association*, 107(500):1599–1609, 2012.

[42] S. Chib and E. Greenberg. Understanding the Metropolis-Hastings algorithm. *The American Statistician*, 49(4):327–335, 1995.

[43] J.-M. Chiou. Dynamical functional prediction and classification, with application to traffic flow prediction. *The Annals of Applied Statistics*, 6:1588–1614, 2016.

[44] J.-M. Chiou, H.-G. Müller, and J.-L. Wang. Functional quasi-likelihood regression models with smooth random effects. *Journal of the Royal Statistical Society. Series B (Statistical Methodology)*, 65(2):405–423, 2003.

[45] J.-M. Chiou, H.-G. Müller, and J.-L. Wang. Functional response models. *Statistica Sinica*, 14(3):675–693, 2004.

[46] J.-M. Chiou, Y.-F. Yang, and Chen Y.-T. Multivariate functional linear regression and prediction. *Journal of Multivariate Analysis*, 146:301–312, 2016.

[47] D.R. Cox. Regression models and life-tables. *Journal of the Royal Statistical Society. Series B (Methodological)*, 34(2):187–220, 1972.

[48] C.M. Crainiceanu and A.J. Goldsmith. Bayesian functional data analysis using Win-BUGS. *Journal of Statistical Software*, 32(11):1–33, 2010.

[49] C.M. Crainiceanu and D. Ruppert. Likelihood ratio tests in linear mixed models with one variance component. *Journal of the Royal Statistical Society. Series B (Statistical Methodology)*, 66(1):165–185, 2004.

[50] C.M. Crainiceanu, D. Ruppert, G. Claeskens, and M.P. Wand. Exact likelihood ratio tests for penalised splines. *Biometrika*, 92(1):91–103, 3 2005.

[51] C.M. Crainiceanu, D. Ruppert, and M.P. Wand. Bayesian analysis for penalized spline regression using winbugs. *Journal of Statistical Software*, 14(14):1–24, 2005.

[52] C.M. Crainiceanu, A.-M. Staicu, and C.Z. Di. Generalized multilevel functional regression. *Journal of the American Statistical Association*, 104(488):1550–1561, 2009.

[53] C.M. Crainiceanu, A.-M. Staicu, S. Ray, and N.M. Punjabi. Bootstrap-based inference on the difference in the means of two correlated functional processes. *Statistics in Medicine*, 31(26):3223–3240, 2012.

[54] P. Craven and G. Wahba. Smoothing noisy data with spline functions. *Numerische Mathematik*, 1:377–403, 1979.

[55] E. Cui. *Functional Data Analysis Methods for Large Scale Physical Activity Studies*. PhD thesis, Johns Hopkins University, Baltimore, MD, June 2023. Available at https://jscholarship.library.jhu.edu/bitstream/handle/1774.2/68330/CUI-DISSERTATION-2023.pdf?sequence=1&isAllowed=y.

[56] E. Cui, C.M. Crainiceanu, and A. Leroux. Additive functional Cox model. *Journal of Computational and Graphical Statistics*, 30(3):780–793, 2021.

[57] E. Cui, A. Leroux, E. Smirnova, and C.M. Crainiceanu. Fast univariate inference for longitudinal functional models. *Journal of Computational and Graphical Statistics*, 31(1):219–230, 2022.

[58] E. Cui, R. Li, C.M. Crainiceanu, and L. Xiao. Fast multilevel functional principal component analysis. *Journal of Computational and Graphical Statistics*, 32(2):366–377, 2023.

[59] E. Cui, E.C. Thompson, R.J. Carroll, and D. Ruppert. A semiparametric risk score for physical activity. *Statistics in Medicine*, 41(7):1191–1204, 2022.

[60] C. de Boor. *A Practical Guide to Splines*. Applied Mathematical Sciences. Springer, 2001.

[61] A.P. Dempster, N.M. Laird, and D.B. Rubin. Maximum likelihood from incomplete data via the EM algorithm. *Journal of the Royal Statistical Society. Series B (Methodological)*, 39(1):1–38, 1977.

[62] C.Z. Di, C.M. Crainiceanu, B. Caffo, and N.M. Punjabi. Multilevel functional principal component analysis. *Annals of Applied Statistics*, 3(1):458–488, 2009.

[63] C.Z. Di, C.M. Crainiceanu, and W.S. Jank. Multilevel sparse functional principal component analysis. *Stat*, 3(1):126–143, 2014.

[64] J. Di, A. Leroux, J. Urbanek, R. Varadhan, A. Spira, and V. Zipunnikov. Patterns of sedentary and active time accumulation are associated with mortality in US: The NHANES study. *bioRxiv*, 08 2017.

[65] K.M. Diaz, V.J. Howard, B. Hutto, N. Colabianchi, J.E. Vena, M.M. Safford, S.N. Blair, and S.P. Hooker. Patterns of sedentary behavior and mortality in U.S. middle-aged and older adults: A national cohort study. *Annals of Internal Medicine*, 167(7):465–475, 2017.

[66] P. Diggle, P. Heagerty, K.Y. Liang, and S. Zeger. *Analysis of Longitudinal Data, 2nd edition*. Oxford, England: Oxford University Press, 2002.

[67] J.D. Dixon. Estimating extremal eigenvalues and condition numbers of matrices. *SIAM Journal on Numerical Analysis*, 20(4):812–814, 1983.

[68] S. Dray and J. Josse. Principal component analysis with missing values: a comparative survey of methods. *Plant Ecology*, 216, 05 2014.

[69] R.M. Dudley. Sample functions of the Gaussian process. *The Annals of Probability*, 1(1):66–103, 1973.

[70] P.H.C. Eilers, B. Li, and B.D. Marx. Multivariate calibration with single-index signal regression. *Chemometrics and Intelligent Laboratory Systems*, 96(2):196–202, 2009.

[71] P.H.C. Eilers and B.D. Marx. Flexible smoothing with B-splines and penalties. *Statistical Science*, 11(2):89–121, 1996.

[72] P.H.C. Eilers and B.D. Marx. Generalized linear additive smooth structures. *Journal of Computational and Graphical Statistics*, 11(4):758–783, 2002.

[73] M. Ester, H.-P. Kriegel, J. Sander, and X. Xu. A density-based algorithm for discovering clusters in large spatial databases with noise. *Proceedings of the Second International Conference on Knowledge Discovery and Data Mining (KDD-96). AAAI Press*, 11:226–231, 1996.

[74] B. Everitt. *Cluster Analysis*. London: Heinemann Educational Books, 1974.

[75] W.F. Fadel, J.K. Urbanek, N.W. Glynn, and J. Harezlak. Use of functional linear models to detect associations between characteristics of walking and continuous responses using accelerometry data. *Sensors (Basel)*, 20(21):6394, 2020.

[76] J. Fan and I. Gijbels. *Local Polynomial Modelling and Its Applications: Monographs on Statistics and Applied Probability 66*, volume 66. CRC Press, 1996.

[77] J. Fan and J.-T. Zhang. Two-step estimation of functional linear models with applications to longitudinal data. *Journal of the Royal Statistical Society: Series B (Statistical Methodology)*, 62(2):303–322, 2000.

[78] Y. Fan, N. Foutz, G.M. James, and W. Jank. Functional response additive model estimation with online virtual stock markets. *The Annals of Applied Statistics*, 8(4):2435–2460, 2014.

[79] J.J. Faraway. Regression analysis for a functional response. *Technometrics*, 39(3):254–261, 1997.

[80] M. Febrero-Bande and M. Oviedo de la Fuente. Statistical computing in functional data analysis: The R package fda.usc. *Journal of Statistical Software*, 51(4):1–28, 2012.

[81] F. Ferraty. *Nonparametric Functional Data Analysis*. Springer, 2006.

[82] F. Ferraty, P. Hall, and P. Vieu. Most-predictive design points for functional data predictors. *Biometrika*, 97(4):807–824, 12 2010.

[83] F. Ferraty, A. Laksaci, and P. Vieu. Functional time series prediction via conditional mode estimation. *Comptes Rendus Mathematique*, 340(5):389–392, 2005.

[84] F. Ferraty, A. Mas, and P. Vieu. Nonparametric regression on functional data: Inference and practical aspects. *Australian & New Zealand Journal of Statistics*, 49(3):267–286, 2007.

[85] F. Ferraty, J. Park, and P. Vieu. Estimation of a functional single index model. In *Recent Advances in Functional Data Analysis and Related Topics*, pages 111–116. Springer, 2011.

[86] F. Ferraty and P. Vieu. *Nonparametric Functional Data Analysis: Theory and Practice*. Springer: New York, NY, USA, 2006.

[87] G. Fitzmaurice, M. Davidian, G. Molenberghs, and G. Verbeke. *Longitudinal Data Analysis*. Boca Raton, FL: Chapman & Hall/CRC, 2008.

[88] E.W. Forgy. Cluster analysis of multivariate data: Efficiency vs interpretability of classifications. *Biometrics*, 21:768–769, 1965.

[89] D. Fourdrinier, W.E. Strawderman, and M.T. Wells. Estimation of a functional single index model. In *Shrinkage Estimation*, pages 127–150. Springer, 2018.

[90] T. Galili. dendextend: An R package for visualizing, adjusting, and comparing trees of hierarchical clustering. *Bioinformatics*, 2015.

[91] M. Gaston, T. Leon, and F. Mallor. Functional data analysis for non-homogeneous Poisson processes. In *2008 Winter Simulation Conference*, pages 337–343, 2008.

[92] I. Gaynanova, N. Punjabi, and C.M. Crainiceanu. Modeling continuous glucose monitoring (CGM) data during sleep. *Biostatistics*, 23(1):223–239, 05 2020.

[93] J.E. Gellar, E. Colantuoni, D.M. Needham, and C.M. Crainiceanu. Variable-domain functional regression for modeling ICU data. *Journal of American Statistical Association*, 109(508):1425–1439, 2014.

[94] J.E. Gellar, E. Colantuoni, D.M. Needham, and C.M. Crainiceanu. Cox regression models with functional covariates for survival data. *Statistical Modelling*, 15(3):256–278, 2015.

[95] S. Geman and D. Geman. Stochastic relaxation, Gibbs distributions, and the Bayesian restoration of images. *IEEE Transactions on Pattern Analysis and Machine Intelligence*, PAMI-6(6):721–741, 1984.

[96] A. Genz and F. Bretz. *Computation of Multivariate Normal and t Probabilities*. Lecture Notes in Statistics. Springer-Verlag, Heidelberg, 2009.

[97] A. Genz, F. Bretz, T. Miwa, X. Mi, F. Leisch, F. Scheipl, and T. Hothorn. *mvtnorm: Multivariate Normal and t Distributions*, 2021. R package version 1.1-3.

[98] J. Gertheiss, J. Goldsmith, C.M. Crainiceanu, and S. Greven. Longitudinal scalar-on-functions regression with application to tractography data. *Biostatistics*, 14(3):447–461, 2013.

[99] J. Gertheiss, A. Maity, and A.M. Staicu. Variable selection in generalized functional linear models. *Stat*, 2(1):86–103, 2013.

[100] J. Gertheiss, E.F. Hessel V. Maier, and A.-M. Staicu. Marginal functional regression models for analyzing the feeding behavior of pigs. *Journal of Agricultural, Biological, and Environmental Statistics*, 20:353–370, 2015.

[101] Y. Goldberg, Y. Ritov, and A. Mandelbaum. Predicting the continuation of a function with applications to call center data. *Journal of Statistical Planning and Inference*, 147:53–65, 2014.

[102] A.J. Goldsmith, J. Bobb, C.M. Crainiceanu, B. Caffo, and D. Reich. Penalized functional regression. *Journal of Computational and Graphical Statistics*, 20(4):830–851, 2011.

[103] A.J. Goldsmith, C.M. Crainiceanu, B. Caffo, and D. Reich. Longitudinal penalized functional regression for cognitive outcomes on neuronal tract measurements. *Journal of the Royal Statistical Society: Series C (Applied Statistics)*, 61(3):453–469, 2012.

[104] A.J. Goldsmith, S. Greven, and C.M. Crainiceanu. Corrected confidence bands for functional data using principal components. *Biometrics*, 69(1):41–51, 2013.

[105] A.J. Goldsmith, F. Scheipl, L. Huang, J. Wrobel, C. Di, J. Gellar, J. Harezlak, M.W. McLean, B. Swihart, L. Xiao, C.M. Crainiceanu, and P.T. Reiss. *refund: Regression with Functional Data*, 2020. R package version 0.1-23.

[106] J. Goldsmith, V. Zipunnikov, and J. Schrack. Generalized multilevel function-on-scalar regression and principal component analysis. *Biometrics*, 71(2):344–353, 2015.

[107] A.D. Gordon. *Classification. Second Edition.* London: Chapman and Hall/CRC, 1999.

[108] C. Goutis. Second-derivative functional regression with applications to near infra-red spectroscopy. *Journal of the Royal Statistical Society. Series B (Statistical Methodology)*, 60(1):103–114, 1998.

[109] S. Greven, C.M. Crainiceanu, B. Caffo, and D.S. Reich. Longitudinal functional principal component analysis. *Electronic Journal of Statistics*, pages 1022–1054, 2010.

[110] S. Greven and F. Scheipl. A general framework for functional regression modelling. *Statistical Modelling*, 17:1–35, 2017.

[111] Z. Gu, R. Eils, and M. Schlesner. Complex heatmaps reveal patterns and correlations in multidimensional genomic data. *Bioinformatics*, 2016.

[112] Z. Gu, L. Gu, R. Eils, M. Schlesner, and B. Brors. circlize implements and enhances circular visualization in r. *Bioinformatics*, 30:2811–2812, 2014.

[113] J.E. Gubernatis. Marshall Rosenbluth and the Metropolis Algorithm. *Physics of Plasmas*, 12(5):57303, 2005.

[114] W. Guo. Functional mixed effects models. *Biometrics*, 58(1):121–128, 2002.

[115] J. Harezlak, B.A. Coull, N.M. Laird, S.R. Magari, and D.C. Christiani. Penalized solutions to functional regression problems. *Computational Statistics & Data Analysis*, 51(10):4911–4925, 2007.

[116] J. Harezlak, D. Ruppert, and M.P. Wand. *Semiparametric Regression with R*. Springer New York, NY, USA, 2018.

[117] J.A. Hartigan. Direct clustering of a data matrix. *Journal of the American Statistical Association*, 67(337):123–129, 1972.

[118] J.A. Hartigan. *Clustering Algorithms*. New York: Wiley, 1975.

[119] J.A. Hartigan and M.A. Wong. Algorithm AS 136: A K-means clustering algorithm. *Applied Statistics*, 28:100–108, 1979.

[120] E. Hartuv and R. Shamir. A clustering algorithm based on graph connectivity. *Information Processing Letters*, 76(4-6):175–181, 2000.

[121] D.A. Harville. Maximum likelihood approaches to variance component estimation and to related problems. *Journal of the American Statistical Association*, 72(358):320–338, 1977.

[122] G. He, H.-G. Müller, J.-L. Wang, and W. Yang. Functional linear regression via canonical analysis. *Bernoulli*, 16(3):705–729, 2010.

[123] L. Horváth and P. Kokoszka. *Inference for Functional Data with Applications*. New York: Springer, 2012.

[124] L. Huang, A.J. Goldsmith, P.T. Reiss, D.S. Reich, and C.M. Crainiceanu. Bayesian scalar-on-image regression with application to association between intracranial DTI and cognitive outcomes. *Neuroimage*, 83:210–223, 2013.

[125] H. Hullait, D.S. Leslie, N.G. Pavlidis, and S. King. Robust function-on-function regression. *Technometrics*, 63(3):396–409, 2021.

[126] R.J. Hyndman and H.L. Shang. Forecasting functional time series (with discussion). *Journal of the Korean Statistical Society*, 38(3):199–221, 2009.

[127] A.E. Ivanescu, C.M. Crainiceanu, and W. Checkley. Dynamic child growth prediction: A comparative methods approach. *Statistical Modelling*, 17(6):468–493, 2017.

[128] A.E. Ivanescu, A.-M. Staicu, F. Scheipl, and S. Greven. Penalized function-on-function regression. *Computational Statistics*, 30(2):539–568, 2015.

[129] Peng J. and D. Paul. A geometric approach to maximum likelihood estimation of the functional principal components from sparse longitudinal data. *Journal of Computational and Graphical Statistics*, 18(4):995–1015, 2009.

[130] J. Jacques and C. Preda. Functional data clustering: A survey. *Advances in Data Analysis and Classification*, 8:231–255, 09 2013.

[131] D. Jaganath, M. Saito, R.H. Gilman, D.M. Queiroz, G.A. Rocha, V. Cama, L. Cabrera, D. Kelleher, H.J. Windle, J.E. Crabtree, and W. Checkley. First detected Helicobacter pylori infection in infancy modifies the association between diarrheal disease and childhood growth in Peru. *Helicobacter*, 19:272–279, 2014.

[132] A.K. Jain and R.C. Dubes. *Algorithms for Clustering Data*. Prentice-Hall, 1988.

[133] P. Jain, C. Jin, S.M. Kakade, P. Netrapalli, and A. Sidford. Streaming PCA: Matching matrix Bernstein and near-optimal finite sample guarantees for Oja's algorithm. In V. Feldman, A. Rakhlin, and O. Shamir, editors, *29th Annual Conference on Learning Theory*, volume 49 of *Proceedings of Machine Learning Research*, pages 1147–1164, Columbia University, New York, New York, USA, 23–26 Jun 2016.

[134] G. James, T. Hastie, and C. Sugar. Principal component models for sparse functional data. *Biometrika*, 87(3):587–602, 2000.

[135] G.M James, J. Wang, and J. Zhu. Functional linear regression that's interpretable. *Annals of Statistics*, 37(5A):2083–2108, 2009.

[136] B.J. Jefferis, T.J. Parsons, C. Sartini, S. Ash, L.T. Lennon, O. Papacosta, R.W. Morris, S.G. Wannamethee, I.-M. Lee, and P.H. Whincup. Objectively measured physical activity, sedentary behaviour and all-cause mortality in older men: Does volume of activity matter more than pattern of accumulation? *British Journal of Sports Medicine*, 53(16):1013–1020, 2019.

[137] H. Jiang and N. Serban. Clustering random curves under spatial interdependence with application to service accessibility. *Technometrics*, 54(2):108–119, 2012.

[138] D. John, Q. Tang, F. Albinali, and S.S. Intille. A monitor-independent movement summary to harmonize accelerometer data processing. *Human Kinetics Journal*, 2(4):268–281, 2018.

[139] I.T. Jolliffe. A note on the use of principal components in regression. *Journal of the Royal Statistical Society, Series C*, 31(3):300–303, 1982.

[140] I.T. Jolliffe. *Principal Component Analysis*. Springer, 2002.

[141] E.L. Kaplan and P. Meier. Nonparametric estimation from incomplete observations. *Journal of the American Statistical Association*, 53(282):457–481, 1958.

[142] M. Karas, J. Muschelli, A. Leroux, J.K. Urbanek, A.A. Wanigatunga, J. Bai, C.M. Crainiceanu, and J.A. Schrack. Comparison of accelerometry-based measures of physical activity: Retrospective observational data analysis study. *JMIR Mhealth Uhealth*, 10(7):e38077, Jul 2022.

[143] K. Karhunen. Über lineare Methoden in der Wahrscheinlichkeitsrechnung. *Annals of the Academy of Science Fennicae. Series A. I. Mathematics-Physics*, 37:1–79, 1947.

[144] R.A. Kaslow, D.G. Ostrow, R. Detels, J.P. Phair, B.F. Polk, and C.R. Rinaldo Jr. The multicenter AIDS cohort study: Rationale, organization, and selected characteristics of the participants. *American Journal of Epidemiology*, 126(2):310–318, 1987.

[145] R.E. Kass and V. Ventura. A spike train probability model. *Neural Computation*, 13:1713–1720, 2001.

[146] L. Kaufman and P.J. Rousseeuw. *Finding Groups in Data: An Introduction to Cluster Analysis*. New York: Wiley, 1990.

[147] R.C. Kelly and R.E. Kass. A framework for evaluating pairwise and multiway synchrony among stimulus-driven neurons. *Neural Computation*, 24:2007–2032, 2012.

[148] J.S. Kim, A.-M. Staicu, A. Maity, R.J. Carroll, and D. Ruppert. Additive function-on-function regression. *Journal of Computational and Graphical Statistics*, 27(1):234–244, 2018.

[149] K. Kim, D. Sentürk, and R. Li. Recent history functional linear models for sparse longitudinal data. *Journal of Statistical Planning and Inference*, 141(4):1554–1566, 2011.

[150] G.S. Kimeldorf and G. Wahba. A correspondence between bayesian estimation on stochastic processes and smoothing by splines. *The Annals of Mathematical Statistics*, 41(2):495–502, 1970.

[151] J.P Klein and M.L. Moeschberger. *Survival Analysis: Techniques for Censored and Truncated Data (2nd ed.)*. Springer, 2003.

[152] G.G. Koch. Some further remarks concerning "A general approach to the estimation of variance components". *Technometrics*, 10(3):551–558, 1968.

[153] P. Kokoszka and M. Reimherr. *Introduction to Functional Data Analysis*. Chapman and Hall/CRC, 2017.

[154] S. Koner and A.-M. Staicu. Second-generation functional data. *Annual Review of Statistics and Its Application*, 10(1):547–572, 2023.

[155] D. Kong, J.G. Ibrahim, E. Lee, and H. Zhu. FLCRM: Functional linear Cox regression model. *Biometrics*, 74(1):109–117, 2018.

[156] E.L. Korn and B.I. Graubard. *Analysis of Health Surveys*, volume 323. John Wiley & Sons, 2011.

[157] D. D. Kosambi. Statistics in function space. *Journal of the Indian Mathematical Society*, 7:76–88, 1943.

[158] J. Kuczyński and H. Woźniakowski. Estimating the largest eigenvalue by the power and Lanczos algorithms with a random start. *SIAM Journal on Matrix Analysis and Applications*, 13(4):1094–1122, 1992.

[159] M.G. Kundu, J. Harezlak, and T.W. Randolph. Longitudinal functional models with structured penalties. *Statistical Modelling*, 16(2):114–139, 2016.

[160] P.A. Lachenbruch and M.R. Mickey. Estimation of error rates in discriminant analysis. *Technometrics*, 10(1):1–11, 1968.

[161] N.M. Laird and J.H. Ware. Random-effects models for longitudinal data. *Biometrics*, 38(4):963–974, 1982.

[162] W. Lee, M.F. Miranda, P. Rausch, V. Baladandayuthapani, M. Fazio, J.C. Downs, and J.S. Morris. Bayesian semiparametric functional mixed models for serially correlated functional data, with application to glaucoma data. *Journal of the American Statistical Association*, 114(526):495–513, 2019.

[163] H. Lennon, M. Sperrin, E. Badrick, and A.G. Renehan. The obesity paradox in cancer: A review. *Current Oncology Reports*, 18(9):56, 2016.

[164] A. Leroux. *Statistical Methods for the Analyis of Functional Data under Models with Complex Association Structures.* PhD thesis, Johns Hopkins University, Baltimore, MD, June 2020. Available at https://jscholarship.library.jhu.edu/bitstream/handle/1774.2/63686/LEROUX-DISSERTATION-2020.pdf?sequence=1.

[165] A. Leroux. *rnhanesdata: NHANES Accelerometry Data Pipeline*, 2022. R package version 1.02.

[166] A. Leroux and C.M. Crainiceanu. Dynamic prediction using landmark historical functional Cox regression. *Under Review*, 2023.

[167] A. Leroux, C.M. Crainiceanu, and J. Wrobel. Fast generalized functional principal component analysis. *Under Review*, 2022.

[168] A. Leroux, J. Di, E. Smirnova, E.J. McGuffey, Q. Cao, E. Bayatmokhtari, L. Tabacu, V. Zipunnikov, J.K. Urbanek, and C.M. Crainiceanu. Organizing and analyzing the activity data in NHANES. *Statistics in Biosciences*, 11(2):262–287, 2019.

[169] A. Leroux, L. Xiao, C.M. Crainiceanu, and W. Checkley. Dynamic prediction in functional concurrent regression with an application to child growth. *Statistics in Medicine*, 37(8):1376–1388, 2018.

[170] A. Leroux, S. Xu, P. Kundu, J. Muschelli, E. Smirnova, N. Chatterjee, and C. Crainiceanu. Quantifying the predictive performance of objectively measured physical activity on mortality in the UK Biobank. *The Journals of Gerontology: Series A*, September 2020.

[171] B. Li and B.D. Marx. Sharpening P-spline signal regression. *Statistical Modelling*, 8(4):367–383, 2008.

[172] C. Li and L. Xiao. *mfaces: Fast Covariance Estimation for Multivariate Sparse Functional Data*, 2021. R package version 0.1-3.

[173] C. Li, L. Xiao, and S. Luo. Fast covariance estimation for multivariate sparse functional data. *Stat*, 9(1):p.e245, 2020.

[174] G. Li, L. Raffield, M. Logue, M.W. Miller, H.P. Santos Jr, T.M. O'Shea, R.C. Fry, and Y. Li. CUE: CpG impUtation ensemble for DNA methylation levels across the human methylation450 (HM450) and EPIC (HM850) BeadChip platforms. *Epigenetics*, 16(8):851–861, 2021.

[175] H. Li, S. Keadle, H. Assaad, J. Huang, and R. Carroll. Methods to assess an exercise intervention trial based on 3-level functional data. *Biostatistics*, 16(4):754–771, 05 2015.

[176] M. Li, A.-M. Staicu, and H.D. Bondell. Incorporating covariates in skewed functional data models. *Biostatistics*, 16(3):413–426, 2015.

[177] T. Li, T. Li, Z. Zhu, and H. Zhu. Regression analysis of asynchronous longitudinal functional and scalar data. *Journal of the American Statistical Association*, 0(0):1–15, 2020.

[178] Y. Li, S. Banerjee, C. Rhee, K. Kalantar-Zadeh, E. Kurum, and D. Şentürk. Multilevel modeling of spatially nested functional data: Spatiotemporal patterns of hospitalization rates in the US dialysis population. *Statistics in Medicine*, 40, 04 2021.

[179] Y. Li, N. Wang, and R.J. Carroll. Generalized functional linear models with semi-parametric single-index interactions. *Journal of the American Statistical Association*, 105(490):621–633, 2010.

[180] Z. Li and S.N. Wood. Faster model matrix crossproducts for large generalized linear models with discretized covariates. *Statistics and Computing*, 30:19–25, 2020.

[181] M.A. Lindquist. The statistical analysis of fMRI data. *Statistical Science*, 23(4):439–464, 2008.

[182] M.A. Lindquist. Functional causal mediation analysis with an application to brain connectivity. *Journal of American Statistica Association*, 107(500):1297–1309, 2012.

[183] S.P. Lloyd. Least squares quantization in PCM. *IEEE Transactions on Information Theory. Technical Note, Bell Laboratories.*, 28:128–137, 1957, 1982.

[184] M. Loève. *Probability Theory, Vol. II, 4th ed. Graduate Texts in Mathematics.* Springer-Verlag, 1978.

[185] T. Lumley. Analysis of complex survey samples. *Journal of Statistical Software*, 9(1):1–19, 2004.

[186] A.R. Lundborg, R.D. Shah, and J. Peters. Conditional Independence Testing in Hilbert Spaces with Applications to Functional Data Analysis. *Journal of the Royal Statistical Society Series B: Statistical Methodology*, 84(5):1821–1850, 11 2022.

[187] D.J. Lunn, A. Thomas, N. Best, and D. Spiegelhalter. WinBUGS - A Bayesian modelling framework: Concepts, structure, and extensibility. *Statistics and Computing*, 10:325–337, 2000.

[188] W. Ma, L. Xiao, B. Liu, and M.A. Lindquist. A functional mixed model for scalar on function regression with application to a functional MRI study. *Biostatistics*, 22(3):439–454, 2019.

[189] J. MacQueen. Some methods for classification and analysis of multivariate observations. In L.M. Le Cam and J. Neyman, editors, *Proceedings of the Fifth Berkeley Symposium on Mathematical Statistics and Probability*, volume 1, pages 1147–1164, Berkeley, CA, 2016. University of California Press.

[190] N. Malfait and J.O. Ramsay. The historical functional linear model. *Canadian Journal of Statistics*, 31(2):115–128, 2003.

[191] E.J. Malloy, J.S. Morris, S.D. Adar, H. Suh, D.R. Gold, and B.A. Coull. Wavelet-based functional linear mixed models: An application to measurement error–corrected distributed lag models. *Biostatistics*, 11(3):432–452, 2010.

[192] B.D. Marx and P.H.C. Eilers. Generalized linear regression on sampled signals and curves: A p-spline approach. *Technometrics*, 41(1):1–13, 1999.

[193] B.D. Marx and P.H.C. Eilers. Multidimensional penalized signal regression. *Technometrics*, 47(1):13–22, 2005.

[194] M. Matabuena, A. Petersen, J.C. Vidal, and F. Gude. Glucodensities: A new representation of glucose profiles using distributional data analysis. *Statistical Methods in Medical Research*, 30(6):1445–1464, 2021.

[195] H. Matsui, S. Kawano, and S. Konishi. Regularized functional regression modeling for functional response and predictors. *Journal of Math-for-Industry*, 1:17–25, 06 2013.

[196] C.E. McCulloch, S.R. Searle, and J.M. Neuhaus. *Generalized, Linear, and Mixed Models, 2nd edition*. New York: Wiley, 2008.

[197] G.J. McLachlan and D. Peel. *Finite Mixture Models*. New York: Wiley, 2000.

[198] M.W. McLean, G. Hooker, A.-M. Staicu, F. Scheipl, and D. Ruppert. Functional generalized additive models. *Journal of Computational and Graphical Statistics*, 23(1):249–269, 2014.

[199] J. Mercer. Functions of positive and negative type and their connection with the theory of integral equations. *Philosophical Transactions of the Royal Society*, 209:4–415, 1909.

[200] N. Metropolis, A.W. Rosenbluth, M.N. Rosenbluth, A.H. Teller, and E. Teller. Equation of state calculations by fast computing machines. *Journal of Chemical Physics*, 21(6):1087–1092, 1953.

[201] M. Meyer, B. Coull, F. Versace, P. Cinciripini, and J. Morris. Bayesian function-on-function regression for multilevel functional data. *Biometrics*, 71:563–574, 03 2015.

[202] B. Mirkin. *Mathematical Classification and Clustering*. Kluwer Academic Publishers, 1996.

[203] I. Mitliagkas, C. Caramanis, and P. Jain. Memory limited, streaming pca. In C.J.C. Burges, L. Bottou, M. Welling, Z. Ghahramani, and K.Q. Weinberger, editors, *Advances in Neural Information Processing Systems*, volume 26. Curran Associates, Inc., 2013.

[204] J.S. Morris. Statistical methods for proteomic biomarker discovery based on feature extraction or functional modeling approaches. *Statistics and Its Interface*, 5 1:117–135, 2012.

[205] J.S. Morris. Functional regression. *Annual Review of Statistics and Its Application*, 2(1):321–359, 2015.

[206] J.S. Morris, P.J. Brown, R.C. Herrick, K.A. Baggerly, and K.R. Coombes. Bayesian analysis of mass spectrometry proteomic data using wavelet-based functional mixed models. *Biometrics*, 64(2):479–489, 2008.

[207] J.S. Morris and R.J. Carroll. Wavelet-based functional mixed models. *Journal of the Royal Statistical Society: Series B (Statistical Methodology)*, 68(2):179–199, 2006.

[208] J.S. Morris, M. Vannucci, P.J. Brown, and R.J. Carroll. Wavelet-based nonparametric modeling of hierarchical functions in colon carcinogenesis. *Journal of the American Statistical Association*, 98(463):573–583, 2003.

[209] F. Mosteller and J.W. Tukey. Data analysis, including statistics. In G. Lindzey and E. Aronson, editors, *Handbook of Social Psychology, Vol. 2*. Addison-Wesley, 1968.

[210] H.-G. Müller, Y. Wu, and F. Yao. Continuously additive models for nonlinear functional regression. *Biometrika*, 100(3):607–622, 2013.

[211] F. Murtagh and P. Legendre. Ward's hierarchical agglomerative clustering method: Which algorithms implement ward's criterion? *Journal of Classification*, 31(3):274–295, 2014.

[212] M. Mutis, U. Beyaztas, G.G. Simsek, and H.L. Shang. A robust scalar-on-function logistic regression for classification. *Communications in Statistics - Theory and Methods*, 52(23):8538–8554, 2023.

[213] H.-G. Müller and U. Stadtmüller. Generalized functional linear models. *The Annals of Statistics*, 33(2):774–805, 2005.

[214] L.T. Nguyen, J. Kim, and B. Shim. Low-rank matrix completion: A contemporary survey. *IEEE Access*, 7:94215–94237, 2019.

[215] J. Niedziela, B. Hudzik, N. Niedziela, M. Gasior, M. Gierlotka, J. Wasilewski, K. Myrda, A. Lekston, L. Polonski, and P. Rozentryt. The obesity paradox in acute coronary syndrome: A meta-analysis. *European Journal of Epidemiology*, 29(11):801–812, 2014.

[216] D. Nychka. Bayesian confidence intervals for smoothing splines. *Journal of the American Statistical Association*, 83(404):1134–1143, 1988.

[217] D. Nychka, R. Furrer, J. Paige, and S. Sain. fields: Tools for spatial data, 2021. R package version 14.1.

[218] R.T. Ogden and E. Greene. Wavelet modeling of functional random effects with application to human vision data. *Journal of Statistical Planning and Inference*, 140(12):3797–3808, 2010. Special Issue in Honor of Emanuel Parzen on the Occasion of his 80th Birthday and Retirement from the Department of Statistics, Texas A&M University.

[219] E. Oja. Simplified neuron model as a principal component analyzer. *Journal of Mathematical Biology*, 15:267–273, 1982.

[220] A. Oreopoulos, R. Padwal, K. Kalantar-Zadeh, G.C. Fonarow, C.M. Norris, and F.A. McAlister. Body mass index and mortality in heart failure: A meta-analysis. *American Heart Journal*, 156(1):13–22, 2008.

[221] F. O'Sullivan. A statistical perspective on ill-posed inverse problems (with discussion). *Statistical Science*, 1(4):505–527, 1986.

[222] H. Park, E. Petkova, T. Tarpey, and R.T. Ogden. Functional additive models for optimizing individualized treatment rules. *Biometrics*, 79(1):113–12, 2023.

[223] S.Y. Park, A.-M. Staicu, L. Xiao, and C.M. Crainiceanu. Simple fixed-effects inference for complex functional models. *Biostatistics*, 19(2):137–152, 06 2017.

[224] Y. Park, B. Li, and Y. Li. Crop yield prediction using Bayesian spatially varying coefficient models with functional predictors. *Journal of the American Statistical Association*, 118(541):70–83, 2023.

[225] Y. Park, L.L. Peterson, and G.A. Colditz. The plausibility of obesity paradox in cancer-point. *Cancer research*, 78(8):1898–1903, 2018.

[226] P.A. Parker and S.H. Holan. A Bayesian functional data model for surveys collected under informative sampling with application to mortality estimation using NHANES. *Biometrics*, 79(2):1397–1408, 2022.

[227] A. Parodi, M. Patriarca, L. Sangalli, P. Secchi, V. Vantini, and V. Vitelli. *fdakma: Functional Data Analysis: K-Mean Alignment*, 2015. R package version 1.2.1.

[228] H.D. Patterson and R. Thompson. Recovery of inter-block information when block sizes are unequal. *Biometrika*, 58(3):545, 1971.

[229] K. Pearson. On lines and planes of closest fit to systems of points in space. *Philosophical Magazine*, 2(11):559–572, 1901.

[230] J. Pinheiro, D. Bates, S. DebRoy, D. Sarkar, and R Core Team. *nlme: Linear and Nonlinear Mixed Effects Models*, 2020. R package version 3.1-149.

[231] J. Pinheiro and D.M. Bates. *Mixed-effects models in S and S-PLUS. Statistics and Computing.* Springer New York, NY, USA, 2006.

[232] M. Plummer. JAGS: A program for analysis of Bayesian graphical models using Gibbs sampling. In *Proceedings of the 3rd International Workshop on Distributed Statistical Computing*, page 1–10, 2003.

[233] C. Preda. Regression models for functional data by reproducing kernel Hilbert spaces methods. *Journal of Statistical Planning and Inference*, 137(3):829–840, 2007.

[234] C. Preda and J. Schiltz. Functional PLS regression with functional response: The basis expansion approach. In *Proceedings of the 14th Applied Stochastic Models and Data Analysis Conference. Universita di Roma La Spienza*, page 1126–1133, 2011.

[235] A. Prelić, S. Bleuler, P. Zimmermann, A. Wille, P. Bühlmann, W. Gruissem, L. Hennig, L. Thiele, and E. Zitzler. A systematic comparison and evaluation of biclustering methods for gene expression data. *Bioinformatics*, 22(9):1122–1129, 2006.

[236] N. Pya. *scam: Shape Constrained Additive Models*, 2021. R package version 1.2-12.

[237] X. Qi and R. Luo. Function-on-function regression with thousands of predictive curves. *Journal of Multivariate Analysis*, 163:51–66, 2018.

[238] S. Qu, J.-L. Wang, and X. Wang. Optimal estimation for the functional Cox model. *The Annals of Statistics*, 44(4):1708–1738, 2016.

[239] S.F. Quan, B.V. Howard, C. Iber, J.P. Kiley, F.J. Nieto, G.T. O'Connor, D.M. Rapoport, S. Redline, J. Robbins, J.M. Samet, and P.W. Wahl. The Sleep Heart Health Study: Design, rationale, and methods. *Sleep*, 20(12):1077–1085, 1997.

[240] R Core Team. *R: A Language and Environment for Statistical Computing.* R Foundation for Statistical Computing, Vienna, Austria, 2020.

[241] J.O. Ramsay and C.J. Dalzell. Some tools for functional data analysis. *Journal of the Royal Statistical Society: Series B (Methodological)*, 53(3):539–561, 1991.

[242] J.O. Ramsay, G. Hooker, and S. Graves. *Functional data analysis with R and MATLAB.* Springer New York, NY, USA, 2009.

[243] J.O. Ramsay, K.G. Munhall, V.L. Gracco, and D.J. Ostry. Functional data analyses of lip motion. *Journal of the Acoustical Society of America*, 99(6):3718–3727, 1996.

[244] J.O. Ramsay and B.W. Silverman. *Functional Data Analysis.* Springer New York, NY, USA, 1997.

[245] J.O. Ramsay and B.W. Silverman. *Functional Data Analysis*. Springer New York, NY, USA, 2005.

[246] J.O. Ramsay, H. Wickham, S. Graves, and G. Hooker. *FDA: Functional Data Analysis*, 2014.

[247] C.R. Rao. Some statistical methods for comparison of growth curves. *Biometrics*, 14(1):1–17, 1958.

[248] S.J. Ratcliffe, G.Z. Heller, and L.R. Leader. Functional data analysis with application to periodically stimulated foetal heart rate data. II: Functional logistic regression. *Statistics in Medicine*, 21(8):1115–1127, 2002.

[249] M. Reimherr, B. Sriperumbudur, and B. Taoufik. Optimal prediction for additive function-on-function regression. *Electronic Journal of Statistics*, 12(2):4571 – 4601, 2018.

[250] P.T. Reiss, A.J. Goldsmith, H.L. Shang, and R.T. Ogden. Methods for scalar-on-function regression. *International Statistical Review*, 85(2):228–249, 2017.

[251] P.T. Reiss, L. Huang, and M. Mennes. Fast function-on-scalar regression with penalized basis expansions. *International Journal of Biostatistics*, 6:1, 2010.

[252] P.T. Reiss and R.T. Ogden. Functional principal component regression and functional partial least squares. *Journal of the American Statistical Association*, 102(479):984–996, 2007.

[253] P.T. Reiss and R.T. Ogden. Functional generalized linear models with images as predictors. *Biometrics*, 66(1):61–69, 2010.

[254] J.A. Rice and B.W. Silverman. Estimating the mean and covariance structure non-parametrically when the data are curves. *Journal of the Royal Statistical Society. Series B (Methodological)*, 53(1):233–243, 1991.

[255] J.A. Rice and C.O. Wu. Nonparametric mixed effects models for unequally sampled noisy curves. *Biometrics*, 57(1):253–259, 2001.

[256] M.N. Rosenbluth. Genesis of the Monte Carlo algorithm for statistical mechanics. In *AIP Conference Proceedings*, volume 690, page 22–30, 2003.

[257] D. Ruppert. Selecting the number of knots for penalized splines. *Journal of Computational and Graphical Statistics*, 11(4):735–757, 2002.

[258] D. Ruppert, M.P. Wand, and R.J. Carroll. *Semiparametric Regression*. Cambridge Series in Statistical and Probabilistic Mathematics. Cambridge University Press, 2003.

[259] P.F. Saint-Maurice, R.P. Troiano, D. Berrigan, W.E. Kraus, and C.E. Matthews. Volume of light versus moderate-to-vigorous physical activity: Similar benefits for all-cause mortality? *Journal of the American Heart Association*, 7(7), 2018.

[260] F. Scheipl, J. Gertheiss, and S. Greven. Generalized functional additive mixed models. *Electronic Journal of Statistics*, 10(1):1455–1492, 2016.

[261] F. Scheipl, A.J. Goldsmith, and J. Wrobel. *tidyfun: Tools for Tidy Functional Data*, 2022. https://github.com/tidyfun/tidyfun, https://tidyfun.github.io/tidyfun/.

[262] F. Scheipl and S. Greven. Identifiability in penalized function-on-function regression models. *Electronic Journal of Statistics*, 10(1):495 – 526, 2016.

[263] F. Scheipl, A.-M. Staicu, and S. Greven. Functional additive mixed models. *Journal of Computational and Graphical Statistics*, 24(2):477–501, 2015.

[264] A. Schmutz, J. Jacques, and C. Bouveyron. *funHDDC: Univariate and Multivariate Model-Based Clustering in Group-Specific Functional Subspaces*, 2021. R package version 2.3.1.

[265] J.A. Schrack, V. Zipunnikov, J. Goldsmith, J. Bai, E.M. Simonsick, C. Crainiceanu, and L. Ferrucci. Assessing the "Physical Cliff": Detailed quantification of age-related differences in daily patterns of physical activity. *The Journals of Gerontology: Series A*, 69(8):973–979, 12 2013.

[266] G. Schwarz. Estimating the dimension of a model. *The Annals of Statistics*, 6(2):461–464, 1978.

[267] E. Sebastián-González, J.A. Sánchez-Zapata, F. Botella, and O. Ovaskainen. Testing the heterospecific attraction hypothesis with time-series data on species co-occurrence. *Proceedings: Biological Sciences*, 277(1696):2983–2990, 2010.

[268] D. Sentürk, L.S. Dalrymple, and D.V. Nguyen. Functional linear models for zero-inflated count data with application to modeling hospitalizations in patients on dialysis. *Statistics in Medicine*, 33(27):4825–4840, 2014.

[269] N. Serban and H. Jiang. Multilevel functional clustering analysis. *Biometrics*, 68(3):805–814, 2012.

[270] R. Sergazinov, A. Leroux, E. Cui, C.M. Crainiceanu, R.N. Aurora, N.M. Punjabi, and I. Gaynanova. A case study of glucose levels during sleep using fast function on scalar regression inference. *Biometrics*, 2023.

[271] J. Shamshoian, D. Şentürk, S. Jeste, and D. Telesca. Bayesian analysis of longitudinal and multidimensional functional data. *Biostatistics*, 23(2):558–573, 10 2020.

[272] H. Shou, A. Eloyan, S. Lee, V. Zipunnikov, A. Crainiceanu, M.-B. Nebel, B. Caffo, M.A. Lindquist, and C.M. Crainiceanu. Quantifying the reliability of image replication studies: The image intraclass correlation coefficient (I2C2). *Cognitive, Affective & Behavioral Neuroscience*, 13(4):714–724, 2013.

[273] H. Shou, V. Zipunnikov, C.M. Crainiceanu, and S. Greven. Structured functional principal component analysis. *Biometrics*, 71(1):247–257, 2015.

[274] C. Skinner and J. Wakefield. Introduction to the design and analysis of complex survey data. *Statistical Science*, 32(2):165–175, 2017.

[275] E. Smirnova, A. Leroux, Q. Cao, L. Tabacu, V. Zipunnikov, C. Crainiceanu, and J.K. Urbanek. The predictive performance of objective measures of physical activity derived from accelerometry data for 5-year all-cause mortality in older adults: National Health and Nutritional Examination Survey 2003-2006. *The Journals of Gerontology Series A Biological Sciences & Medical Sciences*, 75(9):1779–1785, 2020.

[276] Wood S.N., F. Scheipl, and J.J. Faraway. Straightforward intermediate rank tensor product smoothing in mixed models. *Statistics and Computing*, 23(3):341–360, 2013/2011 online.

[277] R.R. Sokal and C.D. Michener. A statistical method for evaluating systematic relationships. *University of Kansas Science Bulletin*, 38, 1958.

[278] A.-M. Staicu, M.N. Islam, R. Dumitru, and E. van Heugten. Longitudinal Dynamic Functional Regression. *Journal of the Royal Statistical Society Series C: Applied Statistics*, 69(1):25–46, 09 2019.

[279] A.M. Staicu, C.M. Crainiceanu, and R.J. Carroll. Fast methods for spatially correlated multilevel functional data. *Biostatistics*, 11(2):177–194, April 2010.

[280] A.M. Staicu, C.M. Crainiceanu, D.S. Reich, and D. Ruppert. Modeling functional data with spatially heterogeneous shape characteristics. *Biometrics*, 68(2):331–343, 2018.

[281] Stan Development Team. *The Stan Core Library*, 2018. Version 2.18.0.

[282] Stan Development Team. *RStan: The R interface to Stan*, 2022. R package version 2.21.7.

[283] J. Staniswalis and J. Lee. Nonparametric regression analysis of longitudinal data. *Journal of the American Statistical Association*, 93(444):1403–1418, 1998.

[284] J.L. Stone and A.H. Norris. Activities and attitudes of participants in the Baltimore Longitudinal Study. *The Journals of Gerontology*, 21:575–580, 1966.

[285] J. Sun, D. Tao, S. Papadimitriou, P.S. Yu, and C. Faloutsos. Incremental tensor analysis: Theory and applications. *ACM Transactions on Knowledge Discovery from Data*, 2(3), October 2008.

[286] B.J. Swihart, B. Caffo, B.D. James, M. Strand, B.S. Schwartz, and N.M. Punjabi. Lasagna plots: A saucy alternative to spaghetti plots. *Epidemiology*, 21(5):621–625, 2010.

[287] B.J. Swihart, A.J. Goldsmith, and C.M. Crainiceanu. Restricted likelihood ratio tests for functional effects in the functional linear model. *Technometrics*, 56(4):483–493, 2014.

[288] B.J. Swihart, N.M. Punjabi, and C.M. Crainiceanu. Modeling sleep fragmentation in sleep hypnograms: An instance of fast, scalable discrete-state, discrete-time analyses. *Computational Statistics and Data Analysis*, 89:1–11, 2015.

[289] M. Taylor, M. Losch, M. Wenzel, and J. Schröter. On the sensitivity of field reconstruction and prediction using empirical orthogonal functions derived from gappy data. *Journal of Climate*, 07 2013.

[290] C.D. Tekwe, R.S. Zoh, F.W. Bazer, G. Wu, and R.J. Carroll. Functional multiple indicators, multiple causes measurement error models. *Biometrics*, 74(1):127–134, 2018.

[291] C.D. Tekwe, R.S. Zoh, M. Yang, R.J. Carroll, G. Honvoh, D.B. Allison, M. Benden, and L. Xue. Instrumental variable approach to estimating the scalar-on-function regression model with measurement error with application to energy expenditure assessment in childhood obesity. *Statistics in medicine*, 38(20):3764–3781, 2019.

[292] O. Theou, J.M. Blodgett, J. Godin, and K. Rockwood. Association between sedentary time and mortality across levels of frailty. *Canadian Medical Association Journal*, 189(33):E1056–E1064, 2017.

[293] T.M. Therneau. *A Package for Survival Analysis in R*, 2020. R package version 3.2-7.

[294] A.N. Tikhonov. Solution of incorrectly formulated problems and the regularization method. *Soviet Mathematics Doklady*, 1963.

[295] T.M. Therneau and P.M. Grambsch. *Modeling Survival Data: Extending the Cox Model*. Springer, New York, 2000.

[296] R.P. Troiano, D. Berrigan, K.W. Dodd, L.C. Mâsse, T. Tilert, and M. McDowell. Physical activity in the united states measured by accelerometer. *Medicine & Science in Sports & Exercise*, 40(1):181–188, 2008.

[297] A.A. Tsiatis. A large sample study of Cox's regression model. *The Annals of Statistics*, 9(1):93–108, 1981.

[298] G. Tutz and J. Gertheiss. Feature extraction in signal regression: A boosting technique for functional data regression. *Journal of Computational and Graphical Statistics*, 19(1):154–174, 2010.

[299] S. Ullah and C.F. Finch. Applications of functional data analysis: A systematic review. *BMC Medical Research Methodology*, 13(43), 2013.

[300] M. Valderrama, F. Ocaña, A. Aguilera, and F. Ocaña-Peinado. Forecasting pollen concentration by a two-step functional model. *Biometrics*, 66:578–585, 08 2010.

[301] G. Verbeke and G. Molenberghs. *Linear Mixed Models for Longitudinal Data*. Springer, New York, 2000.

[302] A. Volkmann, A. Stöcker, F. Scheipl, and S. Greven. Multivariate functional additive mixed models. *Statistical Modelling*, 0(0):1471082X211056158, 2023.

[303] G. Wahba. Bayesian "Confidence Intervals" for the Cross-Validated Smoothing Spline. *Journal of the Royal Statistical Society: Series B*, 45(1):133–150, 1983.

[304] G. Wahba. *Spline Models for Observational Data (CBMS-NSF Regional Conference Series in Applied Mathematics, Series Number 59)*. Society for Industrial and Applied Mathematics, 1990.

[305] G. Wahba, Y. Wang, C. Gu, R. Klein, and B. Klein. Smoothing spline ANOVA for exponential families, with application to the Wisconsin Epidemiological Study of Diabetic Retinopathy: The 1994 Neyman Memorial Lecture. *The Annals of Statistics*, 23(6):1865–1895, 1995.

[306] M. Wand and J. Ormerod. On O'Sullivan penalised splines and semiparametric regression. *Australian & New Zealand Journal of Statistics*, 50:179–198, 06 2008.

[307] J.-L. Wang, J.-M. Chiou, and H.-G. Müller. Functional data analysis. *Annual Review of Statistics and Its Application*, 3(1):257–295, 2016.

[308] W. Wang. Linear mixed function-on-function regression models. *Biometrics*, 70(4):794–801, 2014.

[309] X. Wang, S. Ray, and B.K. Mallick. Bayesian curve classification using wavelets. *Journal of the American Statistical Association*, 102(479):962–973, 2007.

[310] J.H. Ward. Hierarchical grouping to optimize an objective function. *Journal of the American Statistical Association*, 58(301):236–244, 1963.

[311] H. Wickham and G. Grolemund. *R for Data Science: Import, Tidy, Transform, Visualize, and Model Data.* O'Reilly Media, 1 edition, 2017.

[312] R.K.W. Wong and T.C.M. Lee. Matrix completion with noisy entries and outliers. *Journal of Machine Learning Research*, 18(147):1–25, 2017.

[313] S.N. Wood. Thin plate regression splines. *Journal of the Royal Statistical Society: Series B (Statistical Methodology)*, 65(1):95–114, 2003.

[314] S.N. Wood. Stable and efficient multiple smoothing parameter estimation for generalized additive models. *Journal of the American Statistical Association*, 99(467):673–686, 2004.

[315] S.N. Wood. *Generalized Additive Models: An Introduction with R.* Chapman and Hall/CRC, 2006.

[316] S.N. Wood. On confidence intervals for generalized additive models based on penalized regression splines. *Australian & New Zealand Journal of Statistics*, 48(4):445–464, 2006.

[317] S.N. Wood. Fast stable restricted maximum likelihood and marginal likelihood estimation of semiparametric generalized linear models. *Journal of the Royal Statistical Society, Series B*, 73(1):3–36, 2011.

[318] S.N. Wood. On p-values for smooth components of an extended generalized additive model. *Biometrika*, 100(1):221–228, 10 2012.

[319] S.N. Wood. *Generalized Additive Models: An Introduction with R. Second Edition.* Chapman and Hall/CRC, 2017.

[320] S.N. Wood, Y. Goude, and S. Shaw. Generalized additive models for large datasets. *Journal of the Royal Statistical Society, Series C*, 64(1):139–155, 2015.

[321] S.N. Wood, Z. Li, G. Shaddick, and N.H. Augustin. Generalized additive models for gigadata: modelling the UK black smoke network daily data. *Journal of the American Statistical Association*, 112(519):1199–1210, 2017.

[322] S.N. Wood, N. Pya, and B. Safken. Smoothing parameter and model selection for general smooth models (with discussion). *Journal of the American Statistical Association*, 111:1548–1575, 2016.

[323] J. Wrobel. register: Registration for exponential family functional data. *Journal of Open Source Software*, 3(22):557, 2018.

[324] J. Wrobel. *fastGFPCA: Fast generalized principal components analysis*, 2023. https://github.com/julia-wrobel/fastGFPCA.

[325] J. Wrobel, V. Zipunnikov, J. Schrack, and J. Goldsmith. Registration for exponential family functional data. *Biometrics*, 75(1):48–57, 2019.

[326] Y. Wu, J. Fan, and H.G. Müller. Varying-coefficient functional linear regression. *Bernoulli*, 16(3):730–758, 2010.

[327] L. Xiao, L. Huang, J.A. Schrack, L. Ferrucci, V. Zipunnikov, and C.M. Crainiceanu. Quantifying the lifetime circadian rhythm of physical activity: A covariate-dependent functional approach. *Biostatistics*, 16(2):352–367, 10 2014.

[328] L. Xiao, C. Li, W. Checkley, and C.M. Crainiceanu. Fast covariance estimation for sparse functional data. *Statistics and Computing*, 28(3):511–522, 2018.

[329] L. Xiao, C. Li, W. Checkley, and C.M. Crainiceanu. *face: Fast Covariance Estimation for Sparse Functional Data*, 2021. R package version 0.1-6.

[330] L. Xiao, Y. Li, and D. Ruppert. Fast bivariate P-splines: The sandwich smoother. *Journal of the Royal Statistical Society, Series B (Methodological)*, pages 577–599, 2013.

[331] L. Xiao, V. Zipunnikov, D. Ruppert, and C.M. Crainiceanu. Fast covariance estimation for high-dimensional functional data. *Statistics and Computing*, 26(1):409–421, 2016.

[332] M. Xu and P.T. Reiss. Distribution-free pointwise adjusted p-values for functional hypotheses. In G. Aneiros, I. Horová, M. Hušková, and P. Vieu, editors, *Functional and High-Dimensional Statistics and Related Fields*, pages 245–252. Springer, 2020.

[333] Y. Xu, Y. Li, and D. Nettleton. Nested hierarchical functional data modeling and inference for the analysis of functional plant phenotypes. *Journal of the American Statistical Association*, 113(522):593–606, 2018.

[334] F. Yao, H.-G. Müller, and J.-L. Wang. Functional data analysis for sparse longitudinal data. *Journal of the American Statistical Association*, 28(100):577–590, 2005.

[335] F. Yao, H.-G. Müller, and J.-L. Wang. Functional linear regression analysis for longitudinal data. *Annals of Statistics*, 33(6):2873–2903, 2005.

[336] F. Yao, H.G. Müller, A.J. Clifford, S.R. Dueker, J. Follett, Y. Lin, B.A. Buchholz, and J.S. Vogel. Shrinkage estimation for functional principal component scores with application to the population kinetics of plasma folate. *Biometrics*, 59(3):676–685, 2003.

[337] Y. Yuan, J.H. Gilmore, X. Geng, S. Martin, K. Chen, J.-L. Wang, and H. Zhu. Fmem: Functional mixed effects modeling for the analysis of longitudinal white matter tract data. *NeuroImage*, 84:753–764, 2014.

[338] D. Zhang, X. Lin, and M. Sowers. Two-stage functional mixed models for evaluating the effect of longitudinal covariate profiles on a scalar outcome. *Biometrics*, 63(2):351–362, 2007.

[339] Y. Zhang and Y. Wu. Robust hypothesis testing in functional linear models. *Journal of Statistical Computation and Simulation*, 0(0):1–19, 2023.

[340] Y. Zhao, R.T. Ogden, and P.T. Reiss. Wavelet-based lasso in functional linear regression. *Journal of Computational and Graphical Statistics*, 21(3):600–617, 2012.

[341] X. Zhou, J. Wrobel, C.M. Crainiceanu, and A. Leroux. Generalized multilevel functional principal component analysis. *Under Review*, 2023.

[342] H. Zhu, P.J. Brown, and J.S. Morris. Robust, adaptive functional regression in functional mixed model framework. *Journal of the American Statistical Association*, 106(495):1167–1179, 2011.

[343] H. Zhu, P.J. Brown, and J.S. Morris. Robust classification of functional and quantitative image data using functional mixed models. *Biometrics*, 68(4):1260–1268, 2012.

[344] H. Zhu, K. Chen, X. Luo, Y. Yuan, and J.-L. Wang. Fmem: Functional mixed effects models for longitudinal functional responses. *Statistica Sinica*, 29(4):2007, 2019.

[345] V. Zipunnikov, B. Caffo, D.M. Yousem, C. Davatzikos, B. Schwartz, and C.M. Crainiceanu. Multilevel functional principal component analysis for high-dimensional data. *Journal of Computational and Graphical Statistics*, 20(4):852–873, 2011.

[346] V. Zipunnikov, S. Greven, H. Shou, B. Caffo, D.S. Reich, and C.M. Crainiceanu. Longitudinal high-dimensional principal components analysis with application to diffusion tensor imaging of multiple sclerosis. *The Annals of Applied Statistics*, 8(4):2175–2202, 2014.

Index

fields::image.plot 200
filter 147
fit_mims 47
FLAM *see* Functional linear array model
    (FLAM)
FLCM *see* Functional linear Cox model
    (FLCM)
FMEM *see* Functional mixed effects
    modeling (FMEM)
FMM *see* Functional mixed models (FMM)
fMRI studies 253
FoFR *see* Function-on-function regression
    (FoFR)
FoSR *see* Function-on-scalar regression
    (FoSR)
Fourier non-spline bases 38
FPCA *see* Functional principal component
    analysis (FPCA)
fpca.face 57, 58–60, 66, 70–1, 73–4, 162
fpca.sc 57, 58
FPCR *see* Functional principal component
    regression (FPCR)
FPLS *see* Functional partial least squares
    (FPLS)
fREML 162
FSVD *see* Functional SVD (FSVD)
F-tests 53
FUI *see* Fast univariate inference (FUI)
Full conditional distribution 169
Functional additive mixed models (FAMM)
    84ff, 160, 258ff
    in case study 260ff
    not feasible for large data sets 259
    pre-specified spline basis 160
Functional coefficient multiples 127ff
Functional data analysis (FDA)
    assumes infinite-dimensional data
      generation 19
    assumptions for 101
    book by Ramsay and Silverman xi
    computational tricks for
      high-dimensional data 28–9
    conceptual framework 1
    continuous vs. discrete matrix 20
    density and sparseness 15
    exhibiting all properties 6
    as form of linear mixed effects model 19
    framework as guide 2
    history of xi–xiii, 2
    junk 28–9
    multilevel 243ff

non-Gaussian data 76ff
Functional data clustering 265ff
Functional data vs. multivariate data 2
Functional domain 65
Functional effect estimators 53
Functional linear array model (FLAM) 143
Functional linear Cox model (FLCM) 232–4
Functional linear model relation to
    P-splines xi
Functional mixed effects modeling (FMEM)
    143, 256
Functional mixed models (FMM) 256ff
Functional MRI studies 253
Functional partial least squares (FPLS)
    development xi
Functional predictors 109
    for survival analysis 211ff
Functional principal component analysis
    (FPCA) 28, 54ff, 65ff
    for accelerometry data 58–60
    clustering and smoothing 280ff
    comparison of variations 86ff
    in CONTENT child growth study 191ff
    data-driven 160
    defining 65–6
    generalized for non-Gaussian data 76ff
    residual modeling with 161ff
    simulated example 66ff
    smoothing and clustering 280ff
    sparse xiii
    two-level 245ff
    variational Bayes 84
    vs. raw PCA 69–72
Functional principal component regression
    (FPCR) 118ff
    development of xi
    regression on FPC scores 120
Functional process
    assumption 1
    properties 2
Functional regression philosophy and
    practice xii
Functional SVD (FSVD) 28, 54ff
Function describing data set 6
Function-on-function regression (FoFR) 96,
    175ff
    inference for 199ff
    inferential approach xii
    linear mixed effects (LME) 176ff
    model and fit prediction with 180ff